Transactions on Computer Systems and Networks

Transactions on Computer Systems and Networks is a unique series that aims to capture advances in evolution of computer hardware and software systems and progress in computer networks. Computing Systems in present world span from miniature IoT nodes and embedded computing systems to large-scale cloud infrastructures, which necessitates developing systems architecture, storage infrastructure and process management to work at various scales. Present day networking technologies provide pervasive global coverage on a scale and enable multitude of transformative technologies. The new landscape of computing comprises of self-aware autonomous systems, which are built upon a software-hardware collaborative framework. These systems are designed to execute critical and non-critical tasks involving a variety of processing resources like multi-core CPUs, reconfigurable hardware, GPUs and TPUs which are managed through virtualisation, real-time process management and fault-tolerance. While AI, Machine Learning and Deep Learning tasks are predominantly increasing in the application space the computing system research aim towards efficient means of data processing, memory management, real-time task scheduling, scalable, secured and energy aware computing. The paradigm of computer networks also extends it support to this evolving application scenario through various advanced protocols, architectures and services. This series aims to present leading works on advances in theory, design, behaviour and applications in computing systems and networks. The Series accepts research monographs, introductory and advanced textbooks, professional books, reference works, and select conference proceedings.

Ashish Kumar · Divya Singh

Artificial Intelligence in Modern Healthcare System

Springer

Ashish Kumar
School of Computer Science Engineering
and Technology
Bennett University
Greater Noida, Uttar Pradesh, India

Divya Singh
School of Computer Science Engineering
and Technology
Bennett University
Greater Noida, Uttar Pradesh, India

ISSN 2730-7484 ISSN 2730-7492 (electronic)
Transactions on Computer Systems and Networks
ISBN 978-981-96-6702-4 ISBN 978-981-96-6703-1 (eBook)
https://doi.org/10.1007/978-981-96-6703-1

This Springer imprint is published by the registered company Springer Nature Singapore Pte Ltd.
The registered company address is: 152 Beach Road, #21-01/04 Gateway East, Singapore 189721, Singapore

Preface

Artificial intelligence (AI) is transforming healthcare domain by automating and improving diagnosis and treatment. AI is an ever-growing field using multiple subsets, including machine learning, deep learning, natural language processing, and expert systems. When applied in the field of healthcare, AI has the potential to greatly impact diagnostics, treatment, care, operational efficiency, drug discovery and development. Increasingly, artificial intelligence has been used in diagnostics, helping to identify nodules or cancers, predict cardiovascular risk, and provide personalized care for chronic patients.

Various AI models and algorithms, such as convolutional neural networks, decision trees, and support vector machines, among others, have been used to analyze medical imaging data to predict brain tumor biopsy results, cardiovascular disorder, breast cancer and many other critical diseases. AI methods can also be used to analyze high-throughput immunological data for personalized care. These techniques include machine learning, deep learning, and reinforcement learning, based on mathematical, computational, and statistical concepts, and complement immunological research. Benchmark healthcare datasets should encompass at least one million total samples and larger cohorts to yield generalization and interoperability in healthcare. High quality medical datasets should consist of, but are not limited to, imaging data, clinical data, and genomic longitudinal data to mitigate the impact of AI-bias in the predictive outcomes. Apart from clinical datasets, there are certain challenges found in developing robust medical devices, which include the exploitation of AI in medical signal processing, analysis, and interpretation to create smart healthcare system that can utilize expertise-based knowledge and generate solutions. These solutions can accurately refine the identification, diagnosis, as well as therapy of diseases that must be safe, secure, interoperable, and efficient. The early and several benefits of modern healthcare technologies include remote patient monitoring, aiding in early diagnosis, facilitating advanced treatment recommendations for a wider general population, and allowing the unprecedented potential for data sharing, collaboration, and knowledge build-up is possible with the incorporation of AI in healthcare models. To efficiently manage modern clinical needs, many AI algorithms are proposed in dedicated systems including data resource utilities, individual and group support systems, and methods for AI-based classification and predictive modeling.

This book provides a critical overview of AI technologies in the early prediction of chronic diseases and preventing casualties. The book comprises of four parts elaborating on modern healthcare, applications, personalized care and benefits of smart healthcare using AI. Part I contains two chapters. Chapter 1 titled "Introduction to Artificial Intelligence in Modern Healthcare" details the various AI techniques that provide quick, and accurate solutions for transforming treatment using conventional methods of healthcare. Chapter 2 titled "Research Orientation for AI Techniques in Modern Healthcare System" discusses the various aspects and research methodologies currently used in applying AI techniques to realize a smart healthcare-oriented scenario. The great potential of AI depicts the need to reorient and broaden the research interests and focus on modern healthcare systems.

Part II titled "Applications of Artificial Intelligence for Disease Prediction" has six chapters describing various applications and benefits of AI techniques that can be used for diagnosis and prediction of chronic diseases. In this, Chap. 3 titled "Diagnosis and Prediction of Brain Tumor Using Artificial Intelligence" elaborates on the benefits of AI tools and algorithms in diagnosing and predicting brain tumors, particularly in the context of oncology. These tools have the ability to analyze large amounts of brain imaging data, such as magnetic resonance imaging (MRI) and computer tomography (CT) scans, and identify features and patterns typically not detectable by human radiologists, thereby increasing the accuracy of diagnosis and prognosis assessment. Chapter 4 titled "Diagnosis and Prediction of Neurological Disorders Using Artificial Intelligence" details the main idea of diagnosing and predicting a neurological disorder using the latest techniques from imaging, text and signaling data. Chapter 5 titled "Diagnosis and Prediction of Cardiovascular Disorder Using Artificial Intelligence" elaborates on various AI techniques that can predict cardiovascular disorders at an early stage. AI-based data-driven algorithms permit computers to learn gradually and help in the decision-making process for predicting cardiovascular disorders by scrutinizing the diverse range of health data, including electrocardiograms, intervascular ultrasound, genetic, lifestyle, environmental risk factors, and cardiac imaging studies. Chapter 6 titled "Diagnosis and Prediction of Cardiovascular Risk in Retinal Imaging Using Artificial Intelligence" discusses the importance of identifying bioindicators and diabetic retinopathy imaging for the prediction of cardiovascular disease at an early stage. AI systems based on deep neural networks are able to predict the occurrence of some adverse events after the diagnosis of diabetes or hypertension in retinopathy imaging. This has opened up new therapeutic possibilities for ophthalmologists as retinal photography has been found to be a great indicator to predict risk for heart disease, a study that has brought about endless possibilities of collaboration between ophthalmologists and cardiovascular specialists. Chapter 7 titled "Diagnosis and Prediction of Diabetic Foot Ulcer in Modern Healthcare Using Artificial Intelligence" elaborates on the identification of diabetic foot disease (DFU) using AI. DFU data are integrated from foot images and clinical assessments such as ABI, pulses, and vibration perception threshold to carry out diagnosis and determine if the patients are at high risk of developing another foot ulcer or requiring an amputation. A large amount of data is collected from patients in each study to indicate the significant improvement in the stratification of high risk along with increased predictive power. Chapter 8 titled "Diagnosis and Prediction of Breast Cancer Using Artificial Intelligence" elaborates on the severity of breast cancer in high-income countries. These abnormalities can occur in a woman's body gradually and lead to the development of cancer, usually in a few years. Treatment is easier when breast cancer is diagnosed early; early-stage patients have improved outcomes following appropriate treatment. Several AI techniques are reviewed which analyze mammograms and other types of medical imaging in an attempt to detect breast cancer early.

Part III contains four chapters presenting the improvement and benefits offered by AI for personalized care. For this, Chap. 9 titled "Role of Artificial Intelligence in Immunology" details the potential impact of AI in genetic and protein analyses. There has been a shift from single markers to analyzing millions of markers in a comprehensive genome-wide or proteome-wide approach. AI methods are analyzed for processing such complex data sets. Chapter 10 titled "Managing High-Risk Surgery Using Artificial Intelligence" presents the precision and high-ranking decision-making algorithms that are needed for high-risk surgery and make it a natural area for the development and introduction of AI. AI tools for precision surgery are used to predict complications that may arise for a particular patient when decisions are made by the surgery based on information collected by MRI, CT, medical history and examination. There are three main areas where AI can play a beneficial role in precision medicine: (1) preoperative assessment, (2) surgical planning, and (3) postoperative monitoring, which is highlighted in this chapter. AI systems are designed to predict complications that can support experienced surgeons, perioperative physicians, and nursing staff in the assessment of risk and further management of patients. Chapter 11 titled "Benchmark Datasets for Analysis in Medical Systems" compiles the various datasets crucial for the performance review of AI models within medical systems. It emphasizes that benchmark datasets are essential to achieve reproducible and reliable results in modern healthcare. Benchmark datasets consist of training datasets, which are used to develop and train the AI model, and testing datasets, which are used to validate the model before real-time deployment. Chapter 12 titled "Role of AI and Modern Medical Equipment in Smart Healthcare" summarizes AI and medical equipment as evolving complementary tools for delivering smart healthcare services. Advanced medical equipments are analyzed to gather physiological signals for clinical indexes and integrate them to suggest a tentative diagnosis. Based on the accumulated data input, the medical equipment generates a personalized treatment for every patient.

Part IV titled "Artificial Intelligence for Healthcare Digitization" contains three chapters. Chapter 13 titled "Evolution of Traditional Healthcare to Modern Healthcare—Benefits, Opportunities and Challenges" discusses the significant reforms in modern healthcare treatments. It complies the various methodologies the way diseases are diagnosed within the human body, the mode of treatments, medications prescribed, and the research protocols. The conventional labor-intensive treatment procedures are slowly moving to a modern era of personalized precision medicine for the betterment of a human approach, and it would not have been possible without the utility of advanced technologies as a key integrator. Chapter 14 titled "Analysis of AI-Bias in Modern Healthcare Systems" details the development of AI-based models in healthcare and the introduced bias in the outcomes. AI models promise to leverage this data for better diagnoses, more precise treatment decisions, and improved patient outcomes. Despite serving diverse patient populations, existing healthcare data may not be representative of everyone in a given community and can be subject to socioeconomic and cultural biases in healthcare AI models that reflect and propagate existing disparities in care and outcomes. Various bias analyses and suggestions for its mitigation are also highlighted in this chapter. The last Chap. 15 titled "Examining QoS for Modern Healthcare Systems" advances the successful deployment of modern healthcare systems and benefits from advancements in the field of AI, such as using big data technologies or deep learning, which requires awareness of quality of service (QoS) concepts. It discusses the QoS and how effectively AI-based clinical systems perform over a variety of parameters to provide accurate diagnosis and treatment. In many application domains, imperious data requirements and high costs to maintain these standards require additional performance metrics to account for the reliability of digital service at the manipulation layer of these applications.

This book summarizes several changes that are currently taking place or are expected to take place with the potential of AI in the healthcare and medical fields. It will interest the various stakeholders to deeply understand the changes, difficulties, empathy, and insights, and to eventually establish new processes and rules. This book has carefully explained the scientifically proven important facts about the impact of AI in establishing the modern healthcare system and the potential to be the most crucial factor in determining the type of delivery system that healthcare pursues in future.

Greater Noida, India

Ashish Kumar
Divya Singh

Competing Interests The authors have no competing interests to declare that are relevant to the content of this manuscript.

Contents

About the Authors

Ashish Kumar is an associate professor at Bennett University, India, where he has been since 2022. Prior to joining Bennett University, he worked with Bharati Vidyapeeth's College of Engineering. He completed his Ph.D. in Computer Science and Engineering from Delhi Technological University (formerly known as DCE), New Delhi, in 2020. For his contributions to the computer vision domain, he received the Best Researcher Award from Delhi Technological University. Dr. Kumar holds an M.Tech. degree with distinction in Computer Science and Engineering from GGS Indraprastha University, New Delhi. He has published over 35 research papers in various esteemed national and international journals and conferences, as well as more than 15 book chapters in Scopus-indexed books. Additionally, he has authored and edited several books in the fields of artificial intelligence, computer vision, and healthcare. As an active member of various international societies and clubs, Dr. Kumar serves as a reviewer for many reputed journals and participates in the technical program committees of various national and international conferences. He has also chaired sessions at multiple conferences. His current research interests include object tracking, image processing, artificial intelligence, and medical imaging analysis.

Divya Singh is an assistant professor in the School of Computer Science and Engineering at Bennett University, India. She holds a B.Tech. degree in Information Technology and an M.Tech. degree, as well as a Ph.D. in Computer Science Engineering and Technology from IIT BHU, Varanasi. Dr. Singh has two years of teaching experience and has published several research papers in reputable international journals and conferences. Her research interests encompass image and video processing, computer vision, artificial intelligence, machine learning, and deep learning.

Part I
Artificial Intelligence in Modern Healthcare

Chapter 1
Introduction to Artificial Intelligence in Modern Healthcare

Abstract With the advancement in technology, traditional healthcare has evolved to provide improved and accurate diagnosis of critical diseases to save mankind. Traditional healthcare techniques are integrated with artificial intelligence (AI) for early prediction of disease so that proper and personalized treatment can be extended. In modern healthcare system, AI have either machine learning or deep learning techniques which provide quick and automatic solution for prediction of critical diseases such as mental illness, brain tumor, cardiovascular disorder at an early state. The application of machine learning and deep learning models for prediction and prognosis of these diseases has become an irrevocable part of medical treatment aimed at improving the subsequent therapy and management of patients. In this chapter, we have discussed the overview of various machine learning and deep learning techniques which are proposed to address the needs of the patient and provide medical aid at an early stage of the disease.

Keywords Artificial intelligence (AI) · Machine learning (ML) · Deep learning (DL) · Modern healthcare · Smart healthcare · Digital healthcare

1.1 Overview of AI in Modern Healthcare

Artificial intelligence (AI) has diversified applications in various fields such as education [1], sentiment analysis [2], computer vision [3] and many more [4–7]. AI has also revolutionized traditional healthcare by providing early, quick and accurate predictions of critical diseases such as brain tumor [8], cardiovascular disorders (CVD) [9], cervical cancer [10] and many more [11–14]. AI have utilized either machine learning (ML) or deep learning (DL) algorithms for processing medical imaging data such as X-ray, magnetic resonance imaging (MRI), ultrasound (USd), and computed tomography (CT). Apart from imaging data, biomedical signals such as electrocardiogram (ECG), and electroencephalography (EEG) are also investigated for predicting health disorders. In addition, textual data, namely electronic health

A. Kumar and D. Singh, *Artificial Intelligence in Modern Healthcare System*,
Transactions on Computer Systems and Networks,
https://doi.org/10.1007/978-981-96-6703-1_1

records (EHR), tweets, comments, opinions from social media are also very effective in predicting critical disorders in human beings. Generative AI models such as ChatGPT, natural language processing, are also helpful in improving the treatment accuracy and extending personalized treatment to the patients [4, 15]. Figure 1.1 represents the categorization of various AI techniques such as ML, DL or generative AI.

ML based algorithms can be categorized either as supervised learning or unsupervised learning [16, 17]. Treatment accuracy can be alleviated by using labeled medical data in supervised learning algorithms and the commonly adopted algorithms in this category are support vector machine (SVM), random forest (RF), decision trees (DT), XGboost and many others which are used for disease classification. On the other hand, effective representation of medical data from raw unlabeled data is processed using unsupervised learning algorithms which comprise of clustering and dimensionality reduction techniques exploited for disease prediction and reliable clinical diagnoses. Authors have adopted supervised ML algorithms for classification of cells in brain tumors and breast cancer imaging either as malignant or benign [18–21]. On the other hand, clustering techniques and dimensionality reduction also

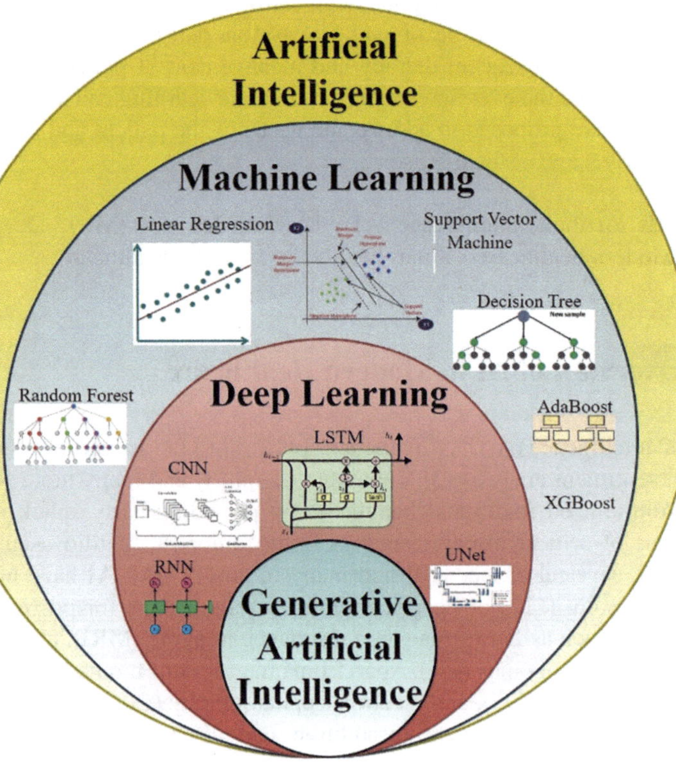

Fig. 1.1 Categorization of AI in medical domain

adopted for providing better results in low resolution imaging [22, 23]. Suppression of noise, contrast and resolution enhancements in medical imaging are adopted during pre-processing process for improving segmentation accuracy of these algorithms. Clinical measurements using ML algorithms focus to minimize the generalizability gaps on different databases and ensure to have better treatment plans with quick and accurate prediction accuracy.

On the other hand, DL algorithms utilize convolutional neural networks (CNN), recurrent neural networks (RNN), U-Net and long short-term memory (LSTM) networks for extracting sensitive and essential information from the various clinical datasets for predicting life-threatening diseases at an early stage [24–28]. In order to prevent the spread of disease such as cancer and tumors, it is crucial to determine its vital markers such as size, shape and location accurately. To understand its appearance and risk factors, the medical imaging is processed using DL models. DL models provide advanced techniques and optimize critical parameters substantial for highly efficient computational models for disease predictions. This is one of the invasive techniques that can identify the criticality of the disorders and save money by eliminating the need for expensive medical tests. It improves the decision making of doctors and medical practitioners by predicting advanced level diseases effectively.

Hybrid approaches are also quite popular in reducing the prediction of critical diseases by processing the large amount of data in less time. These techniques exploited DL techniques for extracting robust features and ML techniques for classification for these features either as benign or malignant [29, 30]. The robust feature extraction and selection is adopted for determining the sensitive features from various medical imaging. These features are processed using either DL models or any other efficient techniques. These features are fed to the ML-based classifier for identification of the advanced diseases. ML-classifiers perform either binary or multi-class classification for computing the criticality of diseases such as foot ulcer, neurological disorder and cancerous cells. In other directions, Attentional mechanisms are also incorporated into the prediction algorithms for extracting the relevant features from the target regions and neglecting the irrelevant features [31–33]. These algorithms are faster in computing the outcomes as only a limited area needs to be processed. Also, the recent algorithms incorporating encoder and decoder networks are widely explored in medical domain for providing enhanced treatment accuracy. In this direction, vision transformer (ViT), Swim transformers are few to name [34, 35]. These algorithms can process the medical data faster to provide accurate and timely information for initiating proper treatment.

Generative AI includes large language models (LLM) and natural language processing (NLP) which are significant in revolutionizing the medical decision-making process [15]. LLM models have a wide range of applications in various radiological specific datasets. These models can automatically generate radiological reports from small sets of keywords, provide diagnosis based on imaging patterns, suggest report summarization for effective treatment and many more. Similarly, NLP has been identified to have potential applications in healthcare management systems along with LLMs [36, 37]. NLP is specifically used with transformer-based algorithms and performs various tasks such as text classification, and extraction. These models can also automate the process of data curation for computing the significant findings from the patient's medical data. The potential of NLP is explored in ChatGPT, a transformer-based method [15]. This chatbot with user-friendly interface provides detailed analysis of medical imaging with quick and accurate diagnosis and segmentation.

This book comprises extensive discussion about the recent, advanced and innovative techniques for revolutionizing medical healthcare. For this, AI-based algorithms are reviewed to emphasize their need for integrating into traditional healthcare methodologies for quick and efficient diagnosis and prognosis of critical disease. AI-based algorithms are widely used for predicting brain tumors, breast cancer, diabetic foot ulcer, neurological disorders and many other life-threatening diseases. The vast availability of medical imaging data reduces the effectiveness of the traditional healthcare system. The outcomes of these medical imaging are highly dependent on the expertise and experience of the radiologist. Secondly, it is time consuming to analyze medical imaging data such as MRI, and CT, which contain numerous images for processing manually. Hence, to provide effective, fast and personalized treatment AI-based techniques are widely explored. In this book, we have summarized the available potential work to predict critical disease and the imaging data publicly available for analysis and prediction.

1.1.1 Challenges in Traditional Healthcare Systems

Traditional healthcare systems have certain limitations which restrict the timely approachability to patients suffering from critical diseases [38, 39]. Traditional healthcare system is incompetent to provide affordable, quick and accurate medical facilities to patients [40]. It has many concerns related to data analysis, trust, reliability, self-reliance and many more. It is difficult to provide early medical aid and healthcare services to the people living in remote areas. Providing the best medical assistance in emergency cases is still challenging as the latest methodology yet to be deployed properly. The limitations of traditional healthcare can be addressed by adopting advanced technologies such as ML, DL, and NLP. These technologies can revolutionize the traditional system by providing smart and innovative healthcare services. Hence, it is essential to understand the limitations of traditional healthcare systems to improve the quality of life significantly. Figure 1.2 represents the limitations of the traditional healthcare systems, and the details are as follows:

Data analysis: A lot of medical data is produced by various medical imaging techniques such as CT, MRI and Ultrasound (USd) scans. It is tedious as well as time consuming to analyze such huge amounts of data manually. Also, the inference from these data highly depends on the expertise and experience of the radiologist.

Fig. 1.2 Limitations of traditional healthcare systems

However, the modern healthcare system addresses this limitation of traditional healthcare systems by exploiting the latest technologies. The analysis is not only fast but also accurate, which is helpful for doctors to make better decisions.

Trustworthiness and reliability: The traditional healthcare system is not very much trustworthy and reliable. The reason could be total dependency on human resources for diagnosis and prognosis. In some cases, the existence of disease in a patient was detected in its last stage as the symptoms were not very clear and precise to be captured manually. The innovation in traditional healthcare is now helpful for capturing and detecting critical disease such as cancer at an early stage to save lives [11].

Efficiency and effectiveness: Traditional techniques in healthcare are not very efficient and effective in early prognosis of disease. The reason could be less accessibility for better medical facilities to the people living in remote areas. The effectiveness of treatment is not efficient either due to the disease detected in its last stage or non-availability of treatment facilities at initial stage of disease.

Availability of expertise: Traditional healthcare system is manual and depends on the proficiency of medical professionals. There is limited availability of specialists which can cure critical diseases. This limitation can be addressed by incorporating the latest technology such as IoT and robotics in medical systems which not only save time but also more accurately treat the disease.

Limitations for personalized treatment: One of the major facilities of smart healthcare is that it provides personalized treatment to the patients by studying the vital parameters such as family history, smoking habits, genetics analysis, and medical biomarkers. Analysis of so many parameters for providing personalized treatment is not possible in traditional healthcare systems. In [41], authors analyzed multidimensional components such as genetic history, functional parameters and influential environmental attributes for providing personalized medical for preventing cardiovascular disorders and cancers.

Physical availability of medical practitioners: Traditional healthcare emphasizes the need for physical examination and medical consultation only. However, due to limited time and non-availability of efficient medical facilities, it is not always possible to visit the doctors physically. To address this challenge, smart healthcare has introduced tele consultations which reduces the number of physical visits to the clinic or hospital and effective in providing primary care to the patients [42]. This facility also helps in preventing the spread of deadly diseases such as COVID 19. Tele consultations proved to be effective for managing people's health in lockdown during COVID19.

Self-reliant healthcare: Traditional healthcare demands all the decisions, measurements and analysis to be taken by medical practitioners only. However, smart healthcare provides self-wearable devices equipped with IoT technology for continuous monitoring of the patient's health. The user can manage its health condition with the help of Apps and information platforms [43]. The data can be understandable to a user and routed to doctor for further analysis.

Geographical barriers: Geographical limitations is one of the biggest barriers in traditional healthcare systems which prevent the accessibility of superior medical services to the people living in remote areas. In smart healthcare systems, various features such as tele consultations, IoT-based wearable devices, personalized healthcare provides reliable treatment in case of emergency.

1.2 Evolution from Traditional Healthcare to Digital Healthcare

Traditional healthcare system is transforming into smart healthcare to make it more intelligent, accessible and accurate. Smart healthcare is not only technological advancement of traditional healthcare system but enhances the medical care, services and experience to the participants.

There are some chronic diseases such as breast cancer, brain tumor, and CVD whose timely management is necessary to prevent casualties. These diseases are curable if predicted at their early stage [11, 34, 44–47]. As the number of patients with such diseases is increasing gradually, traditional healthcare systems which are manual and doctoral centric are incapable to provide proper care and treatment. Hence, smart healthcare emerges with innovative solutions for better management of diseases. Firstly, smart healthcare emphasizes the need for AI-based prediction algorithms and self-monitoring of patients for improved management of such life-threatening diseases. AI-based prediction algorithms can predict the spectrum of these diseases for managing their long-term effects efficiently. Secondly, smart healthcare also suggested on continuous monitoring of patient's vitals and substantial parameters through wearable intelligent devices for timely actions to improve quality of life.

1.2.1 AI and Digitization to Healthcare

Digital innovation of the healthcare sector improves the patient's care and management of critical diseases effectively. Digital transformation offers e-health services that includes the incorporation of tele consultation, management of electronic health record, IoT-based wearable devices for providing health monitoring round the clock [48]. Healthcare digitization offers several advantages such as automatic disease prediction, real-time health monitoring, data driven decision making and self-reliant patients.

Healthcare digitization offers several benefits to manage the patient's health efficiency. The healthcare digitization introduces the concept of tele consultation which allows remote monitoring of the patient's health by providing virtual online appointments. This facility allows the accessibility of the best medical services to the people living in rural areas. This has eliminated the geographical barriers by providing cost effective superior medical facility. Another benefit of digitized healthcare is personalized medical facilities for the patient's suffering from mental disorders, and CVD. Personalized treatment includes a study of patient's genetics and genomics to provide effective medical care. Smart healthcare with recent technologies and IoT makes it possible to provide personalized care for faster recovery.

To summarize, digitization of healthcare offers a wide range of benefits along with continuous health monitoring with the help of smart devices such as watches, mobile apps, and other wearable devices. The data collected will be transferred to the medical practitioners in real-time to take necessary actions. The data can also be further analyzed with the help of AI-based techniques such as ML and DL to predict critical diseases at an early stage.

1.2.2 Smart Healthcare Wearable Devices

Smart healthcare wearables are one of the technological advancements in medical systems which can track and monitor a person's health continuously. These devices can help in providing personalized healthcare along with preventive medicines by understanding the patient's requirements from the data gathered by wearable devices

[49]. The data is streamed to the clinical staff for analysis in real-time. These devices reduce physical visits to the doctor and limit communication between the patient and the clinical staff. The data is helpful in predicting chronic diseases at an early stage and efficient decision-making by the doctors. These devices can track day-to-day activities and recommend exercises for the physical well-being of the patients.

Mainly, wearable devices in healthcare can perform four activities namely, monitoring, screening, detection and prediction [50]. These devices can monitor pulse, heartbeat, and physical activity of the person. Continuous screening of cardiac, and sleep is possible using these devices. These devices can detect and predict clinical risk, less physical activities and early symptoms of any health-related problems. Respiratory rate data, biological age, irregular pulse, biomedical condition and mortality captured from wearable devices can be used for predicting pulmonary diseases and neurological disorders.

The data from wearables has certain limitations and concerns which need to be addressed before its adoption to the public [50]. The concerns are related to its quality, design, technicality, security and privacy. The quality of the product should be high and accurate to ensure the correctness of the data. High quality sensors and accelerometers should be used for capturing error free data. The design of the wearables plays a crucial role in its usage. People prefer to wear healthcare devices in the form of bands and watches if the device design is attractive. Designing these devices should be attractive enough so that individuals prefer to wear them. The interface of such devices should be user-friendly so that individuals with less scientific knowledge and older people can be able to operate it with ease. The available patterns and features should be simple but powerful in screening the potential vitals of the patients. Other concerns are related to security risks and privacy issues associated with the gathered data and its usage. The data protection and security laws must be appropriately deployed in order to achieve the trust and fairness of the users.

To summarize, smart wearables in healthcare are a superior support for monitoring and detecting patient's vital in real-time. Gathered data quality and accuracy should be accessed properly before its usage to the patients. The variability in the sensors, different data collection processes and interpretability of results must be clinically validated to ensure the high-quality standards and interoperability of the devices. Clinical validation is a crucial step to ensure the reliable performance of these devices. In addition, these wearable devices should also be validated for gender equality. It must be ensured that there is no AI-bias in these devices due to missing sociodemographic data, ethnicity, age and nationality. For wider acceptability of wearable devices, it is necessary to consider ethical, legal and social requirements before its deployment to realistic scenarios.

1.3 AI-Bias in Modern Healthcare Systems

AI-based prediction models are high in demand in modern healthcare systems as these models can predict chronic disease accurately at an early stage. However, these models are accessed to be suffered from AI-bias which leads to the non-reliable outcomes and unseen classifications [51]. In AI-based prediction models AI-bias can be introduced either through data or algorithmic design. Broadly, data bias is dependent on dataset selection, sampling and missing historical information. Also, data bias can be introduced due to model training on insufficient, partial and incomplete datasets. The missing values are not handled properly, which creates inconsistencies in the clinical outcomes. The surrogate data is not accessed rightly for its incorporation in the datasets. All these will corrupt the model and biased its output.

Algorithmic bias in AI-models is introduced due to its weak design, missing training-test specifications, hyperparameters selections and interpretability [51]. The algorithmic bias can be introduced during pre-processing, in-processing and post-processing stages of model design and development. The ignorance of sensitive features during feature extraction and selection strategy also leads to algorithmic bias at the pre-processing stage. Also, improper data cleaning and data exploration introduced human bias in the system. During the in-processing step, the black box design of DL architecture with hidden neurons may not be able to interpret the model predictions. The classifier design also influences the model accuracy for diagnosis decision for unbiased and fair outcomes at this step. Bias accountability due to exploitation of less efficient evaluation metrics leads to bias at post-processing step. Insufficient evaluation metrics are utilized for computing the performance of the AI predicting models may leads to its failure after clinical deployment.

There are many strategies suggested to mitigate the impact of AI-bias in prediction models. Strong documentation and auditing of the model design and architecture should be performed to assess the model quality in terms of generalizability and interoperability [52]. Bias assessment tools such as Aequitas, PROBAST and many more must be utilized to compute the bias accountability in the predicted outcomes. Apart from these tools, predicted model quality must be validated by reviewing its technical, clinical and regulatory assessment. Technical validation comprises of evaluating the model quality in terms of its statistical analysis under practical guidance. Clinical validation involves gathering sufficient clinical evidence to ensure the model implementation, design and outcomes follows the designed guidelines. Regulatory assessment performs systematic validation with respect to available guidelines and regulations mandatory for automatic functioning of the model in realistic scenarios.

1.4 Security and Privacy Concerns in Modern Healthcare

The utilization of clinical data in modern healthcare must be checked before its usage for concerns such as ethical, legal, security and privacy in terms of patient, clinical staff and hospital. Ethical concerns in modern healthcare are related to sharing patient's personal details with third parties for analysis without their consent. Ethical issues must be avoided by taking prior consent from the patient [53]. On the other hand, legal issues can be reported when medical data is used across different countries. Data gathered from one country can be analyzed or processed in a different country. There are country specific laws which govern the data collection and usage which must be checked to prevent security and privacy issues [54]. In order to ensure the generalizability and interoperability of the AI-based healthcare prediction model, it is essential to train the model on diverse datasets obtained from different countries. This model trained on diverse datasets has less impact of AI-bias in their outcomes. Hence, it is important to determine the solutions to such usage and modification of dataset cross-countries by avoiding pertaining legal issues [55].

In order to ensure data privacy in healthcare, many rules and regulations are formulated to prevent misuse of patient's data [55]. General data protection regulation (GDPR), and HIPAA are formulated to protect patients' sensitive health information as well as its ethical public usage. These regulations make it mandatory for the written disclosure statement from the patient before sharing their personal information for further analysis. Apart from these regulations, there are AI-based techniques which also ensure to protect the data from unauthorized access. Federated learning, blockchain, and cryptographic techniques are a few techniques which ensure data security in medical domain.

Since a lot of medical data is generated by various medical imaging techniques such as CT, MRI, and USd, the privacy concern regarding big data usage has arisen [56]. The commercial use of this big data in AI-based prediction model has increased the risk of security and privacy breaches. Also, cloud-based E-health systems involve online transfer of critical data digitally such as patient's medical records, radiological reports, billing data, and medical history. High level security schemes must be deployed in cloud to protect them from stealing and its inappropriate usage. Cloud based E-health systems are faster, efficient, robust and effective with minimal human intervention. However, to ensure the trust of user high level security system must be deployed for managing the patient's privacy. Two level security authentication, OTP (One time password) based authentication, attribute based encryption, identity based encryption and many others encryption and authentication techniques were recommended to protect patient's sensitive information from unauthorized access during online data transfer [57].

Generative AI based models such as ChatGPT have offered a wide range of facilities with improved decision-making. To ensure the real-time applicability and acceptability of these facilities, it is essential to address the concerns related to the patient's data confidentiality, availability and privacy [58]. Privacy preserving techniques must be deployed to prevent data poisoning during training and phases.

Sufficient measures must be taken to design models resistant to adversarial attacks. Advanced techniques such as blockchain, and fuzz testing not only to ensure high level data security but also to prevent erroneous and biased medical outcomes.

1.5 Summary

In this chapter, we have summarized the types, features, benefits and limitations of AI-based models either as ML-based approaches or DL-based approaches. The evolution of traditional healthcare to modern healthcare is highlighted to demonstrate the advantages of modern healthcare in today's healthcare. Modern healthcare has transformed the conventional healthcare infrastructure with the help of the latest technologies. It has evolved not only to make patient's self-reliance but also to extend personalized care and treatment to the patients.

The benefits of AI-based prediction models for predicting chronic diseases such as brain tumor, cancer and CVD at an early stage have been discussed. These models are superior in accuracy and efficiency which process the medical imaging quickly. These techniques have reduced the manual intervention of radiologists and process the numerous imaging data in a faster way to provide accurate medical outcomes. Generative AI models-based healthcare models are also efficient in providing a voice-based system for understanding the patient's EHR quickly. These systems can process the patient's multiple medical records to provide a clear understanding of the patient's medical history for the clinical staff.

Since modern e-healthcare systems involve online data collection and transfer, the security and privacy of patient's critical information is foremost concerns. The various recent technologies have been highlighted to provide high-level privacy and security to patient's data. These encryption techniques ensure trust, fairness and transparency to the medical outcomes suitable for their realistic deployment.

References

1. Kumar A, Nayyar A, Sachan RK, Jain R (2023) AI-assisted special education for students with exceptional needs. IGI Global
2. Kumar A, Mangotra A, Ailawadi A, Jain R,. Arora R (2023) Sentiment analysis on multilingual data: Hinglish. In: International conference on data analytics & management. Springer
3. Kumar A (2023) Visual object tracking using deep learning. CRC Press
4. Kumar A, Mamgai R, Jain R (2023) Application of IoT-Enabled CNN for natural language processing. In: IoT-enabled convolutional neural networks: techniques and applications. River Publishers, pp 149–177
5. Kumar A, Vohra R (2023) Impact of deep learning models for technology sustainability in tourism using big data analytics. Deep learning technologies for the sustainable development goals: Issues and solutions in the post-COVID era. Springer, pp 83–96
6. Kumar A, Vohra R, Jain R, Li M, Gan C, Jain DK (2024) Correlation filter based single object tracking: a review. Inf Fusion 102562

7. Kumar A, Walia GS, Sharma K (2020) Real-time multi-cue object tracking: benchmark. In: Proceedings of international conference on IoT inclusive life (ICIIL 2019), NITTTR Chandigarh, India. Springer
8. Bhalla S, Kumar A, Kushwaha R (2023) Comparative analysis on brain stroke prediction using machine learning. In: 2023 14th international conference on computing communication and networking technologies (ICCCNT). IEEE
9. Kumari V, Kumar N, Kumar S, Kumar KA, Skandha SS, Saxena S, Fouda MM (2023) Deep learning paradigm and its bias for coronary artery wall segmentation in intravascular ultrasound scans: a closer look. J Cardiovasc Dev Dis 10(12):485
10. Kumar A, Rai RS, Gheisari M (2021) Prediction of cervical cancer using machine learning. In: Cancer prediction for industrial IoT 4.0. Chapman and Hall/CRC, pp 107–117
11. Kumar A, Ahluwalia R (2021) Breast cancer detection using machine learning and its classification. In: Cancer prediction for industrial IoT 4.0. Chapman and Hall/CRC, pp 65–78
12. Kumar A, Jain R (2021) Behavioral prediction of cancer using machine learning. In: Cancer prediction for industrial IoT 4.0. Chapman and Hall/CRC, pp 91–105
13. Kumar A, Singh D (Eds.) (2025) Revolutionizing Medical Systems Using Artificial Intelligence: A Breakthrough in Healthcare. Elsevier
14. Kumar A, Sareen P, Arora A (2023) Healthcare engineering using AI and distributed technologies. Smart distributed embedded systems for healthcare applications. CRC Press, pp 1–14
15. Rawas S, Samala AD (2024) Revolutionizing brain tumor analysis: a fusion of ChatGPT and multi-modal CNN for unprecedented precision. Int J Online & Biomed Eng. 20(8)
16. Singh D, Srivastava R (2022) Graph neural network with RNNs based trajectory prediction of dynamic agents for autonomous vehicle. Appl Intell 52(11):12801–12816
17. Singh D, Srivastava R (2022) Channel spatial attention based single-shot object detector for autonomous vehicles. Multimedia Tools Appl 81(16):22289–22305
18. Asiri AA, Soomro TA, Shah AA, Pogrebna G, Irfan M, Alqahtani S (2024) Optimized brain tumor detection: a dual-module approach for MRI image enhancement and tumor classification. IEEE Access 12:42868–42887
19. Botlagunta M, Botlagunta MD, Myneni MB, Lakshmi D, Nayyar A, Gullapalli JS, Shah MA (2023) Classification and diagnostic prediction of breast cancer metastasis on clinical data using machine learning algorithms. Sci Rep 13(1):485
20. Deepa B, Murugappan M, Sumithra M, Mahmud M, Al-Rakhami MS (2021) Pattern descriptors orientation and map firefly algorithm based brain pathology classification using hybridized machine learning algorithm. IEEE Access 10:3848–3863
21. Zhang X, Xie W, Huang C, Zhang Y, Chen X, Tian Q, Wang Y (2023) Self-supervised tumor segmentation with sim2real adaptation. IEEE J Biomed Health Inform 27(9):4373–4384
22. Alhassan AM, Zainon WMNW (2020) BAT algorithm with fuzzy C-ordered means (BAFCOM) clustering segmentation and enhanced capsule networks (ECN) for brain cancer MRI images classification. IEEE Access 8:201741–201751
23. Ramachandran M, Patan R, Kumar A, Hosseini S, Gandomi AH (2021) Mutual informative MapReduce and minimum quadrangle classification for brain tumor big data. IEEE Trans Eng Manage 70(8):2644–2655
24. Aamir M, Namoun A, Munir S, Aljohani N, Alanazi MH, Alsahafi Y, Alotibi F (2024) Brain tumor detection and classification using an optimized convolutional neural network. Diagnostics 14(16):1714
25. Gupta BB, Gaurav A, Arya V (2024) Deep CNN based brain tumor detection in intelligent systems. Int J Intell Netw 5:30–37
26. Khushi HMT, Masood T, Jaffar A, Rashid M, Akram S (2023) Improved multiclass brain tumor detection via customized pretrained EfficientNetB7 model. IEEE Access
27. Mustafa E, Jadoon EK, Khaliq-uz-Zaman S, Humayun MA, Maray M (2023) An ensembled framework for human breast cancer survivability prediction using deep learning. Diagnostics 13(10):1688

28. Saleh H, Abd-El Ghany SF, Alyami H, Alosaimi W (2022) Predicting breast cancer based on optimized deep learning approach. Comput Intell Neurosci 2022(1):1820777
29. Agarwal R, Pande SD, Mohanty SN, Panda SK (2023) A novel hybrid system of detecting brain tumors in MRI. IEEE Access 11:118372–118385
30. Rasool A, Bunterngchit C, Tiejian L, Islam MR, Qu Q, Jiang Q (2022) Improved machine learning-based predictive models for breast cancer diagnosis. Int J Environ Res Public Health 19(6):3211
31. Jun W, Liyuan Z (2022) Brain tumor classification based on attention guided deep learning model. Int J Computat Intell Syst 15(1):35
32. Li G, Hui X, Li W, Luo Y (2023) Multitask learning with multiscale residual attention for brain tumor segmentation and classification. Mach Intell Res 20(6):897–908
33. Masoudi B (2024) An optimized dual attention-based network for brain tumor classification. Int J Syst Assur Eng Manage 1–12
34. Hossain S, Chakrabarty A, Gadekallu TR, Alazab M, Piran MJ (2023) Vision transformers, ensemble model, and transfer learning leveraging explainable AI for brain tumor detection and classification. IEEE J Biomed Health Inform 28(3):1261–1272
35. Meshram PA, Joshi S, Mahajan D (2023) Brain tumor detection using Swin transformers. arXiv preprint arXiv:2305.06025
36. Kanzawa J, Yasaka K, Fujita N, Fujiwara S, Abe O (2024) Automated classification of brain MRI reports using fine-tuned large language models. Neuroradiology 1–7
37. Tejani AS (2023) To BERT or not to BERT: advancing non-invasive prediction of tumor biomarkers using transformer-based natural language processing (NLP). Eur Radiol 33(11):8014–8016
38. Singh D, Singh D, Manju, Gupta U (2023) Smart healthcare: a breakthrough in the growth of technologies. In: Artificial intelligence-based healthcare systems. Springer, pp 73–85
39. Tripathy SK, Singh D, Jaiswal A (2023) Multi-Layer feature fusion-based deep multi-layer depth separable convolution neural network for Alzheimer's disease detection. In: 2023 International conference on IoT, communication and automation technology (ICICAT). IEEE
40. Saba Raoof S, Durai MS (2022) A comprehensive review on smart health care: applications, paradigms, and challenges with case studies. Contrast Media & Molecular Imaging 2022(1):4822235
41. Singh M, Kumar A, Khanna NN, Laird JR, Nicolaides A, Faa G, Teji JS, et al (2024) Artificial intelligence for cardiovascular disease risk assessment in personalised framework: a scoping review. EClinicalMedicine 73
42. Carrillo de Albornoz S, Sia K-L, Harris A (2022) The effectiveness of teleconsultations in primary care: systematic review. Family Practice 39(1):168–182
43. Tian S, Yang W, Le Grange JM, Wang P, Huang W, Ye Z (2019) Smart healthcare: making medical care more intelligent. Global Health J 3(3):62–65
44. Arora P, Singh P, Girdhar A, Vijayvergiya R, Chaudhary P (2023) CADNet: an advanced architecture for automatic detection of coronary artery calcification and shadow border in intravascular ultrasound (IVUS) images. Phys Eng Sci Med 46(2):773–786
45. Behera TK, Khan MA, Bakshi S (2022) Brain MR image classification using superpixel-based deep transfer learning. IEEE J Biomed Health Inform 28(3):1218–1227
46. Jabbar A, Naseem S, Mahmood T, Saba T, Alamri FS, Rehman A (2023) Brain tumor detection and multi-grade segmentation through hybrid caps-VGGNet model. IEEE Access 11:72518–72536
47. Shinohara H, Kodera S, Ninomiya K, Nakamoto M, Katsushika S, Saito A, Higashikuni Y, et al (2021) Automatic detection of vessel structure by deep learning using intravascular ultrasound images of the coronary arteries. Plos one 16(8):e0255577
48. Stoumpos AI, Kitsios F, Talias MA (2023) Digital transformation in healthcare: technology acceptance and its applications. Int J Environ Res Public Health 20(4):3407
49. Kang HS, Exworthy M (2022) Wearing the future—wearables to empower users to take greater responsibility for their health and care: scoping review. JMIR Mhealth Uhealth 10(7):e35684

50. Canali S, Schiaffonati V, Aliverti A (2022) Challenges and recommendations for wearable devices in digital health: data quality, interoperability, health equity, fairness. PLOS Digital Health 1(10):e0000104
51. Kumar A, Aelgani V, Vohra R, Gupta SK, Bhagawati M, Paul S, Laird JR (2024) Artificial intelligence bias in medical system designs: a systematic review. Multimedia Tools Appl 83(6):18005–18057
52. Oala L, Murchison AG, Balachandran P, Choudhary S, Fehr J, Leite AW, Nakasi R, et al (2021) Machine learning for health: algorithm auditing & quality control. J Med Syst 45:1–8
53. McGregor C, Dewey C, Luan R (2021) Big data and artificial intelligence in healthcare: ethical and social implications of neonatology. In: 2021 IEEE International Symposium on technology and society (ISTAS). IEEE
54. De Hert P, Czerniawski M (2016) Expanding the European data protection scope beyond territory: article 3 of the general data protection regulation in its wider context. Int Data Privacy Law 6(3):230–243
55. Yadav N, Pandey S, Gupta A, Dudani P, Gupta S, Rangarajan K (2023) Data privacy in healthcare: in the era of artificial intelligence. Indian Dermatol Online J 14(6):788–792
56. Murdoch B (2021) Privacy and artificial intelligence: challenges for protecting health information in a new era. BMC Med Ethics 22:1–5
57. Sivan R, Zukarnain ZA (2021) Security and privacy in cloud-based e-health system. Symmetry 13(5):742
58. Chen Y, Esmaeilzadeh P (2024) Generative AI in medical practice: in-depth exploration of privacy and security challenges. J Med Internet Res 26:e53008

Chapter 2
Research Orientation for AI Techniques in Modern Healthcare System

Abstract Advancement in medical science and health care technology has won increased expected lifetime for people in the twenty-first century. To meet the expectation of health care quality from the public, artificial intelligence (AI) techniques such as rule-based expert systems, fuzzy expert systems, artificial neural networks, genetic algorithms, and hybrid intelligent systems are widely used in medical science and health care services. The main objective is to promote AI applications research to address both theoretical and practical aspects of intelligent medical information, knowledge, and their management. These objectives have to be addressed by closely examining the synergy and the complementary nature of both theoretical and practical solutions of knowledge management and system development of health care intelligent systems. Consequently, current fundamental research results have to be further developed before offering operational procedures for practical robust AI applications in the work process for decision making and knowledge management in the framework of health care level. The medical field has made great strides in medical image processing thanks to recent advancements in deep neural networks (DNNs) and other AI technologies that have found widespread usage in healthcare. A lot of current research is focused on developing automated systems that can evaluate photos and detect acute ailments, including brain tumors, breast cancer, bone fractures, and a host of others. This would greatly benefit medical practitioners. This extensive study summarizes the most current developments in medical imaging that have made use of DNNs. Along with the extensive literature evaluation, there is an overview of publicly available data sources and ideas for future study.

Keywords Artificial intelligence (AI) · Machine learning (ML) · Deep learning (DL) · Modern healthcare · Smart healthcare · AI in medicine · AI-based diagnosis · Medical imaging · Clinical decision support systems

A. Kumar and D. Singh, *Artificial Intelligence in Modern Healthcare System*,
Transactions on Computer Systems and Networks,
https://doi.org/10.1007/978-981-96-6703-1_2

2.1 Introduction

In recent years, there have been significant advancements in the field of artificial intelligence (AI). Machine learning (ML) is a subfield of AI that has achieved practical use in real-world scenarios [1, 2]. Here, some significant advancements revolve around neural networks (NN). The evolution of artificial neural networks (ANNs) may be described as progressing in a sinusoidal manner. Following an initial fascination in the late 1950s and early 1960s, there was a period of inactivity until 1986, when James McClelland and David Rumelhart released their renowned book. This rekindled enthusiasm for neural network research. Nevertheless, toward the end of the twentieth century, the enthusiasm for neural networks waned again. A contributing factor to the lack of progress was the insufficiency of computer hardware with the capacity to process the extensive data required to implement neural network models [3] effectively. Only over the past decade has there been a resurgence of interest in neural networks, leading to the creation of effective applications that can tackle real-world challenges. Several neural network topologies have garnered significant attention in the field of DNNs. These techniques have been applied in several fields, including medical picture classification, electromyography recognition, illness recognition and segmentation. Nevertheless, our focus in this study is to present a comprehensive analysis of the application of DNNs in the field of medical imaging.

DNNs have greatly enhanced the process of diagnosing, arranging therapy, and providing care to patients by completely transforming several aspects of medical image processing. Their ability to identify significant features and patterns from medical images utilizing large-scale datasets has shown to be highly successful, leading to more accurate and efficient analysis [4]. DNNs demonstrate exceptional efficacy in tasks like as image classification and segmentation, which are crucial in the field of medical imaging. DNNs have the ability to acquire the knowledge of identifying and classifying various anatomical characteristics, abnormalities, or lesions in medical images through extensive training on large datasets that have been labeled and annotated. In addition, they have the ability to accurately partition organs or other regions of interest, enabling precise measurements and quantitative analysis. DNNs have been extensively integrated into computer-aided diagnostic (CAD) systems [5]. By using large amounts of labeled data, these networks may learn to detect subtle patterns or abnormalities in medical pictures that may be challenging for human observers to see. DNN-based computer-aided diagnosis (CAD) systems provide radiologists and clinicians with more information and improve the accuracy of diagnoses. Deep learning (DL) algorithms have proven to be highly successful in the application of picture repair and enhancement tasks within the medical industry [6]. For example, DNNs have the ability to generate high-quality images from defective or noisy data in computed tomography (CT) and magnetic resonance imaging (MRI). This process reduces artifacts and improves the overall picture quality [7].

The key contributions of this chapter are as follows:

- We have categorized the pattern recognition task, image modality and abdominal region by using deep-learning-based medical imaging. The salient features of each category are highlighted and elaborated to determine its benefits and limitations.
- We have highlighted the salient features of high-risk surgery and modern medical equipment assisted by AI. We have also mentioned the benefits and challenges.
- A summary of the paradigm shift from traditional to smart healthcare, AI biased medical systems and analysis

These advancements lead to faster scans, reduced radiation doses, and improved visibility of anatomical structures. DNNs have been utilized to achieve precise and resilient image registration, a process that includes matching numerous medical pictures obtained from various modalities or time points. Through the acquisition of spatial changes, DNNs are able to automatically align pictures and enable comparisons, such as monitoring the advancement of diseases or strategizing solutions. Disease detection and prediction: DNNs have demonstrated potential in automating the identification and prediction of diseases through the analysis of medical imagery. Through the utilization of extensive datasets, these networks are capable of discerning precise imaging biomarkers that are linked to various illnesses [8]. For instance, in the field of cancer imaging, DNNs have the capability to detect tumor properties, forecast tumor malignancy, and evaluate the effectiveness of treatment by analyzing radiological pictures. DNNs have the capability to produce synthetic medical pictures, which may be used to enhance data augmentation, expand the training set, and create authentic simulations for training and testing. The amalgamation of photos is especially advantageous in scenarios with a scarcity of labeled data or uncommon circumstances, whereby DNNs can provide a wide range of instances to augment the efficacy of models. The DL models can utilize medical imaging along with other clinical data, such as genetics or electronic health records, to facilitate the practice of tailored care. DNNs can aid in treatment planning, prognostication, and therapy selection by incorporating patient-specific data and analyzing image-derived characteristics and patterns [8].

The rest of the chapter is organized as follows. Section 2.2 analyzed the recent deep-learning applications using medical imaging like classification, segmentation, registration and detection for different diseases. In addition, AI assisting high risk surgery and AI and modern medical equipment, from traditional to smart healthcare of paradigm shift, AI-bias in medical system are highlighted in Sect. 2.3. In Sect. 2.4, Quality of Services for AI based Healthcare system is discussed. Applications of AI-based Healthcare system are mentioned in Sect. 2.5. Lastly, the concluding remarks and future directions, i.e., summary is sketched in Sect. 2.6 (Fig. 2.1).

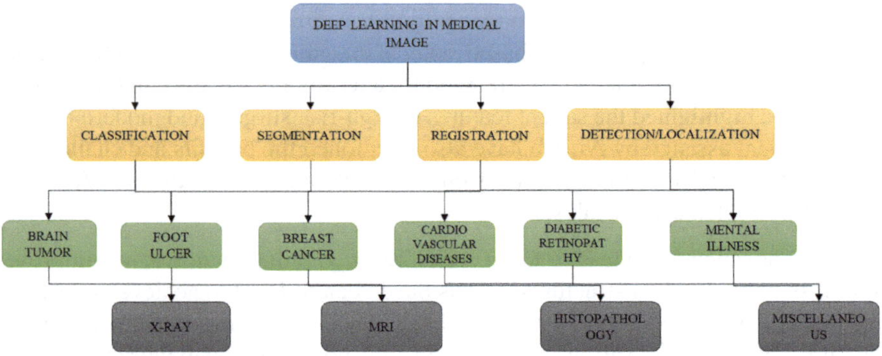

Fig. 2.1 Classifications and methods for disease detection using medical imaging

2.2 Classification of Diseases

The integration of machine learning (ML) and DL in the healthcare sector is a recent development that has not yet been thoroughly explored [7]. The medical healthcare business is a promising area of study, as current research trends indicate [9, 10]. In the following sections, we will explore some of the most significant recent literature on the methodologies, contributions, and applications of machine learning and DL in several disciplines within this industry. Table 2.1 presents a comprehensive analysis of different diseases that cover the use of DL and ML technologies in conjunction with smart healthcare integration. The methods and contributions of each organization depicted in this table have been thoroughly examined.

2.2.1 Mental Illness

People's susceptibility to mental diseases is heightened by sudden shifts in living standards, economic volatility, and excessive utilization of social media platforms. Mental diseases result in elevated levels of stress, which in turn give rise to significant neurological troubles in individuals, including depression, suicidal inclinations, and various mental issues [11, 12]. AI has facilitated the prediction, monitoring, and planning of mental health disorders in the population with its sophisticated capabilities. AI has developed prediction models capable of analyzing health information, brain imaging, and clinical notes to accurately detect mental diseases [13]. AI is widely used to analyze social media sites like Twitter and Facebook to diagnose depression in users. This is done by extracting crucial information from the tweets and comments they publish. This chapter has covered the prominent characteristics and restrictions of several AI-driven prediction models that are helpful in identifying mental disease problems at an early stage [14, 15].

Table 2.1 Similarities and differences between various AI-based disease detection

Disease	Mental illness	Brain tumor	Diabetic retinopathy	CVD	Breast cancer	Diabetic foot ulcer
ML models	SVM, Decision Tree, Naïve Bayes Classifier, Logistic Regression	RF, SVM, KNN	Ensemble classifier combining random forest, K-NN, LR, NB, gradient boosting, and AdaBoost	RF, decision tree classifier, multilayer perceptron, XGBoost, K-mode clustering and GridSearchCV	KNN, SGD, AdaBoost, SVM, RF	RF, SVM, SVD, PCA, Naïve Bayes Classifier, LDA, KNN
DL Models	RNN, MLP	CNN, RNN	CNN, Inception + ResNet, VGG16, VGG19, DenseNet	CNN, RNN		CNN (RCNN, Faster RCNN), RNN(LSTM, GRU)
Evaluation Metrices	AUC, ROC, Accuracy	DSC, IoU, ACC, Computation time, Hausdorff metrics and p-value	F1-score, Precision, Recall, Accuracy	AUC, Accuracy,	ACC, PR, RE, AUC-ROC, and F1-score	Accuracy, Precision, Recall, sensitivity, specificity, ROC Curve
Data Type	EEG, ECG	MRI, CT images	Retrospective data Coronary prediction Dataset	IVUS	Numerical Dataset, Textual EHR dataset, Serial mammography	CT scan, X-ray, MRI, and ultrasound

2.2.2 Brain Tumor

Medical imaging, or radiology, is the branch of medicine where healthcare experts produce pictures of different bodily areas for the goal of diagnosis or therapy. Medical imaging treatments encompass non-invasive diagnostic tests that enable physicians to identify injuries and diseases without causing intrusion or discomfort (TechTarget, n.d.). Several tools and strategies are employed to automate the interpretation of medical pictures obtained through different image-processing technologies [16]. The brain is a very intricate and sizable organ within the human body. Detecting anomalies from brain pictures, such as MRI, CT, PET scans, etc., is a significant field of study in medical image analysis. Brain image analysis utilizes a range of image processing techniques, including filtering, thresholding, geometry models, graph models, region-based analysis, connected component analysis, machine learning (ML) models, DL models, and hybrid models [17]. Brain tumors are a prevalent kind of brain illness that has a significant mortality rate. Analyzing brain pictures to identify tumors is challenging due to the diverse nature of their shape, location, size, texture, and other features [18]. This paper provides a thorough examination of brain tumor image analysis, covering the fundamental concepts of brain tumors, brain imaging, tasks involved in brain image analysis, models used for brain image analysis, features of brain tumor images, performance metrics for evaluating the models, and available datasets for brain tumor and medical images. The text discusses several issues associated with brain tumor analysis, as well as providing ideas for future research approaches [19].

2.2.3 Diabetic Retinopathy Using Retinal Imaging

Diabetic retinopathy (DR) is a serious eye disorder and a leading cause of permanent blindness globally. The condition is a result of injury to the blood vessels in the retina. The symptoms of Diabetic Retinopathy (DR) include the presence of black strings or spots that appear to float in the individual's field of vision, the occurrence of empty regions within their visual field, a decline in their ability to perceive colors accurately, and the experience of hazy and inconsistent vision [20]. In severe instances, the individual experiences permanent visual impairment. Historically, the process of DR screening involved manually examining fundus photographs. Nevertheless, this procedure, apart from being laborious, also requires meticulousness during large-scale screening to prevent any diagnostic errors. These constraints can be surmounted by an automated computer-aided diagnostic system (CAD) for DR. A DR-CAD system refers to the automated analysis of fundus pictures for the purpose of classifying diabetic retinopathy and identifying related retinal diseases. A DR-CAD system can aid medical professionals in accurately interpreting medical pictures [21]. Furthermore, it can also aid in the identification and highlighting of prominent structures in the retina, which can then be utilized for more accurate examination of their severity. This chapter showcases a carefully chosen assortment of machine learning and DL models that are used for detecting diabetic retinopathy. The study encompasses models for binary and multistage diabetic retinopathy (DR) categorization, as well as the identification and delineation of four primary lesions—namely, microaneurysms, hemorrhages, cotton wool spots, and hard exudates. DR-CAD systems enable the automated identification of DR in its first phase, facilitating the management of the gradual deterioration of the retina [22].

2.2.4 CVD (Cardio Vascular Diseases) Risk

Cardiovascular diseases (CVD) are the primary cause of mortality worldwide and are seeing a concerning upward trend, as reported by the American Heart Association's Heart Attack and Stroke Statistics 2021. This surge has been intensified due to the ongoing coronavirus (COVID-19) epidemic, consequently augmenting the strain on existing healthcare services. Smart and Connected Health (SCH) offers a practical and effective answer for the current healthcare difficulties [23]. It has the ability to transform the direction of healthcare to become more strategic, preventative, and tailored, hence enhancing its effectiveness with additional services that provide value. This research aims to categorize the most advanced SCH (Smart City Hub) technologies through a detailed examination of existing literature and analysis. The goal is to provide a comprehensive definition of SCH characteristics and highlight the technological difficulties that need to be addressed for widespread adoption of SCH. Additionally, we present an architectural model that encompasses the technology element of the SCH solution, its context, and the key players involved [24]. It functions as a benchmark for the adoption and implementation of SCH. We analyzed a case study on COVID-19, which demonstrated how several nations have approached the pandemic by utilizing diverse technologies in the field of public health, such as big data, cloud computing, Internet of Things, AI, robots, blockchain, and mobile apps. SCH has been effectively utilized at several phases, including illness diagnosis, viral identification, individual monitoring, tracking, managing, and resource allocation, in the fight against the pandemic. Moreover, this analysis emphasizes the obstacles to the acceptability of SCH (Smart Connected Health) and suggests prospective research avenues to improve patient-centric healthcare [25, 26].

2.2.5 Breast Cancer Prediction

Since 2020, breast cancer has attained the highest global incidence rate among all types of malignancies. Early detection and intervention by breast imaging greatly contribute to improving the prognosis of breast cancer patients. Over the last ten years, DL has made significant advancements in the analysis of breast cancer imaging.

It has the potential to effectively comprehend the abundant information and intricate context of several breast imaging techniques. Given the fast advancements in DL technology and the growing severity of breast cancer, it is crucial to evaluate previous achievements and pinpoint potential obstacles that need to be tackled [25]. This work presents a comprehensive analysis of breast cancer imaging research that utilizes DL techniques. It encompasses studies conducted in the previous 10 years, focusing on mammograms, ultrasound, MRI, and digital pathology pictures. This text elaborates and discusses the main techniques and uses of DL in imaging-based screening, diagnosis, treatment response prediction, and prognosis. Based on the results of this survey, we provide a thorough analysis of the difficulties and possible directions for further investigation in DL-based breast cancer imaging [23, 24].

2.2.6 Detection of Diabetic Foot Ulcer

Diabetes is a persistent medical disorder resulting from unregulated amounts of glucose in the human body. Early detection of this condition can help prevent serious consequences, such as the development of diabetic foot ulcers (DFUs). A Diabetic Foot Ulcer (DFU) is a severe medical ailment that has the potential to result in the surgical removal of a diabetic patient's lower extremity. The diagnosis of DFU poses significant challenges for medical professionals due to its complex nature, frequently requiring many expensive and time-consuming clinical tests. In the era of excessive data, the utilization of advanced techniques such as DL, machine learning, and computer vision has offered several ways to aid physicians in reaching more accurate and expedient diagnostic judgments [27]. Consequently, the scientific community has recently shown increased interest in automatically identifying DFU. The attributes of the wound and the way they are perceived visually in the context of computer vision and DL, namely convolutional neural network (CNN) methods, have shown promising options for diagnosing diabetic foot ulcers (DFU). These methodologies possess the capacity to be really beneficial in contemporary medical procedures. Hence, it was necessary to conduct a thorough and extensive examination of these current methods [28]. The publication sought to furnish scholars with an elaborate account of the present state of automated DFU detection tasks. Existing works have shown that the utilization of both classic machine learning (ML) and sophisticated DL approaches is essential in assisting doctors to produce quicker and more dependable diagnostic conclusions. Image features in standard machine learning (ML) methods play a crucial role in providing meaningful information regarding diabetic foot ulcer (DFU) wounds, aiding in their correct diagnosis. Nevertheless, sophisticated DL methods have demonstrated more potential compared to machine learning (ML) approaches. The issue domain has been predominantly dominated by CNN-based solutions put forth by several authors [29]. A diligent researcher will effectively discern the main concept in the DFU identification task, and this article will assist them in solidifying their future study objective [30].

2.2.7 *AI in Immunology*

The human immune system has a high level of intricacy. Traditionally, comprehending it necessitated specific knowledge and experience acquired through years of study. Recently, the implementation of technology like AIoMT (Artificial Intelligence of Medical Things), genetic intelligence algorithms, and smart immunological techniques has simplified this procedure. These technologies have the ability to observe and identify relationships and patterns that are also perceivable by people, as well as patterns that are not detectable by humans [31]. Moreover, these technologies have also facilitated our comprehension of the many cellular components inside the immune system, including their compositions, significance, and influence on human immune response, particularly in devastating conditions like cancer. This paper examines the current AI approaches used in the field of immunology. This study begins by elucidating the incorporation of AI in the healthcare sector and its transformative impact on the medical field. Additionally, it provides an overview of the present uses of AI in various healthcare sectors, as well as the primary obstacles encountered when attempting to incorporate AI into healthcare. It also highlights the recent advancements and contributions made by other researchers in this subject [32]. The primary objective of this study is to investigate the prevailing categorizations of health ailments, immunology, and its principal subfields. The latter portion of the paper provides a statistical analysis of the advancements made in AI within various areas of immunology. It also includes a comprehensive examination of the machine learning and DL techniques and algorithms that have been utilized in the field of immunology. In addition, we have examined a compilation of machine learning and DL datasets pertaining to several subdomains within the field of immunology. Ultimately, the paper concludes by examining the potential avenues for future research in the subject of AI in immunology and offering potential remedies for the identified issues [33].

2.3 AI Tools for Automated Medical Systems

The use of IoMT (Internet of Medical Things) and its associated technologies has successfully addressed several challenges in the fields of remote monitoring, telemedicine, robotics, and sensors. Nevertheless, achieving widespread acceptance presents difficulties stemming from considerations like as data privacy and security, the handling of vast quantities of data, scalability, and the need for upgrades [34, 35]. This organized systematic review will enhance the efficiency of healthcare practitioners, policymakers/decision-makers, scientists, and researchers in assessing the application of IoMT in healthcare, despite the already existing abundance of knowledge and information sharing.

2.3.1 Modern Medical Equipment

The abrupt outbreak of the Coronavirus illness (COVID-19) has placed the whole healthcare system in a state of heightened vigilance. The Internet of Medical Things (IoMT) has significantly alleviated the situation. Additionally, the COVID-19 pandemic has spurred scientists to develop a new "Smart" healthcare system that prioritizes early diagnosis, prevention of transmission, education, treatment, and adaptation to the new normal. This review seeks to determine the role of Internet of Medical Things (IoMT) applications in enhancing the healthcare system. It also aims to assess the current state of research that demonstrates the effectiveness of IoMT benefits for patients and the healthcare system. Additionally, it provides a brief overview of the technologies that support IoMT and the challenges encountered in developing a smart healthcare system [36].

Biomedical research progress produces a wide range of healthcare-related data, such as medical records and information on the maintenance of medical devices. The COVID-19 pandemic has a substantial impact on the worldwide death rate, leading to a tremendous need for medical technologies. With the advancement of information technology, the idea of intelligent healthcare has increasingly become more important [37, 38]. Smart healthcare employs advanced information technologies, including the Internet of Things (IoT), big data, cloud computing, and AI, to revolutionize the conventional medical system [39]. A predictive model is presented to forecast medical equipment failure in order to intelligently manage healthcare services and introduce the notion of smart healthcare. The Internet of Things (IoT) has had a significant influence on the progress of the healthcare business. The advent of Medicine 4.0 has led to a greater focus on the development of platforms, encompassing both hardware and software components. This concept has resulted in the creation of Healthcare Internet of Things (H-IoT) solutions [40]. The fundamental technologies that facilitate the functioning of a system include the communication networks that facilitate the exchange of information between the sensing nodes and the processors, as well as the processing algorithms that are responsible for creating an output based on the data acquired by the sensors [41]. Currently, these facilitating technologies are also backed by several emerging technologies. AI has revolutionized the H-IoT systems across several levels. The fog/edge concept involves putting processing capacity in close proximity to the deployed network, therefore addressing several issues in the process. Big data enables the management of vast quantities of data. In addition, Software Defined Networks (SDNs) provide system flexibility, while blockchains are being utilized for innovative purposes in H-IoT systems.

2.3.2 AI-Assisted High-Risk Surgery

Global surgery encompasses a fast-growing interdisciplinary subject that focuses on improving and ensuring fair access to surgical care in worldwide healthcare systems.

Global surgical programs typically prioritize enhancing capacity, advocating for, educating, researching, and developing policies in low- and middle-income countries (LMICs) [42]. The current lack of sufficient surgical, anesthetic, and obstetric treatment is responsible for causing 18 million fatalities annually that might have been prevented. Consequently, there is an increasing fascination in the fast expansion of AI which presents a unique chance to improve surgical services in LMICs. AI modalities have been utilized to customize surgical education, automate administrative procedures, and create practical and cost-efficient simulation training programs that cater to individuals with specific requirements [43]. In addition, AI may contribute to offering valuable information for governance, infrastructure development, and the monitoring and prediction of stock take or logistics failure, which can enhance the foundations of global surgery. AI-powered telemedicine platforms have enabled healthcare providers to remotely assist in intricate procedures, potentially enhancing surgical accessibility in LMICs. One of the challenges in integrating AI technology is the misrepresentation of minority groups in the datasets, which might result in discriminatory bias. Further research is needed to better understand human reluctance, employment insecurity, automation bias, and the impact of confounding factors in order to ensure fair and effective use of AI. By employing a concentrated and empirically-supported strategy, AI has the potential to assist several LMICs in overcoming administrative inefficiencies and enhancing the effectiveness of their surgical systems [44].

2.3.3 Traditional to Smart Healthcare

An overview of the fundamental aspects, such as the evolution of healthcare systems from traditional to smart and innovative healthcare, is important. Awareness of the obstacles posed by several healthcare paradigms to sophisticated health solutions helps to overcome them. Healthcare paradigms have experienced continual advancements, intertwined with sophistication added to each [42]. Proper judgment or assessment and service of modern healthcare composite solutions are indeed essential to entail the most efficient strategies to tackle health issues. It is important to understand the deluge of health problems people are likely to face in the coming years and the healthcare needs that will cater to people's health. Smart healthcare, driven by the latest technological advancements, is also increasing in demand. Technological revolution in recent years has changed the way healthcare has been traditionally viewed. Traditional healthcare generally refers to the medical care conducted by general practitioners, nurses, and others along similar lines [45]. Smart healthcare mainly leverages the use of advanced technology to help simplify diagnosis and bring services closer to patients. Even though there is an increase in smart and innovative healthcare, some traditional setups are also being used. The only problem with these traditional healthcare systems is that they are unable to cope with the rapid increase in sickness brought on by novel bacteria and focus on trying to bring medicine to the people instead of the people to medicine [46].

2.4 Short Notes on QoS in Smart Healthcare

An essential demand in the field of medical healthcare is the effective calculation of quality of service (QoS) during the processing of medical data, achieved via the use of intelligent measurement methods. Emergency medical services frequently need the transfer of vital data, which means that they have strict demands for network quality of service (QoS). This study makes three different contributions [34]. The proposed system, called Adaptive QoS Computation system (AQCA), aims to monitor performance indicators such as transmission power, duty cycle, and route selection during medical data processing in healthcare applications. The technique is designed to ensure fairness and efficiency in this monitoring process. Furthermore, a QoS computing framework for medical applications is provided, including the physical, medium access control (MAC), and network levels. Furthermore, a QoS computation method is constructed using the suggested AQCA, together with the consideration of quality of experience (QoE). In addition, the assessment of QoS computation for medical healthcare applications is conducted using user terminal (UT) devices with big screens ranging from 4 to 10 inches, such as LCD panels with certain sizes and resolutions. These devices prioritize good visualization, long battery life, and power optimization for ECG service in emergency situations [39]. These UT gadgets are utilized to attain the utmost level of pleasure in terms of reduced power consumption, prolonged battery life, and best route selection. The analysis focuses on determining the extent to which each QoS parameter influences the processing of medical data, taking into account the calculation of QoE perception. The experimental findings suggest that Quality of Service (QoS) is determined at the physical, MAC, and network levels using specific parameters such as transmission power (-15 dBm), latency (100 ms), jitter (40 ms), throughput (200 Bytes), duty cycle (10%), and route selection (optimal). Therefore, it can be concluded that the suggested AQCA is a more suitable option for QoS computation in medical healthcare applications compared to the Baseline [47].

2.5 Artificial Intelligence for Healthcare: Applications

Artificial intelligence (AI) technologies have, for several years, slowly but assuredly been penetrating several facets of our lives ranging from industries to entertainment. Similarly, the healthcare sector is not exempt from this advancement. In recent years, healthcare providers have started to adopt AI technologies to enhance the efficiency and outcomes of the services being provided. By looking at the current trajectory of AI deployment across healthcare systems, this transformation seems to continue at a higher pace in the forthcoming years. Likewise, the adoption of artificial intelligence in healthcare is being motivated by several key drivers, including improved patient care, enhanced operational efficiencies, and the reduced burden on healthcare professionals. Consequently, the subsequent sections seek to elucidate the discussion

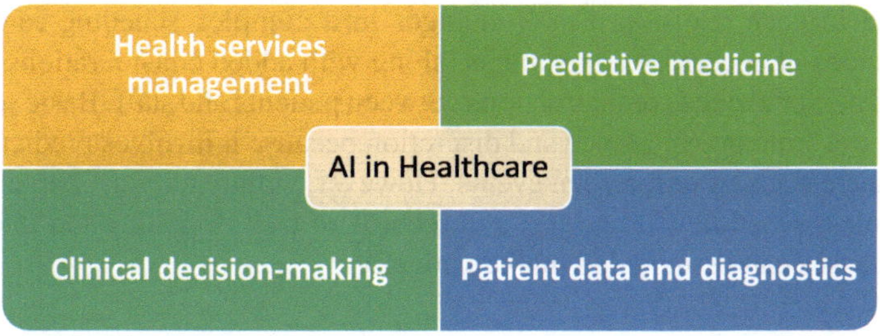

Fig. 2.2 Dominant variables for AI in healthcare

around the utilization of AI approaches in healthcare. The shown elements may be observed in Fig. 2.2.

2.5.1 Health Services Management

Artificial Intelligence (AI) is rapidly reshaping the management of health services within hospitals and healthcare facilities. Health services management encompasses the planning, organizing, coordination, and evaluation of health services to ensure efficient delivery of care. For many years, AI technologies have been implemented to streamline the management of health services, particularly administrative processes and tasks that consume substantial manual effort, and are error-prone due to human involvement. As a result, a variety of AI-driven solutions have emerged to automate administrative tasks, allowing healthcare professionals to devote their time to priorities that directly concern patient welfare. The implementation of AI technologies in health services management thus brings a dual benefit: operational improvement and enhanced patient care. AI applications enable hospitals and healthcare services to operate more efficiently due to the following reasons:

- Clinicians can instantly access data as and when required.
- Nurses can enhance patient safety while administering medication.
- Patients can remain informed and actively participate in their healthcare by communicating with their medical teams during hospital stays.

In addition, in healthcare, a large proportion of everyday tasks consists of routine actions that do not require extensive experience or deep knowledge. Rather, these tasks often involve simple choices based on easily identifiable grounds, such as checking a set of data against known parameters and taking a standardized action if they fall outside the accepted range. Other common routine tasks include scheduling actions based on temporal relationships, transferring data from one place to another, adding entries to records, and looking up data based on specific tags. AI technologies can automate many such tasks. For example, scheduling is a very widespread everyday task that can be quite complicated when multiple parameters must be considered in each decision. Small changes in a complex schedule can result in large cascades of necessary adjustments, along with added considerations related to health and personal needs or interactions between patients and staff. Basic scheduling often requires human experience and discretion because it involves predicting priorities and the effects of unforeseen events. However, with readily available data such as fixed rules, schedules, time logs, and records of past decisions and interactions, these processes can be modeled mathematically.

2.5.2 Predictive Medicine

Predictive medicine is a recently blossomed field seeking to boost patients' life quality and longevity through tailored preemptive actions instead of conventional therapies that move reactively post disease emergence. Artificial Intelligence (AI) enhances patients' results by predicting upcoming health issues and recommending intervention suggestions. These models can be used to refine the classic preventive healthcare strategy of periodic risk screenings into a more efficient On-Demand screening strategy, prompting the analysis of selected individuals flagged as risky by the model. Numerous successful predictive medicine implementations exist across specialties, illustrating the wide applicability of AI predictive algorithms. A notable achievement of predictive medicine is in Early Diagnosis systems, where models predict future health events and suggest timely intervention. The key to predictive medicine efficacy is the quantification of data-driven insights from retrospective data in the form of numerical indices that capture relevant public health aspects. Early-stage interventions are vital for systems regarding disease proliferation and irreversible damaging effects. Recently emerged healthcare systems focused on patients' pivotal feature: proactivity in disease management. Ailing prevention is based on periodic screening of the entire population, potentially overlooking many endangered individuals. Instead, proactive healthcare strategies aim to boost currently applied predictive screening techniques.

2.5.3 Clinical Decision-Making

Healthcare organizations are under constant pressure to enhance clinical decision-making processes and improve patient outcomes. Artificial intelligence (AI) technologies can assist healthcare professionals in making better-informed decisions. Clinical decision support systems based on AI algorithms can analyze patient data and suggest the most effective treatment options. The suggestions made by the decision support systems are typically in the form of evidence-based recommendations that take the form of a ranked list of possible interventions. Several studies explore the emerging use of AI systems in enhancing decision-making capabilities in clinical contexts where efficiency plays a critical role, such as disease diagnostic processes and epidemiological surveillance. Integrating AI within electronic health records (EHRs) could augment clinicians' capabilities, enabling them to spend less time collecting and analyzing data and more time providing care. However, concerns exist regarding the interpretability of AI recommendations. Resolution of these concerns will require research into the balance of power within collaborative decision-making between healthcare providers and AI systems. Responsibility for the correctness of a decision made with the involvement of an AI system should lie with the healthcare provider who interprets the system's output, thus reinforcing their decision-making role.

2.5.4 Patient Data and Diagnostics

Health worldwide has recently been significantly affected by Covid-19. Artificial Intelligence (AI) could improve patient care quality and hospital performance, making it highly attractive for healthcare recovery. An essential part of the clinical environment is patient data, which healthcare organizations collect, maintain, and utilize daily during clinical operations. Clinical records include every patient-related activity in the hospital or healthcare organization, right from admission to discharge. Healthcare facilities maintain diverse clinical records of patients such as demographic attributes, medical history, progress notes, vitals, lab reports, treatments, medications, medical images, and outcomes. These clinical records are beneficial for creating patients' profiles and performing diagnoses. With the rapid growth of digitization in healthcare, patient data management has become a major challenge, but at the same time, a key technology determinant for better diagnostics and improved healthcare outcomes.

The patient information files in hospitals or healthcare organizations are a rich source of structured and unstructured data related to each patient's clinical history. AI-operated systems can utilize this data to prepare a comprehensive profile for each patient and assist the physician in making better clinical decisions. Most hospitals maintain health information systems (HIS), which are either standalone or integrated systems to manage different clinical applications such as patient record management, pharmacy, lab, radiology, billing, and so on. These applications capture, store, and maintain a large amount of data related to patients, diseases, and clinical activities. The advent of huge clinical data repositories has the potential to improve the accuracy of disease diagnosis by revealing hidden patterns in the data, which is too complex for human analysis. Data mining techniques provide the means to discover knowledge from massive datasets and use such knowledge to automate decision-making. Predictive analytics and machine learning models can enhance the diagnostic capability of the healthcare system by discovering data-driven insights from historical records of patients, diseases, and treatments. Unfortunately, most healthcare organizations still rely on traditional data handling techniques and expert knowledge for disease diagnostics, leading to inaccurate diagnoses in many cases. Along with this, improper data recording and a lack of interoperability between different health information systems restrict the application of data mining in healthcare. Creating a comprehensive profile for each patient and mechanism implementation for protecting patient data privacy are important concerns for healthcare organizations planning to adopt artificial intelligence.

2.6 Summary

In medical imaging, there are several artificial intelligence (AI) models and techniques available to perform the task of segmentation, which refers to demarcating target structures, organs, and areas in medical images. We summarize the AI models for this domain in this section. These models can be divided into two groups based on the approaches used to develop them: classical computer vision and deep learning. In addition, deep learning models can be further divided on the basis of their respective architectures. Convolutional neural networks (CNNs) are widely applied in the tasks of object segmentation, detection, classification, and are considered the state-of-the-art architecture. We further summarize the AI models as follows:— Classical image processing-based models—Deep learning-based models—CNN-based models—Segmentation by integration of domain knowledge models—U-Net-based models—Attention-based models Each of these AI models has distinct advantages and disadvantages. Some are very powerful and can segment very specific structures like airway walls, lung nodules, lung lesions, and colon polyps, among others. Depending on the AI model selected and the nature of the dataset being explored, one or more of the following common processes need to be performed: pre-processing, model training, and model evaluation. In terms of AI model selection, knowledge about the dataset and the problem at hand is an essential factor that needs to be taken into account while developing an AI model. The performance of any developed AI model will essentially be a function of the dataset from which the model was trained. High-quality labeled image datasets are key to developing accurate and precise AI models. Therefore, it is almost impossible to develop domain-agnostic gold standard models. Domain knowledge should be involved while devising and training AI models that are to be deployed in a clinical or research setting.

References

1. Kumar A (2023) Visual object tracking using deep learning. CRC Press
2. Kumar A, et al (2023) Object tracking technology: trends, challenges and applications
3. Bowers JS et al (2023) Deep problems with neural network models of human vision. Behav Brain Sci 46:e385
4. Hua C et al (2023) Dynamic neural network models for time-varying problem solving: a survey on model structures. IEEE Access 11:65991–66008
5. Wahab Sait AR, Dutta AK (2023) Developing a deep-learning-based coronary artery disease detection technique using computer tomography images. Diagnostics 13(7):1312

6. Magboo VPC, Magboo MSA (2023) Diagnosis of coronary artery disease from myocardial perfusion imaging using convolutional neural networks. Procedia Comput Sci 218:810–817
7. Woźniak M, Siłka J, Wieczorek M (2023) Deep neural network correlation learning mechanism for CT brain tumor detection. Neural Comput Appl 35(20):14611–14626
8. Raffy P et al (2023) Deep learning body region classification of MRI and CT examinations. J Digit Imaging 36(4):1291–1301
9. Kumar A, Bhalla S (2025) Exploring brain tumor detection through artificial intelligence. In Revolutionizing Medical Systems using Artificial Intelligence, pp 93–120. Academic Press
10. Kumar A, Sareen P, Arora A (2023) Healthcare engineering using AI and distributed technologies. Smart distributed embedded systems for healthcare applications. CRC Press, pp 1–14
11. Tripathy SK, Singh D, Jaiswal A (2023) Multi-Layer feature fusion-based deep multi-layer depth separable convolution neural network for Alzheimer's disease detection. In: 2023 International conference on IoT, communication and automation technology (ICICAT)
12. Tripathy SK et al (2023) Multiscale low-level feature fused multilayer convolution neural network for Alzheimer's disease detection. IET Conf Proc 2023(5):466–470
13. Sachan RK, et al (2023) Diagnosis of mental health from social networking posts: an improved ml-based approach. In: International conference on emergent converging technologies and biomedical systems. Springer
14. Kumar A, Singh D (Eds.) (2025) Revolutionizing Medical Systems Using Artificial Intelligence: A Breakthrough in Healthcare. Elsevier
15. Kamra V, Kumar P, Mohammadian M (2023) An intelligent disease prediction system for psychological diseases by implementing hybrid hopfield recurrent neural network approach. Intell Syst Appl 18:200208
16. Anagun Y (2023) Smart brain tumor diagnosis system utilizing deep convolutional neural networks. Multimedia Tools Appl 82(28):44527–44553
17. Bhalla S, Kumar A, Kushwaha R (2023) Comparative analysis on brain stroke prediction using machine learning. In: 2023 14th International conference on computing communication and networking technologies (ICCCNT). IEEE
18. Aggarwal M et al (2023) An early detection and segmentation of Brain Tumor using deep neural network. BMC Med Inform Decis Mak 23(1):78
19. Gupta BB, Gaurav A, Arya V (2024) Deep CNN based brain tumor detection in intelligent systems. Int J Intell Netw 5:30–37
20. Gadekallu TR, et al (2023) Deep neural networks to predict diabetic retinopathy. J Ambient Intell Humanized Comput 1–14
21. Ali G et al (2023) A hybrid convolutional neural network model for automatic diabetic retinopathy classification from fundus images. IEEE J Trans Eng Health Med 11:341–350
22. Naik S, et al (2023) Efficient diabetic retinopathy detection using convolutional neural network and data augmentation. Soft Comput 1–12
23. Li Y et al (2023) A deep learning approach to cardiovascular disease classification using empirical mode decomposition for ECG feature extraction. Biomed Sig Process Control 79:104188
24. Zhang D et al (2023) A physics-guided deep learning approach for functional assessment of cardiovascular disease in IoT-based smart health. IEEE Internet Things J 10(21):18505–18516
25. Ahmad S et al (2023) Diagnosis of cardiovascular disease using deep learning technique. Soft Comput 27(13):8971–8990
26. Kavatlawar A, et al (2024) CardioVascular disease (CVD) recognition using convolutional neural networks. Grenze Int J Eng Technol (GIJET) 10
27. Thotad PN, Bharamagoudar GR, Anami BS (2023) Diabetic foot ulcer detection using deep learning approaches. Sens Int 4:100210
28. Dhar MK et al (2024) FUSegNet: a deep convolutional neural network for foot ulcer segmentation. Biomed Signal Process Control 92:106057
29. Biswas S et al (2023) DFU_MultiNet: a deep neural network approach for detecting diabetic foot ulcers through multi-scale feature fusion using the DFU dataset. Intell-Based Med 8:100128

30. Reyes-Luévano J et al (2023) DFU_VIRNet: a novel Visible-InfraRed CNN to improve diabetic foot ulcer classification and early detection of ulcer risk zones. Biomed Signal Process Control 86:105341
31. MacMath D, Chen M, Khoury P (2023) Artificial intelligence: exploring the future of innovation in allergy immunology. Curr Allergy Asthma Rep 23(6):351–362
32. Kalyakulina A et al (2023) Small immunological clocks identified by deep learning and gradient boosting. Front Immunol 14:1177611
33. Li L-S et al (2023) From immunology to artificial intelligence: revolutionizing latent tuberculosis infection diagnosis with machine learning. Mil Med Res 10(1):58
34. Pradhan B, et al (2023) An AI-assisted smart healthcare system using 5G communication. IEEE Access
35. Kothamali PR et al (2023) Smart healthcare: enhancing remote patient monitoring with AI and IoT. Revista de Inteligencia Artificial en Medicina 14(1):113–146
36. Jayachitra S et al (2023) AI enabled internet of medical things in smart healthcare. AI models for blockchain-based intelligent networks in IoT systems: concepts, methodologies, tools, and applications. Springer, pp 141–161
37. Kumar A et al (2023) Security and privacy issues in smart healthcare using machine-learning perspectives. 6G-Enabled IoT and AI for smart healthcare. CRC Press, pp 41–56
38. Kumar A, et al (2023) AI-assisted special education for students with exceptional needs. IGI Global
39. Suleiman TA, Adinoyi A (2023) Telemedicine and smart healthcare—the role of artificial intelligence, 5G, cloud services, and other enabling technologies. Int J Commun Netw Syst Sci 16(3):31–51
40. Ramakrishnan S, Jijitha S, Amudha T (2024) Roadmap of AI and IoMT in smart healthcare: current applications and future perspectives. Internet Med Things Smart Healthcare 137–161
41. Awotunde JB et al (2023) Internet of medical things for enhanced smart healthcare systems. Implementation of smart healthcare systems using AI, IoT, and blockchain. Elsevier, pp 1–28
42. Guni A et al (2024) Artificial intelligence in surgery: the future is now. Eur Surg Res 65(1):22–39
43. Spinelli A et al (2023) Artificial intelligence in colorectal surgery: an AI-powered systematic review. Tech Coloproctol 27(8):615–629
44. Shao S et al (2023) Artificial intelligence assists surgeons' decision-making of temporary ileostomy in patients with rectal cancer who have received anterior resection. Eur J Surg Oncol 49(2):433–439
45. Ardeti VA et al (2023) An overview on state-of-the-art electrocardiogram signal processing methods: traditional to AI-based approaches. Expert Syst Appl 217:119561
46. Renugadevi N, Saravanan S, Sudha CN (2023) Revolution of smart healthcare materials in big data analytics. Mater Today: Proc 81:834–841
47. Singh PD et al (2023) BOSS: a new QoS aware blockchain assisted framework for secure and smart healthcare as a service. Expert Syst 40(4):e12838

Part II
Applications of Artificial Intelligence for Disease Prediction

Chapter 3
Diagnosis and Prediction of Brain Tumor Using Artificial Intelligence

Abstract Brain tumor is a disorder which occurs due to unconditional and uncontrolled growth of the brain cells. Sometimes, the growth of these cells is malignant and leads to brain cancer. AI with conventional and advanced algorithms analyze various imaging data such as CT scans and MRI scans of the brain to identify the growth of cancerous cells in brain at an early stage. Early diagnosis of brain tumor is helpful in extending the accurate treatment and preventing the mortality to a great extent. In this chapter, we have exhaustively analyzed and reviewed the various conventional and advanced techniques which can detect brain tumor cells at an early stage. Also, the performance of these algorithms is explored to identify the limitations and suggest future solutions.

Keywords Brain tumor · Machine learning (ML) · Deep learning (DL) · Transfer learning (TL) · Brain imaging · Generalizability

3.1 Introduction

Brain is a part of central nervous system which controls the movement and actions of other body organs [1, 2]. Brain is an incredible organ of the body which sends instructions to other organs for processing of the decision taken by them. Brain has a very complex structure and require specialized skills to understand its disorders and abnormalities [3, 4].

Artificial Intelligence (AI) has provided many applications in the field of tracking [5, 6], tourism [7], education [8, 9] and many more [10, 11]. In the field of medical image analysis, AI has provided many machine learning (ML) [12, 13] and deep learning (DL) [14, 15] algorithms for early diagnosis and prognosis of critical diseases. AI has provided many solutions to segment the tumorous cells from the healthy cells in the brain imaging. AI-based algorithms diagnose the severity of brain cancer and saves lives.

Mainly, brain imaging data acquire using either CT (Computer tomography) [16] or MRI (Magnetic resonance imaging) [2, 17] scans to process them for determining

A. Kumar and D. Singh, *Artificial Intelligence in Modern Healthcare System*,
Transactions on Computer Systems and Networks,
https://doi.org/10.1007/978-981-96-6703-1_3

brain abnormalities. Both of them are non-invasive techniques and quite popular among neurologists to perform initial stage of test for predicting the existence of brain tumor. Numerous algorithms have been proposed to process these imaging data for determining the type, stage and location of tumor in brain [4, 18]. The captured images are poor in contrast and resolution. Hence, proper preprocessing steps must be taken to improve contrast and suppress noise for accurate prediction [19]. After this, features are extracted, and feature selection is performed to determine the crucial and sensitive features. This step is essential for determining the better computational efficiency of the algorithm. Next, AI-based algorithms were exploited to provide discriminative information for segmenting the brain imaging for tumor cell classification.

Earlier, brain imaging data can be analyzed manually. It includes examining the patient's physical appearance, checking the medical and family history. However, manual prediction of cancerous cells in brain tumor is highly dependent on the expertise of the radiologists, medical practitioners and is time consuming too. In modern medical imaging analysis, conventional and advanced techniques are quite innovative, popular, fast, and accurate. These techniques utilized various features such as intensity, texture, gradient, Gabor and deep features for predicting various brain diseases [20–23]. Depending on the requirements, medical experts can utilize either of the techniques for prediction.

In this chapter, various AI-based brain tumor prediction algorithms are discussed and reviewed. These algorithms can predict the various categories of brain tumor and analyze the severity so that proper treatment can be extended. The key contributions of the chapter are as follows:

- We have analyzed the brain anatomy to determine the various abnormalities by evaluating the various brain regions.
- The challenges of AI-based predictive algorithms for segmenting the brain imaging to predict the category of tumor are highlighted to provide suitable solutions to address each limitation.
- AI-based predictive algorithms for brain tumor segmentation are broadly reviewed into two categories as conventional approaches and advanced approaches. The potential work under each category is exhaustively examined to determine the accuracy and efficacy of each method for brain tumor prediction.
- The various categories of AI-based predictive algorithms are compared and elaborated. The salient features of each category are highlighted to determine the benefits and limitations for their applicability to clinical deployment.

The details about the brain structure and its abnormalities are as follows.

3.1.1 Brain Structure and Abnormalities

The brain is a complex organ of the central nervous system which controls all the other body parts. It responds and takes decisions to instruct the various body parts to take necessary actions. The human brain has three main components namely, cerebrum, cerebellum and brain stem. There are two matters, namely, gray matters and white matters are present in brain which control its activity by forming neuronal and cells of brain. Figure 3.1 illustrates the structure of the brain along with its main components.

Brain anatomy describes cerebrum as its largest part and known for controlling the major activities of the human that includes thinking, body parts movements, reactions, and feelings. It is divided into two hemispheres known as right and left parts. They contain four lobes namely, temporal, frontal, parietal, and occipital lobe. Each lobe is characterized for a set of functions taken by brain. Frontal lobe located in forward part of the brain and is responsible for reasoning, emotion and language. Behind the frontal lobe, parietal lobe exists and is responsible for sensation movements that include touch, pressure, taste, smell and pain. Occipital lobes are known for processing visuals in the brain. This lobe is crucial in memory management, face recognition, color identification and determining depth and distance. Temporal lobe is separated from frontal lobe by later fissure. It is involved in processing sensory and auditory information. It helps in recognizing voice, faces and creating memories [24].

As per brain anatomy, the second largest part is cerebellum. Cerebellum is located in head backside and interconnected with brain stems. Primarily, it is responsible for walking, posture, hand movements and other body activities. Another important component in brain anatomy is brain stem. It is present in the bottom part of the brain connected with the spinal cord. Brain stem controls many crucial body functions such as breathing, digestion, heart rate, cough, vomiting, sleep cycles, yawning and many more. Apart from this, cerebrospinal fluid also flows within, and around the brain. The glucose, white blood cells and body salts are formed in this fluid. It also prevents brain tissues from severe injury [24].

Quality of life is impacted by the life-threatening diseases that occur in various body parts. Brain is also a sensitive body organ which is prone to many abnormalities that may lead to patients' death if not diagnosed and treated properly. Brain tumor

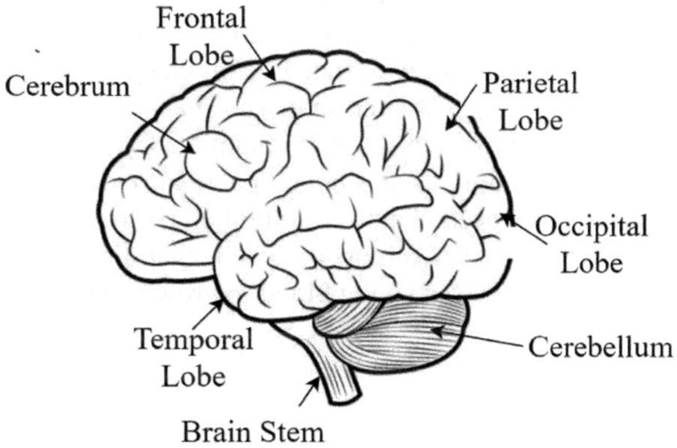

Fig. 3.1 Anatomy of healthy brain and its main components

is one of the serious life threating diseases that occur due to uncontrolled growth and mutation of normal cells in the brain. These cells absorb body blood to grow at a rapid rate. This increases the pressure, causes blockage and swelling in the brain nerves and impacts the other brain parts. This situation further leads to other severe neurological conditions such as dementia [25], stroke [26], and depression [27]. The severity of these diseases increases with time and causes another disease known as Alzheimer, in which patients lose memory, ability of learn, visual perception and ability of recognize people [28–30]. Hence, early diagnosis of brain abnormalities is crucial to provide effective treatment to the patient to prevent the condition from worsening.

3.1.2 Description of Brain Tumor

Brain tumor is a condition which causes abnormal growth of cells in the human brain. This unconditional growth of cells in brain causes abnormalities and may also lead to death of the person without proper and timely treatment. Brain tumors are broadly classified either as malignant or non-malignant. Malignant are cancerous tumors and require proper treatment and therapies to prevent their growth. On the other hand, non-malignant tumors are benign which are non-cancerous and do not impact the other parts of the body if left untreated. Hence, tumor concentrated in brain is not essentially cancerous. Its accurate diagnosis and prognosis are essential to determine it spread and type.

To prevent criticality and spread to other organs, the early detection and prognosis of brain tumor is paramount. Benign tumors do not grow quickly and impact the other body parts. However, malignant tumors grow and spread quickly. Brain tumors are also categorized either as primary tumors or secondary tumors. The abnormalities of cells in the brain are known as primary tumors. On the other hand, metastatic brain cancer is known as a secondary tumor which originated in other body parts and then spread to the brains. Based on the growth, size, appearance and position, brain tumors are also graded in four categories [31]. Grade I: It is the initial stage of brain tumor which develops and grows slowly. Their timely diagnosis can cure them and prevent causalities. Grade II: This category of tumor can impact the neighboring tissues and can grow over time. The chances of recurrence of this stage of tumor are high and require time-to-time evaluation. Grade III: This stage of tumor spread faster to its surrounding tissues in comparison to grade II. Apart from surgical treatment, these tumors need chemo or radiotherapy treatment to prevent their growth. Grade IV: This category of tumors is most dangerous, faster growing and spreading quickly. These tumors absorb body blood to grow and spread aggressively.

3.1.3 Challenges in Brain Tumor Detection

Brain tumor detection in the brain imaging is challenging to the complex brain structure, and the limitations of adopted methodology for acquiring the training datasets. There are many reasons that needs to be addressed for effective and efficient brain tumor prediction and detection [32, 33]. The details are as follows:

Location uncertainty: The distribution and mutation of glioma in the brain is from gluey cells which are widely spread in the brain. Gluey cells are kind of support to the nerves in the brain which can occur from low to high grade glioma. Due to widespread distribution of gluey cells, the precise localization of tumorous cells in brain is really challenging.

Morphological ambiguity: Brain tumors vary in shape, size and structure. The morphological uncertainty occurs due to variations in shape and size in images. The outer layer of brain known as edema, is also different in different sub regions and locations. Hence, the variations in tumor shape and size make its detection and segmentation from the neighboring tissues tedious.

Low contrast imaging: The captured brain images either from CT or MRI are of poor quality with low resolution and contrast. To design efficient methods for accurate segmentation of tumor from the surroundings, the imaging data needs to contain high quality diverse information. The blurry images make the tumor boundaries hard to be classified from the nearby regions.

Noise in images: Apart from the low quality, brain imaging data contains noise which make the segmentation process hard and difficulty. It has been observed that during image projection and acquisition process, images contain the artifacts, and details about motion of external equipment along with the tumor's cells. The presence of such noise in brain imaging restricts accurate localization of the tumor.

Handling of multimodal information in datasets: MRI brain imaging acquire data from multiple channels which varied in contrast and resolution [34]. This multi-modal information causes scattering effect due to which boundaries of tumorous cells in the brain imaging became blurry and hard to detect. To ensure the accuracy of tumor detection, the multi-modal information in the imaging data must be handled appropriately.

Manual labeling: Labeling of tumorous cells in the brain is done manually by the medical experts, trainers and practitioners. The manual annotation of brain imaging can be varied from small region to larger region. The manual preparation of ground truth for the localization of tumor in brain is time consuming and required high end skills and expertise. Sometimes experts varied in their decision about data labeling that will lead to annotation bias. The ground truth should be free from annotation bias for accurate and efficient segmentation results.

Data imbalance: Public datasets for brain tumor detection are highly unbalanced in terms of size of brain imaging [35, 36]. These datasets vary in number of voxels in different tumor regions. These datasets have many cases for one region while very few for the other regions. This imbalance in datasets impacts algorithm learning which do not segment the small tumor accurately in comparison to the larger tumors.

Generalizability: Many AI-based segmentation algorithms have been proposed for predicting the brain tumor in brain imaging [18, 35, 37]. These algorithms are highly accurate and efficient in predicting various categories of tumor. But the performance of these algorithms is evaluated on specific datasets. The performance is not tested on variety of datasets to ensure the generalizability of the model. In addition, model hyperparameters settings and other details are model specific which restricts its applicability to real-time environments.

Paper to practice: A lot of research papers are published to predict brain cancer by accurate segmentation of brain imaging. However, most of the work is theoretical and real-time deployment is either not possible or time-consuming. The overall deployment cost and accuracy in real-time environments is also one of the main reasons that restricts the clinical deployment of these works. There is a requirement of robust algorithms that will predict efficient results and meet the needs of medical personnel when applied to real-time situations.

The rest of the chapter is organized as follows. Section 3.2 categories the AI-based brain tumor predictive models into various categories. Section 3.3 classifies the conventional approaches for brain tumor segmentation algorithms. The salient features of each methodology under each category are investigated in detail. The advanced approaches are classified and salient details for each category are highlighted in Sect. 3.4. Section 3.5 compares and evaluates the various AI-based brain tumor detection methods. Lastly, the concluding remarks and future directions are sketched in Sect. 3.6.

3.2 AI-Based Predictive Models for Brain Tumor Prediction

Classification of tumorous cells from its neighboring tissues requires skilled and expert physician with detailed knowledge of brain anatomy and its illnesses. Similarly, MRI and CT images of brain also possess significant challenges for processing image, eliminating noise, tumor recognition and explanation. To address these challenges, AI-based predictive models for brain tumor detection have proved highly accurate and efficient. These are non-invasive which can automatically segment the tumor so that proper treatment can be provided to the patient. Broadly, AI-based brain tumor detection models are categorized either as conventional approaches or advanced approaches. Figure 3.2 illustrates the categorization of AI-based brain tumor prediction model. The various categories are represented in a tree-like structure to provide the clarity of various methodologies under each classification.

Conventional approach consists of two categories of algorithms. These techniques utilized either ML [4, 38, 39] or non-ML [16, 37, 40] algorithms for proposing efficient tumor segmentation frameworks. ML algorithms exploited techniques such

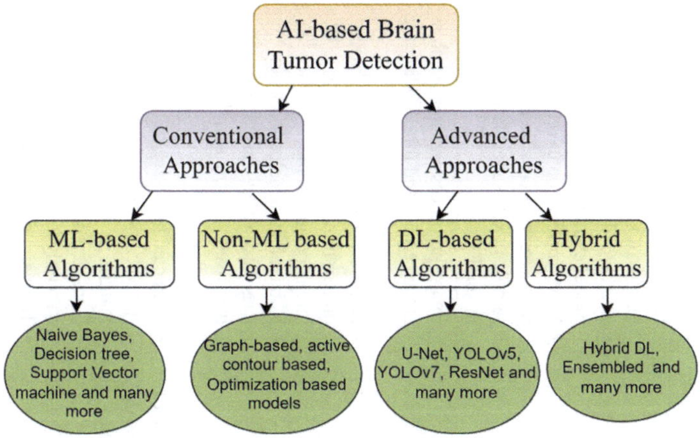

Fig. 3.2 Categorization of AI-based predictive models for brain tumor detection

as logistic regression (LR), multi-layer perceptron (MLP), support vector machine (SVM), random forest (RF), and decision tree (DT). On the other hand, non-ML techniques investigated fuzzy c-means [41], dynamic graph learning [37], active contouring [40], extreme learning machine [23, 42] and many more [16, 43] for developing robust model for brain cancer prediction. These approaches segmented the tumor from MRI or CT scans efficiently to a great extent.

Advanced approaches contain either DL-based networks [2, 17, 34] or hybrid algorithms [44–46] for segmenting brain tumor from brain imaging. Generally, DL-based networks utilize CNN, UNet and its variants for generating efficient results. But hybrid algorithms combined ML-techniques along with DL-techniques to integrate the potential of both the methodologies. The techniques under advanced approaches focus to improve the image processing capabilities along with enhanced computational power. The next section will elaborate on the conventional approaches for brain tumor detection in detail.

3.3 Conventional Approaches for Brain Tumor Prediction

Brain scan interpretation and understanding is really crucial for identifying the regions infected with cancerous cells. For this, conventional approaches have provided robust algorithms for classifying the brain imaging, lesion and cancerous cells from the MRI and CT scans. These algorithms follow step by step procedure that involves image preprocessing, feature extraction, feature selection and classification. Under this category, brain tumor detection algorithms are discussed either as ML-based algorithms or non-ML based algorithms. Table 3.1 tabulates the salient features of the representative work categorized in the domain of conventional approaches. The details about the ML-based algorithms are as follows.

Table 3.1 Representative work in conventional approaches for brain tumor prediction

Reference	Dataset description	Model utilized/proposed	Modality	Performance measures	Summary
Zhang et al. [39]	• OASIS DB, BTCV, CHAOs, KiTS, Pancreas-CT, and two subsets from MSD challenge	Unsupervised Sim2Real training regime	Brain MRI and Abdomen CT imaging	DSC, SEN, SPE	• Classified varying size tumors based on their intensity and the texture of surroundings tissues • Synthetic data generation to train model efficiently • Mask verification to reduce reconstruction error in each sample in self-supervised model in stage I
Yu et al. [16]	• Hemorrhage-K (DB1), Hemorrhage-NEH (DB2), Hemorrhage-RSNA (DB3), Hemorrhage-CQ (DB4), Hemorrhage-MH (DB5), Tumor-AI (DB6) • Tumor/Healthy: 100/100 (DB1), 17/27 (DB2), 50/50 (DB3), 40/30 (DB4), 36/46 (DB5), 100/100 (DB6)	Improved sparrow search algorithm	CT Scan Image	ACC, CC	• Image preprocessing steps such as enhancement and thresholding for noise removal • Feature extraction strategy based on haralick, Gabor, Tamura, LBP, intensity, and shape • Robust feature selection from large feature sets based on improved sparrow search algorithm
Ma et al. [37]	• ABIDE dataset from New York university • Total/Normal/Disorder: 184/105/79	Multi-scale dynamic graph learning	fMRI	ACC, PR, RE, AUC, F1-score and p-value	• Extracted multi-scale spatiotemporal dynamic fMRI features at whole graph level • Considered resting-state fMRI data only • Trained model on small fMRI datasets

(continued)

Table 3.1 (continued)

Reference	Dataset description	Model utilized/proposed	Modality	Performance measures	Summary
Asiri et al. [4]	• Nanfang Hospital, Guangzhou, China and General Hospital, Tianjin Medical University, China • Total: 3064 from 233 patients • Pituitary/Meningiomas/Gliomas: 930/708/1426	SVM	Enhanced-contrast MRI scans	ACC, SEN, SPE, DSC, CM and computational time	• Utilized image enhancement techniques such as adaptive Wiener filtering, neural networks, and independent component analysis • SVM for segmentation and tumor classification • Classified tumor into multiple classes
Xu et al. [40]	• MRI DB • Total/Utilized/Healthy: 7023/2000/50	Active contouring mechanism	MRI	ROC, FPR, TPR, and F1-score	• Utilized active contour model to identify brain tumors of different sizes, and shape • Texture analysis to eliminate false tumor regions • Correct segmentation area ratio for evaluating tumor performance
Deepa et al. [38]	• BraTs 2013 and ISLES DB • Total/tumor/stroke: 1100/600/500	Hybridized SVM, RF	MRI	FPR, F1-score, ACC, PR, RE, JC and CM	• Extracted texture, intensity, and shape features from brain imaging • Feature selection strategy using firefly algorithm based on maximum A priori • Tumor categorization into four classes as high-grade, low grade, acute and sub-acute stroke • Identified tumor regions as edema, necrotic and non-enhancing

(continued)

Table 3.1 (continued)

Reference	Dataset description	Model utilized/ proposed	Modality	Performance measures	Summary
Alhassan and Zainon [41]	• MRI DB • Total/Infected/Healthy: 253/155/98 • Training/Validation/Test: 70/15/15	Enhanced capsule networks and Bat algorithm with fuzzy c-ordered means	Multimodal MRI	ACC, PR, RE, and F1-score	• Utilized clustering techniques for automatic segmentation of brain imaging • Enhanced capsule networks for classification into healthy and infected tumor • Data augmentation techniques to increase dataset size to avoid overfitting
Gumaei et al. [42]	• Public DB • Total: 3064 from 233 patients	Regularized extreme learning machine	MRI	ACC, and CM	• Utilized min–max normalization for enhancing image contrast • Hybrid feature extraction using normalized GIST descriptor along with PCA • Introduced regularizer to prevent overfitting and increase training speed
Ramachandran et al. [43]	• CBTRUS DB	MapReduce and minimum quadrangle classification	Imaging	ACC, Complexity and Computational time	• MapReduce to handle large scale big data • Improved classification ACC using Lagrange multipliers and radial basis kernel function • Minimum quadrangle SVM to remove unwanted features from brain DB

(continued)

Table 3.1 (continued)

Reference	Dataset description	Model utilized/proposed	Modality	Performance measures	Summary
Amin et al. [20]	• BraTS2012 (DB1), BraTS2013 (DB2), BraTS2014 (DB3), BraTS2015 (DB4), and ISLES (DB5) • Total: 80 (DB1), 30 (DB2), 191 (DB3), 274 (DB4), and 64 (DB5)	RF	MRI	SPE, SEN, PPV, NPV, ACC, AUC, DSC, and segmentation quality	• Extracted LBP, HOG, SFTA and GWF features • Categorized into complete, enhancing, and non-enhancing tumor • Feature fusion for improving the classification ACC
Sharif et al. [23]	• BraTS2012 (DB1), BraTS2013 (DB2), BraTS2014 (DB3), and BraTS2015 (DB4) • Total: 50 (DB1), 40 (DB2), 300 (DB3), 384 (DB4)	Extreme learning machine	MRI	JS, DSC, SEN, SPE, FNR, FPR, PPV, CM and ACC	• Extracted Gabor and similar texture features • Triangular fuzzy median filtering for image enhancement • Compared results with fuzzy and non-fuzzy filters

DB: Database, CBTURS: Central brain tumor registry of the United States, ISLES: Ischemic stroke lesion segmentation, CT: Computed tomography, MRI: Magnetic imaging resonance, LBP: Local binary pattern, HOG: Histogram of gradient, GWF: Gabor wavelet features, SFTA: Segmentation based fractal texture analysis, MSD: Medical segmentation decathlon, PCA: Principal component analysis, ACC: Accuracy, CC: Convergence curve, DSC: Dice score coefficient, SPE: Specificity, SEN: Sensitivity, PR: Precision, RE: Recall, AUC: Area under the curve, ROC: Receiver operating characteristics, FPR: false positive rate, TPR: True positive rate, PPV: Positive predicted value, NPV: Negative predicted value

3.3.1 Machine-Learning Based Algorithm for Brain Tumor Prediction

In this section, we will discuss the ML algorithms exploited for brain tumor detection. In this direction, Zhang et al. [39] exploited self-supervised ML approach for brain tumor segmentation. Synthetic data was generated for model pretraining and developed scalable pipeline for layer decomposition to perform tumor segmentation tasks. Candidate sourcing was done for validating the potential candidate mask. Mask verification for each sample was done to determine low reconstruction error sample with high precision. Authors introduced two module computerized method as highly accurate and fast method for tumor detection [4]. Initially, image enhancement techniques were utilized for improving image contrast and reducing noise in MRI imaging. After this, SVM algorithm classified the tumor as meningiomas and pituitary tumors. The method handled MRI images limitations of low contrast, resolution, coherence and noise efficiently to improve segmentation accuracy.

Tumor segmentation in brain imaging is challenging due to variations in tumor shape, size and texture. Lesion localization in the brain is also tedious due to complex brain structure. To address this, authors utilized RF based unsupervised clustering approach along with fused feature vector to classify tumoral region as complete, enhancing and non-enhancing [20]. Clustering algorithm utilizes five clusters for segmenting the lesion region by calculating values such as min, max, range and interval for these clusters. Multiple features were extracted, and fused feature vector was generated which fed to the classification algorithm to recognize the labels. On the other hand, authors proposed hybridized ML algorithms to classify the brain illnesses as stroke or tumor [38]. Feature extraction methods such as neoteric directional based quantized extrema pattern, clustering-based wavelet transforms and conventional shape descriptors were used for extracting texture, intensity and shaper, respectively. The segmented image was subjected to SVM based RF for classify the various classes of tumor.

To summarize, ML-based methods for tumor classification can handle the brain imaging limitations by using various image enhancement techniques. These techniques enhance the classification accuracy by handling the low contrast, resolution and noise in the images. ML-techniques are also efficient to classify the lesion areas into healthy and tumors regions. It can segment the exact boundaries of the abnormal regions so that necessary treatment can be extended before it impacts the neighboring cells. The next section will detail the non-ML based algorithms for brain tumor detection.

3.3.2 *Non-machine Learning Based Algorithms for Brain Tumor Prediction*

Brain diagnostic images require efficient methods to segment so that brain diseases can be detected at an early stage. For this, authors introduced improved sparrow search algorithms for brain disease classification [16]. Tent chaotic initialization and adaptive crossover operation was used as local search strategy to enhance search algorithm capabilities. Features were selected using binary operators and fed to classification algorithms such as KNN, SVM, DT and RF. KNN achieved highest accuracy of 85% in comparison to other classification algorithms. On the other hand, Ma et al. [37] exploited multi-scale dynamic graph for capturing spatiotemporal information from functional MRI features for detecting brain disorders. Node-level features were learned using K graph isomorphism network and spatial attentional learning was done suing Sero readout operation. Multiple dynamic spatiotemporal features captured at various spatial levels and fused using multi-scale fusion. In [41], authors recommended clustering algorithm with fuzzy c-ordered means for accurate segmentation of tumor. Bat algorithms was exploited as clustering method that calculated the initial centroid and within pixel distance for determining the distance among tumor region and non-tumor region. Further, the enhanced capsule networks analyzed and categorized the tumor into healthy or cancerous.

The different size and shape of tumors are challenging to detect due to similar contour regions for tumors and the background. To address this, authors utilized active contour models to find location, measure shape and size to detect tumor [40]. False tumor areas from the suspected tumor areas were eliminated using area ratio scheme. False larger sizes were identified using texture analysis. Segmentation area ratio localized the tumor in brain accurately. However, Ramachandran et al. [43] proposed mutual informative MapReduce and minimum quadrangle classification to address the concerns related with big data in brain tumor classification. Mutual informative MapReduce is used for eliminating the redundant features in brain tumor detection dataset using mutual information at preprocessing step. Minimum quadrangle was created using SVM with Lagrange multipliers to improve the classification accuracy. Radial basis kernel function and MAXMIN values were compared to detect the tumor at an early stage.

Brain tumor localizes near to brain surface are hard to detect using standard detection techniques. To improve the detection accuracy for such kind of tumors, Sharif et al. [23] proposed extreme learning machine for tumor classification. For accurate segmentation results, triangular fuzzy median filtering and fuzzy based unsupervised clustering was applied as an image enhancement technique. Median values with different padding and window size were selected for handling noise in the images. Similar texture feature obtained from Gabor filter response was used for discriminating between pathological and normal brain images. On the other hand, authors exploited regularized extreme learning along with hybrid feature for classifying brain tumor [42]. Images were preprocessed using min–max normalization to enhance quality by improving contract of brain edges and regions. Hybrid features include normalized GIST with PCA (Principal component analysis) normalizing GIST with L2-norm. GIST feature represented features using spatial envelope by computing spatial structure of the image. Regularized extreme learning not only prevents algorithms overfitting but also improves training speed.

To summarize, accurate identification of tumors is difficult due to variations in their size, shape, location and textures. To improve the survival rate, early detection of tumor is essential. Conventional techniques have the potential to predict the cancerous cells from the neighboring regions and help medical practitioners to provide timely diagnosis so that appropriate treatment can be given to the patients at initial stage.

3.4 Advanced Approaches for Brain Tumor Detection

In this section, advanced approaches for brain tumor detection are elaborated in two categories namely, DL-based algorithms [14, 15, 18, 21] and hybrid algorithms [3, 19, 44, 47]. DL-models can process large amounts of imaging data efficiently to improve model accuracy. On the other hand, hybrid models are powerful with high levels of effectiveness and superior efficiency in terms of classification accuracy and computation complexity. The details about the DL-based algorithms for brain tumor segmentation are follow in turn.

3.4.1 Deep Learning-Based Algorithms for Brain Tumor Prediction

Table 3.2 tabulates the salient features of the representative work proposed in the category of DL-based algorithms for brain tumor detection. The exploited datasets, methodology and image modality are also extracted to provide valuable insights about the recent advances in the field.

DL-model requires a large amount of training data for producing effective results. However, high quality medical data is limitedly available for processing due to various environmental issues. To address this issue, authors generated realistic synthetic MRI data using generative adversarial networks (GANs) [48]. UNet and Swin transformers are utilized for segmenting the tumor from brain imaging. Four 2D GANS and 2D diffusion model evaluated comprehensively for brain tumor images and annotations. Models were trained using both real and synthetic images which improved the model's classification and segmentation accuracy. Generative model results were qualitatively evaluated by neuroradiologist for proving the model's efficiency.

Typically, DL networks are computationally expensive in comparison to UNet models. UNet models can be easily modified to provide efficient segmentation results.

Table 3.2 Representative work in DL-based brain tumor prediction

Reference	Dataset description	Model utilized/proposed	Modality	Performance measures	Summary
Akbar et al. [48]	BraTS 2020 (DB1) and 2021(DB2) Total: 369 (DB1), 1251 (DB1) Training/Testing: 313/56 (DB1), 1195/56 (DB2)	UNet and Swin transformer	MRI Imaging	FID, and IS	• Two methods for data augmentation namely, geometric and intensity • UNet model with extra depth layer, instance normalization, and cross entropy loss • Trained and implemented swim transformer using MMSegmentation, and 3-channel RGB images
Almufareh et al. [14]	DB from Southern Medical University, Guangzhou Total: 3064 from 233 patients Meningioma/Glioma/ Pituitary: 708/1426/930	YOLOv5 and YOLOv7	MRI Imaging	CM, PR, RE, mean and average PR	• Brain imaging segmented to categorize tumor into three distinct classes • Advanced mask alignment as preprocessing step for better precision • Different YOLO algorithms to test the capability for model for accurate segmentation results
Anaya-Isaza et al. [18]	BraTS2020 challenge DB Total: 369 images	UNet	MRI imaging	DSC, IoU, ACC, Computation time, Hausdorff metrics and p-value	• Designed 4-stage encoder-decoder transformers • Cross-attentional model with separable convolutional layer for light weight computation • Modified attention model by modifying transition layers, encoder and decoder blocks

(continued)

Table 3.2 (continued)

Reference	Dataset description	Model utilized/proposed	Modality	Performance measures	Summary
Chauhan et al. [15]	BraTS2018 dataset Training/Validation/Test: 285/66/191	UNet, PSPNet, DeepLabV3 + and ResNet50	MRI imaging	DSC, SEN, SPE	• Utilized preprocessing steps such as cropping, resizing, and normalization • Tumor segmentation using 9-layer UNet like design • 7-layer UNet for augmenting tumor segmentation results as input
Khushi et al. [22]	Kaggle dataset Total/Training/Test: 3264/2870/394	AlexNet, VGG16, VGG19, ResNet50, InceptionV3, DenseNet121, variants of EfficientNet and Customized EfficientNetB7	MR Images	ACC, loss, PR, SEN, SPE, RE, F1-score, mean IoU and CM	• Applied preprocessing to improve the image quality and increase training data size • Multiple CNN networks along with TL for efficient outcomes • Additional layers including global average pooling, dense layer and softmax classifier for modified EfficientNetB7 model
Lee et al. [2]	Public datasets: Figshare, Bhuvaji on Kaggle and Br35H by Hamada Total/Training/Validation/Test: 4800/1920/480/2400	ViT-B/16, MaxViT-B, TresNet-M, and EfficientNetV2-M	MR images	ACC, and F1-score	• Reduced noise in MR images by applying Gaussian filters • Patterned GridMask for generalized performance of deep neural networks • Computer aided diagnosis for improved results and early detection of tumor

(continued)

Table 3.2 (continued)

Reference	Dataset description	Model utilized/proposed	Modality	Performance measures	Summary
Li et al. [17]	BraTs2019, BraTs 2020 and Jun Cheng	Vector quantized variational autoencoder	MR images	DSC, SEN, SPE, and HD95	• Corrected brain tumor segmentation errors using corrective diffusion • Reduced model dimensionality to improve model stability • Enhanced segmented performance by using multi-fusion attention mechanism
Liu et al. [34]	BraTs 2022 and self-generated DB	DL-framework	Multi-modal MR images	DSC, JC, SEN, SPE, PR, inference time and Hausdroff distance	• Highlighted tumor related features using multi-modal brain images • Feature alignment module to learn from robust fused features • Global correlation block to obtain fused features from fully connected layers
Metlek and Cetiner [35]	BraTS2018 (DB1), BraTS 2019 (DB2), BraTS 2020 (DB3) Total/Training/Valdiation/ Testing: 660/369/125/166 (DB1), 626/335/125/166 (DB2), 542/285/66/181 (DB3)	ResUNet +	MRI	DSC, JC, ACC, SPE, SEN, PR and loss graph	• Modified encoder stage to obtain low-level features • Incorporated residual blocks in encoder to address vanishing gradient problem • Added nodes in encoder and decoder layer to obtain the lost features

(continued)

Table 3.2 (continued)

Reference	Dataset description	Model utilized/proposed	Modality	Performance measures	Summary
Ren et al. [49]	BraTS 2023 Training/Test: 9/1	3D UNet	MRI	DSC, and HD95	• Rescaled voxel intensities in images to identify important features • Histogram contrast matching to improve intensity distribution in the images • Incorporated edge loss component to obtain correct boundary information and reduce intensity dependency
Shah et al. [50]	BraTS 2015 subset DB from Kaggle Total/Utilized/Infected/Healthy: 3762/3060/1500/1500 Training/Validation/Test: 2400/600/60	EfficientNet-B0	MRI	ACC, PR, SEN, SPE, Loss, CM, ROC and F1-score	• Followed three steps preprocessing strategy to improve images low quality • Additional layers in EfficientNet-B0 to improve classification accuracy • Data augmentation to increase training dataset size for preventing model overfitting

(continued)

Table 3.2 (continued)

Reference	Dataset description	Model utilized/proposed	Modality	Performance measures	Summary
Zaitoon and Syed [51]	BraTS17 (DB1), BraTS18 (DB2), BraTS19 (DB3), BraTS20 (DB4) Total/patients: 885/285 (DB1), 798/266 (DB2), 855/285 (DB3), 1005/335 (DB4) Total/HGG/LGG: 3515/2345/1168	RU-Net2 +	MRI	ACC, PR, RE, F1-score, DSC, CM and ROC-AUC	• Cox multivariate model to analyze the survivability prediction • Integrated detection, segmentation, classification, and risk prediction for early prognosis • Convolutional normalized mean filter for image enhancement for better classification accuracy
Aggarwal et al. [52]	BraTS 2020 from Kaggle Total/Training/Test: 369/125/169	Enhanced ResNet	MRI	ACC, RE, MSE, PSNR, computational time, JC, DSC, SEN, SPE and F1-Measure	• Jump relationship combined with convolution input for robustness • Identification function to identify significant layer for better bottom level processing • Utilized long skip connection in the residual network

(continued)

Table 3.2 (continued)

Reference	Dataset description	Model utilized/proposed	Modality	Performance measures	Summary
Behera et al. [21]	NITR-DHH (DB1), DS-75 (DB2), DS-160 (DB3) Total: 356(DB1), 75(DB2), 160 (DB3) Augmented size: 2932 (DB1), 1504 (DB2), 2100 (DB3) Training/Validation/Test: 8/1/1	Simple linear iterative clustering based super pixel with CNN	MRI	ACC, SEN, SPE, PR, F1-score, computational time and CM	• Extracted texture features and performed CNN based classification • Different superpixels inputs from multiple datasets to test generalizable performance • Binary classification of brain tumor either as normal or abnormal
Wang et al. [53]	In Vivo Hyperspectral human brain (DB1) and Multidimensional Choledoch (DB2) MedHSIs: 36 (DB1), 174 (DB2)	Deep autoencoder	Microscopy hyperspectral imaging	Overall ACC, average ACC, and Kappa coefficient	• Classifier and feature extraction integrated to robust feature extraction • Extra cosine margin into embedded soft-max for compact and separable feature extraction • Utilized two-stage training strategy for better ACC

DB: Database, FID: Frechet inception distance, IS: Inception score, MRI: Magnetic resonance imaging, GLCM: Gray level co-occurrence matrix, HGG: High-grade glioma, LGG: Low grade glioma, CM: Confusion matrix, ACC: Accuracy, SEN: Sensitivity, SPE: Specificity, PR: Precision, RE: Recall, DSC: Dice score coefficient, JC: Jaccard Coefficient, IoU: Intersection-over-union, AUC: Area under the curve, ROC: Receiver operating characteristics, PSNR: Peak signal to noise ratio, MSE: Mean square error

In this direction, authors designed the UNet into 4-stage deep encoder-decoder structure with cross-attention model and separable convolutional layers [18]. Separable convolutional layers were low cost and improve the computational efficiency of the model. Dice coefficient loss function was used to compute the difference between the predicted values and the ground truth values to improve model's performance. Zaitoon and Syed [51] employed RUNet2+ (Residual UNet 2+) for precise detection of brain tumor. Survivability rate was also predicted by incorporating the Cox multivariate model on the extracted features. Convolutional normalized mean filter was used in preprocessing step for noise removal while preserving the edges. DBT-CNN was adopted for tumor multi-class classification including high-grade glioma and low-grade glioma. But authors developed hybrid UNet as ResUNet+ based on the residual block [35]. HighHat and lowHat transformations were applied at preprocessing to reduce the impact of illumination variations in MRI images. Model was trained with random weights without pretrained weights for better segmentation accuracy. On the other hand, authors constructed segmentation network for multimodal MRI images based on 3D UNet [49]. Z-Score normalization is utilized to cater intensity variations in images. Feature visibility was enhanced by rescaling voxel intensities to ease important feature identification. Histogram contrast matching for aligning intensity distributions between images with different contrast.

Early tumor detection with accuracy is crucial to provide sufficient treatment for improving the chances of life survival. For this, authors utilized multiple pretrained models along with different variants of EfficientNet [22]. Images were resized and cropped to highlight the salient features. FastNIMeans denoising colored filter removed the noise from images and augmentation technique to prevent model overfitting. Among all the networks EfficientNetB7 was highly accurate as additional layers and fine tuning was done to improve the overall accuracy. Similarly, authors fine-tuned the base model of EfficientNet-B0 for detecting brain tumor efficiently [50]. Three step preprocessing strategy was followed to improve the brain images for segmentation process. Adam optimizer was utilized for optimizing the network hyperparameters such as learning rate, and loss function to improve segmentation accuracy. Similarly, authors compared segmentation accuracy for various DL-models such as UNet, PSPNet, DeepLabV3+ and ResNet50 [15]. Out of these, 3D UNet obtained the highest segmentation accuracy. Transfer learning of pre-trained weights was used for fine tuning of model to improve computational efficiency of the model. Also, Lee et al. [2] demonstrated the segmentation performance for brain tumor detection using four DL-models. Image enhancement such as noise removal and generalization strategies were adopted to improve early detection accuracy. Model performed the multiclass classification of brain tumor into glioma, meningioma, pituitary and healthy tissues.

In another line of work, authors utilized hyperspectral images for diagnosis of tumor by utilizing deep margin cosine autoencoder [53]. Basic architecture of MedHSI was used for feature extraction and soft-max classifier for predicting the labels from the output layer. Extra cosine margin was used in soft-max classifier to maximize angular space for extracting compact and separable features for obtaining great results. Behera et al. [21] presented an ensembled approach based on transfer learning by utilizing linear iterative superclustering-based superpixel with CNN. Superclustering segmented the image into clusters based on the similarity measures determined by perceptual feature space. In [17], authors developed corrective diffusion model for enhancing segmentation performance. Vector quantized variational autoencoder compressed images for improving model stability by reducing dimensionality of training data. Multi-fusion attentional mechanism enhanced segmentation model reliability and flexibility.

Segmenting the brain tumor without ignoring the brain appearance information is needed for accurate tumor segmentation. For this, authors proposed multimodal DL framework with variational autoencoder to present brain images based on latent distribution [34]. In the decoder layer, feature alignment module was presented to resolve the feature compatibility issues between the multimodal brain tumor and monomodal normal brain images. Fusion module based on global correlation block concatenated the features from same channel to generate fused feature. On the other hand, authors compared YOLOv5 and YOLOv7, object detection algorithms for classification of brain tumor as meningiomas, gliomas and pituitary [14]. Mask alignment scheme for standardizing the dataset images for better training outcome. Techniques were applied to identify the tumor boundaries for accurate segmentation of the tumor.

To summarize, timely tumor brain tumor detection is challenging and necessary to increase the survival rate. The evolution in DL-based algorithms have improve the segmentation performance and shown promising results in tumor detection. These algorithms are capable of multi-class classification of brain abnormalities for better analysis and appropriate treatment. Pre-processing steps in brain tumor imaging not only enhances the image quality but also improves the training computational power of the algorithm.

3.4.2 Hybrid Algorithms for Brain Tumor Prediction

Brain tumors are becoming the major cause of death globally because of inaccurate and late diagnosis. The reason could be the time taken for model training by large datasets which delayed the outcome. Due to this, there is a requirement of hybrid models which not only process the training data quickly but also provide effective and efficient results. Table 3.3 tabulates the salient features of representative work utilizing hybrid approach in their algorithms. The dataset description, performance metrics and exploited methodology are also elaborated to understand the innovation and advancement in the field.

Authors introduced hybrid model by integrating CapsNet model with VGGNet model for automatic and accurate segmentation of brain tumor [44]. The problems of model training with large datasets were addressed by automatic extraction of radiomic features for brain cancer classification. Transfer learning models were optimized for extracting complex features for brain tumor classification into four classes as normal, pituitary, meningioma and glioma. Aggarwal et al. [54] employed hybrid model in which features were extracted using CNN and multiple ML models for brain tumor

Table 3.3 Representative work in hybrid brain tumor prediction

Reference	Dataset description	Model utilized/proposed	Modality	Performance measures	Summary
Agarwal et al. [54]	Three public DB Training/Test: 75/25 Total/Training/Test: 4850/3637/1213	CNN, DT, NB, AdaBoost, KNN and SVM	MRI imaging	F1-score, PR, RE, CM, AUC-ROC, and ACC	• Extracted features from max pooling layer using CNN with Adam and cross entropy optimizer • Improved and optimized results using five ML techniques from the extracted features • Classified tumor into five classes namely, normal, meningioma, pituitary, and glioma
Anantharajan et al. [3]	DB from Kaggle Total/Healthy/Tumor: 255/98/155	Ensemble deep neural SVM	MRI imaging	ACC, SEN, SPE, Computational time. JC, PSNR	• Utilized ACEA and median filtering for noise removal as preprocessing step • Fuzzy C-means for segmentation of MRI brain imaging • Extracted features using GLCM for better identification of abnormal tissues
Hossain et al. [19]	Total: 3264 Training/validation/ Test: 8/1/1	IVX16 ensemble model, Vision transformers namely, SWIN, CCT and EANet	Imaging dataset	PR, RE, F1-score, support, CM and ACC	• Classified tumor into multi-class such as glioma, meningioma, pituitary and no tumor • Ensembled six CNN models namely, InceptionV3, VGG16, Xception, VGG19, ResNet50, and Inception ResNetV2 • Implemented three vision-based transformer namely, SWIN, CCT and EANet
Jabbar et al. [44]	BraTS 2020 (DB1) Training and BraTS 2019 (DB2) Testing DB1: Total (4575), patients (36), Training/ Testing (3146/1419)	Caps-VGGNet	3D MRI imaging	ACC, SPE, SEN, F1-score, DSC, PPV, AUC and CM	• Utilized TL-based five pretrained and fine-tuned algorithms for tumor classification into four classes • Applied image preprocessing steps such as normalization, crossing, resizing and augmentation • Extracted features using GLCM

(continued)

Table 3.3 (continued)

Reference	Dataset description	Model utilized/proposed	Modality	Performance measures	Summary
Lamba et al. [47]	Public DB Total/Normal/Infected: 3264/500/2764	Deep neural network and SVM	MRI	ACC, PR, SEN, SPE, F1-score and CM	• Automated brain tumor segmentation into healthy and infected from MR images • Efficiently handled class imbalance, overfitting, computational requirements, and generalization on unseen data • Utilized VGG-16 for robust feature extraction
Mohsen et al. [45]	Kaggle DB Total/Glioma/Pituitary: 1800/900/900	ResNext101_32 × 8d and VGG19	MRI images	RE, PR, ACC, ROC-AUC, CM and ACC	• Applied single-image super-resolution to enhance feature extraction in MRI images • Classified brain tumor into two categories namely, glioma and pituitary • Applied hyper parameter optimization and k-fold cross validation to improve accuracy
Shah et al. [55]	Contrast enhanced brain neoplasm Figshare DB Total/Training/Test: 3064/2800/264	Voting based semi-supervised Bayesian Ensemble attention mechanism	MRI	ACC, SEN, SPE,	• Computer aided algorithm to diagnose brain tumor • Ensemble architecture along with squeeze and excitation attention blocks for improved efficiency and accuracy • Dense classifier for obtaining the multi-class classification of brain tumor
Vinod et al. [56]	BraTS2020 Training/Validation/Test: 70/15/15	UNet, CNN and modified Self organizing feature map	Multi modal MRI	ACC, mean IoU, DCS	• Applied pixel intensity normalization and resizing of images at preprocessing step • Determined survival prediction rate using modified self-organizing feature map • Modified UNet encoder and decoder for enhancing classification accuracy

(continued)

Table 3.3 (continued)

Reference	Dataset description	Model utilized/proposed	Modality	Performance measures	Summary
Vinu et al. [57]	MRI DB Total: 3290 images Training/Test: 7/3	Ensembled approach: CNN, SVM, RNN, KNN and RF	MRI	ACC, SEN, SPE, CM and DSC	• Extracted shape, model and intensity features from the images • Image resizing and rotation as preprocessing steps • Utilized ensembled approach for tumor classification
Mallampati et al. [58]	RSNA-MICCAI DB from Kaggle	Hybrid model with KNN and gradient boosting classifier	MRI	ACC, PR, RE, F1-score, ROC-AUC T-test and CM	• Extracted features using 3D-UNet and 2D-Unet • Employed ensembled ML algorithms for brain tumor classification • Performed statistical test to understand the statistical difference between results
Rajendran et al. [36]	Figshare DB	UNet and 3D CNN	MRI	ACC, PR, F1-score, SEN, and DSC	• Utilized GLCM to eliminate noise from the MRI images • Obtained similarity score using Gaussian Kernel function • Applied mirroring, rotation, and gamma correction during training to improve accuracy and reduce overfitting

(continued)

Table 3.3 (continued)

Reference	Dataset description	Model utilized/proposed	Modality	Performance measures	Summary
Saeedi et al. [59]	Contrast enhanced MRI Total/Infected/Healthy: 3264/2764/500	2D CNN, auto encoder, MLP, SVM, RF, LR, SGD and KNN	MRI	PR, RE, F-Measure, loss, CM, ROC and ACC	• Number of layers in the DL network adjusted to improve classification accuracy • Extracted shape, intensity and model-based features for better accuracy • Multi-class classification of brain tumor as glioma, pituitary, meningioma and no tumor
Renugadevi et al. [46]	BraTS2020 Training/Validation/ Test: 8/1/1	UNet++, SGD, DT, RF, XGBoost and SVM	MRI	ACC, PR, RE, MSE, R-squared and F1-score	• Extracted radiomics features from the segmented image • Synthetic minority oversampling and adaptive synthetic technique for handling class imbalance and dataset sampling size • Categorized brain tumor into high-grade and low-grade glioma

DB: Database, CNN: Convolutional neural network, DT: Decision tree, NB: Naïve Bayes, KNN: K Nearest Neighbor, SVM: Support vector machine, RNN: Reccurrent neural network, RF: Random forest, MLP: Multi-layer perceptron, LR: Logistics regression, SGD: Stochastic gradient descent, ACEA: Adaptive contrast enhancement algorithm, GLCM: Gray-level co-occurrence matrix, MRI: Magnetic resonance imaging, TL: Transfer learning, PR: Precision, RE: Recall, CM: Confusion matrix, AUC-ROC: Area under the curve-receiver operating characteristics, ACC: Accuracy, JC: Jaccard Coefficient, DCS: Dice score coefficient, PSNR: Peak signal to noise ratio, SPE: Specificity, SEN: Sensitivity, PPV: Positive predicted value classification. CNN model utilized pooling layer to reduce the image dimensionality by preserving the significant data for faster computations. Automatic optimization of the model prevented overfitting by maintaining the low learning rates. But authors recommended to preprocess the images using adaptive contrast enhancement, median filter and fuzzy c-means based segmentation for eliminating the noise from images and improve their quality [3]. Multiple features were extracted using Gray-level co-occurrence matrix and abnormal tissues were classified from the healthy tissues using ensembled DL SVM. Hossain et al. [19] proposed multi-class classification of tumor using ensemble models with transfer learning approach. Several DL models were investigated with three best performing transfer learning models for detecting brain tumors. In addition, model explainability model local interpretable model-agnostic explanations (LIME) was used to generate interpretable model for result validity.

Automatic segmentation of brain tumors is significantly effective to save significant time and cost. Pretrained hybrid DL-models can efficiently classify tumors from brain imaging. In this direction, authors proposed ResNext101_32 $\times$ 8d and VGG19 models for brain tumor classification into two classes [45]. Single image super-resolution method was applied to enhance image resolution so that crucial features can be captured easily. Data augmentation techniques were used by both the models to prevent overfitting. ResNext101_32 $\times$ 8d utilized rotation, horizontal and vertical flip, whereas VGG19 exploited rotation, width and height shift for data augmentation. However, authors integrated UNet, CNN and modified self-organizing feature map (mSOFM) for precise segmentation of tumor [56]. UNet was used for image segmentation and mSOFM for capturing complex data patterns. mSOFM also predicted patient survivability by analyzing the segmented images. UNet encoder section captured contextual information and decoder section recovered the spatial information to generate segmentation mask. Shah et al. [55] proposed voting system based semi supervised Bayesian ensemble attention mechanism for multiclass brain tumor classification. Voting technique used for identifying the final abnormality. Squeeze and excitation attention network integrated into CNN for selecting efficient features by scaling each feature with a weight parameter.

To improve medical aid for brain tumor patients, hybrid technique proposed by handling the issues of class imbalance and time consumption [47]. Image resizing and augmentation techniques were applied during preprocessing to improve image quality and prevent model overfitting. Pre-trained deep neural network such as InceptionV3, DenseNet121 and ResNet50 were used for feature extraction and SVM for classifying the MRI images either as infected or healthy. On the other hand, authors utilized 3D-UNet and 2D-UNet DL models for feature extraction [58]. For tumor classification, ML models KNN and gradient boosting classifier were combined using soft voting. Vinu et al. ensembled CNN, RNN, KNN and RF for achieving high segmentation accuracy [57]. High end intricate tumor features such as shape, depth and model were extracted for knowing the tumor in-depth. Resizing and rotation were applied during preprocessing for enhancing dataset capabilities for robust feature extraction. In [59], authors developed 2D CNN and convolutional auto-encoder network for multi-class classification of brain tumor. Model consisted of two parts namely, convolutional auto encoder for feature extraction and CNN along with six ML algorithm for tumor classification. Encoder network removed the critical feature and output layer provided the classification outcome. Similarly, Renugadevi et al. [46] investigated encoder-decoder UNet++ architecture along with ML techniques for robust feature extraction and tumor classification. Regression techniques such as SGD, linear, ridge and extreme gradient boosting were also utilized predicting the lifetime of high glioma patients. Synthetic minority oversampling and adaptive synthetic approaches increased the class size for handling dataset imbalance. PCA and tree-based feature selection methodology for determining the robust features.

To summarize, hybrid models integrated either multiple DL models or DL models with multiple ML models for feature extraction and tumor classification. Automatic segmentation of brain tumor for investigating the level and stage of tumor for extending proper treatment was done effectively by hybrid models. Various preprocessing techniques were utilized for handling the limitation associated with the quality, resolution and contrast of the brain imaging. Also, techniques were adopted for catering data set imbalance using data augmentation methodology for preventing model overfitting. These models are effective and efficient in automatic brain tumor segmentation for providing disease management at an early stage.

3.5 Comparison of Various Artificial Intelligence-Based Brain Tumor Prediction Algorithms

In this section, we have reviewed AI-based tumor prediction algorithms as conventional approaches and advanced approaches. The salient features of each of the techniques are tabulated in Table 3.4. The various parameters are discussed and reviewed, and performance is analyzed to provide the future perspective for each technique for brain tumor classification and segmentation.

It has been observed that methods under each category have their advantages and limitations. These approaches are automatic, effective and accurate to detect brain tumors at an early stage. In addition, these approaches have capabilities for multi-class classification to predict the severity of cancerous cells in the brain. Preprocessing techniques are also investigated to cater the limitations of MRI brain imaging such as low contrast, resolution, and noise. Preprocessing techniques not only improve the image quality but also ease the tumor segmentation by preserving its edges and highlighting its boundaries. Further, data augmentation techniques are also helpful in preventing the model from overfitting and improving its accuracy.

Most of these techniques are published by various researchers. These methods are highly accurate, and efficient in terms of various performance metrics proved by these researchers. However, very little efforts are undertaken to evaluate these methods and determine their validity for real-time deployments. The efficacy of these methods should be determined for automatic and early detection of brain tumors for clinical deployments.

Table 3.4 Similarities/differences of various brain tumor prediction techniques

Attributes	Segmentation/classification techniques			
Approach	Conventional approach		Advanced approach	
Methodology	Non-ML	ML	DL	Hybrid
Feature extraction	✓	✓	✓	✓
Feature selection	Limited	Limited	High	High
Preprocessing techniques	✗	✓	✓	✓
Dataset	Imaging	Imaging	Imaging	Imaging
Model complexity	Less	Less	Moderate	High
Automation	✗	✗	✓	✓
Computational resources	Less	Less	Moderate	High
Generalizability	✗	✗	Limited	Limited
Interpretability	✗	✗	✓	✓
Performance	Moderately accurate	Moderately accurate	Highly accurate	Highly accurate

3.6 Summary

In this chapter, we have exhaustively reviewed the various AI-based brain tumor prediction algorithms. For better understanding of the advancements in the field, we have categorized the AI-based brain tumor segmentation algorithms as conventional techniques and advanced learning techniques. It has been evident that multimodal MRI and CT scans are quite popular among medical practitioners to detect the brain tumor. However, the images collected by these methodologies are quite poor in contrast and resolution. These images are containing noise either due to artefacts used in collection or the varying illumination levels. In order to address imaging limitations, preprocessing techniques such as image enhancements are applied by various researchers. Also, to mitigate the impact of model overfitting, data augmentation techniques are utilized during preprocessing steps only. These initial steps of data preparation are really crucial to ensure the high accuracy and efficiency of the models.

AI-based brain tumor detection techniques extract various features from the brain imaging to perform the tumor segmentation. Some of the techniques adopt feature selection methodology by enhancing the tumor boundaries and edges not only to improve the classification accuracy but also to maintain the computational complexity. Among all the discussed approaches, hybrid algorithms are highly superior in terms of effectiveness and efficiency for tumor segmentation. These algorithms consider ensembled approaches in which multiple DL-models are utilized for feature extraction and multiple ML-models are exploited for tumor classification.

The algorithms not only perform multi-class classification but also can predict the survivability rate in patients diagnosed with brain tumor.

In future, good quality multi-modal brain imaging data is required for producing accurate segmentation results. The complexity concerns while using auto-encoder transformer networks for robust feature extraction need to be addressed. The clinical deployment of these methods along with accurate survivability rate prediction is accessed for better usability and generalizability of these methods.

References

1. Karim S, Tong G, Yu Y, Laghari AA, Khan AA, Ibrar M, Mehmood F (2024) Developments in brain tumor segmentation using MRI: deep learning insights and future perspectives. IEEE Access
2. Lee J-h, Chae J-W, Cho H-C (2024) Improved classification of different brain tumors in MRI scans using patterned-GridMask. IEEE Access
3. Anantharajan S, Gunasekaran S, Subramanian T, Venkatesh R (2024) MRI brain tumor detection using deep learning and machine learning approaches. Measurement: Sensors 31:101026
4. Asiri AA, Soomro TA, Shah AA, Pogrebna G, Irfan M, Alqahtani S (2024) Optimized brain tumor detection: a dual-module approach for mri image enhancement and tumor classification. IEEE Access 12:42868–42887
5. Kumar A (2023) Visual object tracking using deep learning. CRC Press
6. Kumar A, Walia GS, Sharma K (2020) Recent trends in multicue based visual tracking: a review. Expert Syst Appl 162:113711
7. Kumar A, Vohra R (2023) Impact of deep learning models for technology sustainability in tourism using big data analytics. Deep learning technologies for the sustainable development goals: Issues and solutions in the post-COVID era. Springer, pp 83–96
8. Kumar A, Nayyar A, Sachan RK, Jain R (2023) AI-assisted special education for students with exceptional needs. IGI Global
9. Kumar A, Singh D, Vohra R (2023) Improving learning abilities using AI-Based education systems. In: AI-Assisted special education for students with exceptional needs. IGI Global. pp 137–155
10. Kumar A, Mamgai R, Jain R (2023) Application of IoT-Enabled CNN for natural language processing. In: IoT-enabled convolutional neural networks: techniques and applications. River Publishers, pp 149–177
11. Kumar A, Sarren P (2023) Raja, deep learning-based multi-object tracking. In: Object tracking technology: trends, challenges and applications. Springer, pp 183–199
12. Kumar A, Ahluwalia R (2021) Breast cancer detection using machine learning and its classification. In: Cancer prediction for industrial IoT 4.0. Chapman and Hall/CRC, pp 65–78
13. Kumar A, Jain R (2021) Behavioral prediction of cancer using machine learning. In: Cancer prediction for industrial IoT 4.0. Chapman and Hall/CRC, pp 91–105
14. Almufareh MF, Imran M, Khan A, Humayun M, Asim M (2024) Automated brain tumor segmentation and classification in MRI using YOLO-based deep learning. IEEE Access
15. Chauhan AS, Singh J, Kumar S, Saxena N, Gupta M, Verma P (2024) Design and assessment of improved convolutional neural network based brain tumor segmentation and classification system. J Integr Sci Technol 12(4):793–793
16. Yu W, Kang H, Sun G, Liang S, Li J (2022) Bio-inspired feature selection in brain disease detection via an improved sparrow search algorithm. IEEE Trans Instrum Meas 72:1–15
17. Li W, Huang W, Zheng Y (2024) CorrDiff: corrective diffusion model for accurate MRI brain tumor segmentation. IEEE J Biomed Health Inf

18. Anaya-Isaza A, Mera-Jiménez L, Fernandez-Quilez A (2023) CrossTransUnet: a new computationally inexpensive tumor segmentation model for brain MRI. IEEE Access 11:27066–27085
19. Hossain S, Chakrabarty A, Gadekallu TR, Alazab M, Piran MJ (2023) Vision transformers, ensemble model, and transfer learning leveraging explainable AI for brain tumor detection and classification. IEEE J Biomed Health Inf
20. Amin J, Sharif M, Raza M, Yasmin M (2018) Detection of brain tumor based on features fusion and machine learning. J Amb Intell Humanized Comput, pp 1–17
21. Behera TK, Khan MA, Bakshi S (2022) Brain MR image classification using superpixel-based deep transfer learning. IEEE J Biomed Health Inf
22. Khushi HMT, Masood T, Jaffar A, Rashid M, Akram S (2023) Improved multiclass brain tumor detection via customized pretrained EfficientNetB7 Model. IEEE Access
23. Sharif M, Amin J, Raza M, Anjum MA, Afzal H, Shad SA (2020) Brain tumor detection based on extreme learning. Neural Comput Appl 32:15975–15987
24. Noback CR, Ruggiero DA, Strominger NL, Demarest RJ (2005) The human nervous system: structure and function. Springer Science & Business Media
25. Kumar A, Singh D (Eds.) (2025) Revolutionizing Medical Systems Using Artificial Intelligence: A Breakthrough in Healthcare. Elsevier
26. Bhalla S, Kumar A, Kushwaha R (2023) Comparative analysis on brain stroke prediction using machine learning. In: 2023 14th International conference on computing communication and networking technologies (ICCCNT). IEEE
27. Sachan RK, Kumar A, Shukla D, Sharma A, Kumar S (2023) Diagnosis of mental health from social networking posts: an improved ml-based approach. In: International conference on emergent converging technologies and biomedical systems. Springer
28. Singh D, Singh D, Manju, Gupta U (2023) Smart healthcare: a breakthrough in the growth of technologies. In: Artificial intelligence-based healthcare systems. Springer, pp 73–85
29. Tripathy SK, Singh D, Jaiswal A (2023) Multi-Layer feature fusion-based deep multi-layer depth separable convolution neural network for Alzheimer's disease detection. In: 2023 International conference on IoT, communication and automation technology (ICICAT). IEEE
30. Tripathy SK, Singh D, Srivastava S Srivastava R (2023) Multiscale low-level feature fused multilayer convolution neural network for Alzheimer's disease detection
31. Solanki S, Singh UP, Chouhan SS, Jain S (2023) Brain tumor detection and classification using intelligence techniques: an overview. IEEE Access
32. Liu Z, Tong L, Chen L, Jiang Z, Zhou F, Zhang Q, Zhou H (2023) Deep learning based brain tumor segmentation: a survey. Complex & intelligent systems 9(1):1001–1026
33. Soomro TA, Zheng L, Afifi AJ, Ali A, Soomro S, Yin M, Gao J (2022) Image segmentation for MR brain tumor detection using machine learning: a review. IEEE Rev Biomed Eng 16:70–90
34. Liu H, Ni Z, Nie D, Shen D, Wang J, Tang Z (2024) Multimodal brain tumor segmentation boosted by monomodal normal brain images. IEEE Trans Image Process
35. Metlek S, Çetıner H (2023) ResUNet+: A new convolutional and attention block-based approach for brain tumor segmentation. IEEE Access
36. Rajendran S, Rajagopal SK, Thanarajan T, Shankar K, Kumar S, Alsubaie N, Mostafa SM (2023) Automated segmentation of brain tumor MRI images using deep learning. IEEE Access
37. Ma Y, Wang Q, Cao L, Li L, Zhang C, Qiao L, Liu M (2023) Multi-Scale dynamic graph learning for brain disorder detection with functional MRI. IEEE Trans Neural Syst Rehabil Eng
38. Deepa B, Murugappan M, Sumithra M, Mahmud M, Al-Rakhami MS (2021) Pattern descriptors orientation and map firefly algorithm based brain pathology classification using hybridized machine learning algorithm. IEEE Access 10:3848–3863
39. Zhang X, Xie W, Huang C, Zhang Y, Chen X, Tian Q, Wang Y (2023) Self-supervised tumor segmentation with sim2real adaptation. IEEE J Biomed Health Inf
40. Xu M, Guo L, Wu H-C (2024) Novel robust automatic brain-tumor detection and segmentation using magnetic resonance imaging. IEEE Sens J
41. Alhassan AM, Zainon WMNW (2020) BAT algorithm with fuzzy C-ordered means (BAFCOM) clustering segmentation and enhanced capsule networks (ECN) for brain cancer MRI images classification. IEEE Access 8:201741–201751

42. Gumaei A, Hassan MM, Hassan MR, Alelaiwi A, Fortino G (2019) A hybrid feature extraction method with regularized extreme learning machine for brain tumor classification. IEEE Access 7:36266–36273
43. Ramachandran M, Patan R, Kumar A, Hosseini S, Gandomi AH (2021) Mutual informative MapReduce and minimum quadrangle classification for brain tumor big data. IEEE Trans Eng Manage 70(8):2644–2655
44. Jabbar A, Naseem S, Mahmood T, Saba T, Alamri FS, Rehman A (2023) Brain tumor detection and multi-grade segmentation through hybrid caps-VGGNet model. IEEE Access
45. Mohsen S, Ali AM, El-Rabaie E-SM, ElKaseer A, Scholz SG, Hassan AMA (2023) Brain tumor classification using hybrid single image super-resolution technique with ResNext101_32x8d and VGG19 Pre-Trained Models. IEEE Access
46. Renugadevi M, Narasimhan K, Ravikumar C, Anbazhagan R, Pau G, Ramkumar K, Prabu S, et al (2023) Machine learning empowered brain tumor segmentation and grading model for lifetime prediction. IEEE Access
47. Lamba K, Rani S, Anand M, Maguluri LP (2024) An integrated deep learning and supervised learning approach for early detection of brain tumor using magnetic resonance imaging. Healthcare Analytics, p 100336
48. Usman Akbar M, Larsson M, Blystad I, Eklund A (2024) Brain tumor segmentation using synthetic MR images—a comparison of GANs and diffusion models. Scientific Data 11(1):259
49. Ren T, Honey E, Rebala H, Sharma A, Chopra A, Kurt M (2024) An optimization framework for processing and transfer learning for the brain tumor segmentation. arXiv preprint arXiv:2402.07008
50. Shah HA, Saeed F, Yun S, Park J-H, Paul A, Kang J-M (2022) A robust approach for brain tumor detection in magnetic resonance images using finetuned efficientnet. IEEE Access 10:65426–65438
51. Zaitoon R, Syed H (2023) RU-Net2+: a deep learning algorithm for accurate brain tumor segmentation and survival rate prediction. IEEE Access
52. Aggarwal M, Tiwari AK, Sarathi MP, Bijalwan A (2023) An early detection and segmentation of brain tumor using deep neural network. BMC Med Inform Decis Mak 23(1):78
53. Wang M, Xu Y, Wang Z, Xing C (2023) Deep margin cosine autoencoder based medical hyperspectral image classification for tumor diagnosis. IEEE Trans Instrum Measur
54. Agarwal R, Pande SD, Mohanty SN, Panda SK (2023) A novel hybrid system of detecting brain tumors in MRI. IEEE Access 11:118372–118385
55. Shah SMAH, Ullah A, Iqbal J, Bourouis S, Ullah SS, Hussain S, Mustafa G et al (2023) Classifying and localizing abnormalities in brain MRI using channel attention based semi-Bayesian ensemble voting mechanism and convolutional auto-encoder. IEEE Access
56. Vinod D, Prakash SS, AlSalmanH, Muaad AY, Heyat MBB (2024) Ensemble technique for brain tumour patient survival prediction. IEEE Access
57. Vinu M, Pasupathy V, Senthilkumar K, Balasubramanian K, Dhanaselvam J, Kumar K (2024) Accurate brain tumour segmentation in MRI images using enhanced CNN and machine learning methods. Int J Intell Syst Appl Eng 12(11s):547–556
58. Mallampati B, Ishaq A, Rustam F, Kuthala V, Alfarhood S, Ashraf I (2023) Brain tumor detection using 3D-UNet segmentation features and hybrid machine learning model. IEEE Access 11:135020–135034
59. Saeedi S, Rezayi S, Keshavarz H, Niakan Kalhori SR (2023) MRI-based brain tumor detection using convolutional deep learning methods and chosen machine learning techniques. BMC Med Inf Decis Making 23(1):16

Chapter 4
Diagnosis and Prediction of Neurological Disorders Using Artificial Intelligence

Abstract With abrupt changes in living standards, economic instability, and excessive use of social media platforms, people are observed to be vulnerable to mental disorders. Mental disorders lead to an increase in stress levels which causes severe neurological complications in humans such as depression, suicidal tendencies, and other psychiatric problems. AI with its advanced tools has provided support for the prediction, monitoring, and planning of mental health illnesses in the population. AI has provided predictive models that can analyze health records, brain imaging, and clinical notes to identify mental disorders. AI is prevalent in analyzing social media platforms such as Twitter, Facebook, and many more for diagnosing depression in the user by extracting critical information from the tweets and comments posted by them. In this chapter, we have discussed the salient features and limitations of various AI-based predictive models useful in addressing mental disorder complications at an early stage.

Keywords Artificial intelligence (AI) · Clinical diagnosis · Social media · Signalling data · COVID-19 · Personalised treatment

4.1 Introduction

One of the most common reasons for mental illness and neurological complications is depression. Depression causes an increase in stress levels and is considered the foremost reason for mental disability, suicidal ideation, anxiety, schizophrenia, bipolar disorder, and psychological impairments [1]. Depression can impact any individual regardless of their sex, age, and ethnicity. Depression is a psychic condition whose early diagnosis is very crucial to prevent its negative impact on human lives. Mental illness impacts the health, remembering ability, cognitive skills, and societal well-being of an individual [2].

Generally, mental disorders in a human can be diagnosed either clinically or by analyzing social media posts. The symptoms such as changes in sleep patterns, mood swings, difficulty in making decisions, and variations in concentration are a few

A. Kumar and D. Singh, *Artificial Intelligence in Modern Healthcare System*, Transactions on Computer Systems and Networks,
https://doi.org/10.1007/978-981-96-6703-1_4

parameters that help in the clinical diagnosis of mental illness. Clinical depression is a neurological disorder that can be identified by analyzing textual data, imaging data, and signaling data. For this, electronic health records (EHR) [3–5], Magnetic resonance imaging (MRI) scans [6], voice data [7, 8] and Electroencephalography (EEG) [9–11] signals are investigated for clinical diagnosis of various mental disorders. On the other hand, social media data is also explored by various researchers for the diagnosis of mental illness at an early stage [12–15]. Social media is very prevalent among all age groups people and has worldwide connectivity. People's profiles, posts, comments, and suggestions can be examined for predicting depression, suicidal tendencies, anxiety, and changes in a person's behavior and attitude [16–18]. Figure 4.1 represents various categories for analysis of depression in human beings.

To diagnose mental illness at an early stage, AI-based algorithms have gained popularity in the automatic detection of depression, anxiety, and suicidal ideation. AI-based algorithms can predict critical diseases such as cancer [19], brain stroke [20], and many others [21, 22]. AI-based algorithms can be broadly categorized as machine learning (ML) based algorithms [12, 18, 23, 24], deep learning (DL) based algorithms [25–28], and hybrid algorithms [29–33] for either clinical or social media-based diagnosis of mental health. ML-based algorithms utilized Logistic regression (LR), Naïve Bayes (NB), Support vector machine (SVM), K-nearest neighbor (KNN), Decision tree (DT), and many other popular approaches. DL-based algorithms exploit convolutional neural networks (CNN), recurrent neural networks (RNN), Long short-term memory (LSTM), transformers, and many other extended networks to provide efficient results. Hybrid algorithms integrated ML and DL approaches along with

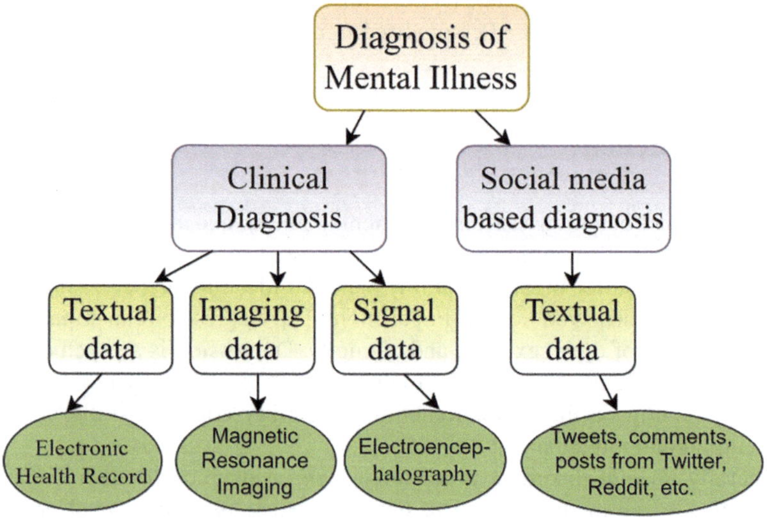

Fig. 4.1 Various methods for diagnosis of mental illness in an individual

natural language processing (NLP) and clustering techniques for the detection of depression.

In this chapter, we have analyzed the recent AI-based algorithms exploiting ML, DL, and a combination of both for predicting mental health. These algorithms have the potential to predict mental illness at an early stage along with its severity level. On the other hand, traditional approaches are quite slow and are known to suffer from human bias in comparison to AI-based approaches. AI-based is not only fast but also efficient and effective in mental health diagnosis. The key contributions of this chapter are as follows:

- We have categorized the diagnosis of mental health either clinically or by analyzing social media data. The salient features of each category are highlighted and elaborated to determine its benefits and limitations.
- AI-based predictive models for mental healthcare are reviewed into three categories namely, ML-based, DL-based, and hybrid approaches. The potential work in each category is elaborated and tabulated to examine the accuracy and efficiency of each method.
- Various AI-based solutions are detailed for improving the teaching–learning experience of students with special needs. AI-based tools are also analyzed to enhance the learning ability of specialized students.
- Mental health during the COVID-19 pandemic is analyzed to determine the role of social media platforms in predicting mental disorders using AI-based algorithms.

The rest of the chapter is organized as follows. Section 4.2 details the clinical diagnosis and social media-based prediction of various mental disorders. Section 4.3 classifies AI-based prediction models into various categories. The salient features of various methodologies under each category are elaborated to investigate their efficiency and accuracy. Various AI-based methods for students with special needs are discussed in Sect. 4.4. In addition, various AI-based tools adopted for enhancing the learning abilities of students with special needs are also discussed. Section 4.5 examines the impact of COVID-19 on people's mental health and the increase in depression among people due to isolation and lockdown situations. Lastly, the concluding remarks and future directions are sketched in Sect. 4.6.

4.2 Diagnosis of Mental Disorders

Mental disorders cause changes in the thoughts, behavior, and personality of a person suffering from distress and physiological impairments. Mental disorders can occur in humans in the form of depression, anxiety, suicidal ideation, and many other major depressive disorders [1]. These disorders can be treated if diagnosed accurately. There are two ways to predict depression in human beings, either clinical diagnosis or social media-based diagnosis. The details about the clinical diagnosis of depression are as follows.

4.2.1 Clinical Diagnosis of Mental Disorders

Early prediction and diagnosis of mental disorders can be done through clinical intervention. Clinical diagnosis of mental disorders can be done by investigating various medical data namely, textual, imaging, voice, and signaling. Textual data involves a patient's prescription, medical notes, or EHR data [3–5]. Imaging data considers MRI scans and signaling data analyzed EER signal [6, 34]. Audiovisual recordings are analyzed to predict mental-health-related problems so that appropriate measures are taken for their treatment [7, 28].

In [3], authors have reviewed the work utilizing EHR data, and brain imaging data to predict a person's mental status. EHR data is the subjective and written form of analysis of a patient's mental health. This data has the potential for prediagnosis of mental illness with the help of AI-based screening tools. Authors have utilized patient's medical history documented in the form of EHR data for predicting depression [4]. This data contains information about the patient's procedure and demographic information for predicting chronic diseases like depression. Similarly, Msosa et al. [5] have exploited unstructured data from EHR to diagnose mental illness. EHR records are easily accessible, and their processing requirements are flexible and simple. These records are considered to be the cheapest and richest source of health information crucial for predicting mental wellness.

Further, MRI data can be categorized as structural MRI (sMRI) and functional MRI (fMRI) for analysis of a person's mental health [24]. sMRI contains anatomical details about the human brain whereas fMRI contains underlying brain functioning. In [6], authors have processed sMRI to predict various psychiatric disorders [6]. sMRI measures the alterations in brain data by analyzing its anatomical structure to diagnose critical mental disorders such as schizophrenia and bipolar disorder. Mousavian et al. [24] have exploited fMRI for extracting information from various brain regions to provide accurate diagnosis. fMRI data of a person can be captured either when he is performing any activity or idle. Capturing either of the MRI scans is quite expensive. On the other hand, EEG signaling is a non-invasive and effective technique that captures electrical signals from the brain for detecting mental disorders [10]. These signals are complex in nature but easy and cheap to record. These signals act as a potential biomarker for predicting depressive disorders in a person. Figure 4.2 displays the difference in EEG signals recorded for a healthy and depressed person. Authors have recorded EEG signals by capturing brain waves for a specified duration to predict neurological disorders [11]. Captured EEG signals are analyzed for predicting the complex clinical depression in a person. On the other hand, Sarkar et al. [34] exploited the EEG dataset from the Kaggle website for predicting clinical depression. The datasets consist of various dependent and independent variables for determining mental illness.

Audio or speech analysis is a robust technique used for depression estimation along with text records and videos [28]. Speech has acoustic features that assess depression efficiently. The change in speaking patterns can be observed in a person suffering from mental diseases. Audio-visual capture by taking personal interviews

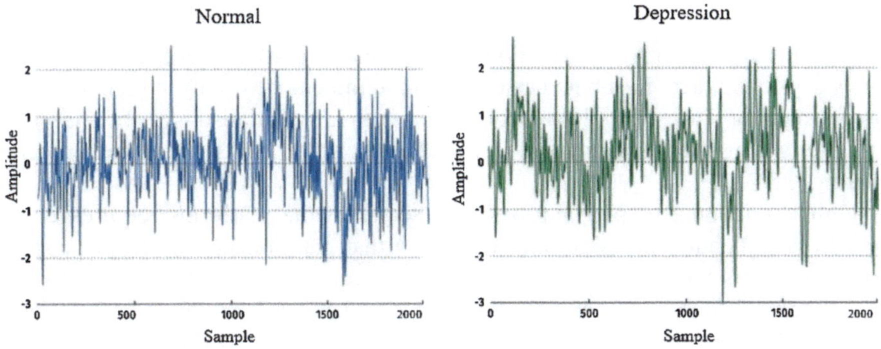

Fig. 4.2 Difference in EEG signal recorded for a normal and depressed person [10]

can be used for predicting depression [7]. These multi-modal data are a strong descriptor for diagnosing mental health. Speech analysis is cheap, non-invasive, and efficient in predicting neurological disorders.

To summarize, unstructured data such as clinical notes and EHR records predict mental illness efficiently. Careful analysis of these records can recognize changes in patient's behavior and attitude. However, this data is different for different professions, and extracting relevant information is a complex task. Another potential biomarker in determining mental health is MRI scans. These scans are non-invasive and provide 3D view of the brain anatomy for better visualization of depression. However, MRI scans are of high dimensionality and require a lot of processing power to extract desired results. Signaling data such as EER is easy to capture and requires less computation for predictions. Multi-modal data such as audio and text information are fused efficiently to study depression in a person. Text information comprises a set of questionnaires to assess depression and non-depression whereas audio data analyses the variation in speech of a person suffering from mental disorders.

4.2.2 Social Media-Based Diagnosis of Mental Disorders

Clinical diagnosis of mental depression is fast, accurate, and effective. However, due to societal barriers and unawareness people neither have understanding nor have acceptability for their mental illness. Due to this proper treatment cannot be provided to those people and that will impact their quality of life and peace of mind. Hence, it became essential to analyze people's day-to-day activities and behavior patterns to predict the status of their mental health. Textual data from various social networking websites can be examined to predict mental disorders at an early stage [2, 12, 13, 35].

Recently, social media platforms such as Twitter [13, 15], Facebook [16], and Reddit [14, 17] are quite popular among people of all age groups. People create their profiles on these websites and share their thoughts, opinions, and emotions which can be investigated to detect various mental issues such as depression, anxiety, and suicidal ideation. Sharing information on social media platforms not only represents one's day-to-day activities but is also a potential indicator to analyze mental health. Social media are investigated to track mental health in a naturalistic way by analyzing and interpreting one's behavior and feelings based on their profile, likes, and thoughts sharing [35].

Online screening of people's posts on social media platforms requires a tool or API (Application programming interface) that can collect data from people's social accounts with their consent and agreement [13]. These tools or APIs can access people's public data and check for specific keywords or phrases. The related keywords such as "depression", "anxiety", "suicide", "kill" and other similar expressions are retrieved from various social media accounts. After this, these gathered posts are analyzed using NLP, ML, and/or DL approaches to predict the state of mind as depressed or healthy. Sentimental analysis determines linguistic patterns along with lexicon evaluation to determine stress, suicidal tendencies, and other bipolar disorders.

4.3 Artificial Intelligence-Based Models for Predicting Mental Healthcare

AI has provided many predictive models for analyzing mental healthcare exploiting clinical data such as EHR [4, 25], MRI [24], EER [10, 11], and multi-modality or social media posts and profiles. Based on the utilized methodology, AI-based predictive models are broadly categorized into ML-based algorithms, DL-based algorithms, and hybrid algorithms. ML-based algorithms utilized various ML techniques such as LR [14, 16], SVM [17, 36], KNN [24, 37], and RF [13], along with lexicon analysis [24] to detect depression. On the other hand, CNN, RNN, and LSTM [2, 10, 11] are exploited in DL-based predictive models for prediction. Hybrid models integrated ML and DL models along with NLP techniques to design a robust model with better efficiency and accuracy [31, 32]. Table 4.1 tabulates the similarities and differences between AI-based predictive models for predicting various mental disorders.

4.3.1 Machine Learning Based Models for Predicting Mental Disorders

Mental disorders are one of the known reasons for disability and other critical diseases such as diabetes, hypertension, and many others [39–41]. It is essential to diagnose mental disorders at an early stage to prevent personal and societal loss. For this, ML-based algorithms are investigated by various researchers for early diagnosis of depressive mental disorders. These algorithms utilize either clinical data or social

Table 4.1 Similarities and differences between various AI-based predictive models

SN	Attributes	ML-based algorithms	DL-based algorithms	Hybrid algorithms
1	Techniques	LR [14, 16], SVM [17, 36], KNN [24, 37], cluster analysis [15]	CNN, RNN, LSTM [2, 10, 11], Transformers [4, 38]	NLP and a combination of both ML and DL algorithms [3, 5]
2	Feature selection	✗	✓	✓
3	Data types	Clinical data and social-media data	Clinical data and social-media data	Clinical data and social-media data
4	Computational efficiency	Less	Moderate	High
5	Model generalizability	Limited	High	High
6	Performance	Less	Moderate to High	Relatively high
7	Advantages	Simple and easy to implement	Interpretable and explainable	Analyzed pool of data with better accuracy
8	Limitations	Less accurate	Complex in implementation	Highly complex

media data to predict the extent of mental illness. Table 4.2 tabulates the representative work that employs ML-based predictive algorithms to predict the status of mental health.

ML-based depression diagnosis analyzes the clinical data namely, text, audio, imaging, and signals. In this direction, Chao et al. [23] captured brain responses of depressed and healthy patients using fNIRS (functional near-infrared spectroscopy) devices. Statistical-based features and vector-based features were extracted from brain responses and processed using four variants of neural networks. ReliefF was used for selecting robust features and specifying critical brain regions for predicting depressive disorder. Authors extracted multiple behavioral features such as speech behavior, speech prosody, eye movements, and head pose [36]. Thirty-eight feature selection algorithms were utilized for interpreting depression. On the other hand, Hao et al. [42] analyzed depression and anxiety in undergraduate students. Questionnaire data was collected from selected students and clustered into two distributions namely, anxiety and depression using a random sampling method. The correlation between these clusters was identified using the correlation analysis method. Similarly, the authors surveyed to prepare a dataset from 21 questionnaire details selected using the Hamilton tool and psychiatrist suggestions [37]. These details were then analyzed using various ML-based algorithms to detect a person suffering from depression.

Recently, social media has been explored for analyzing depression by studying emotional expression and sentiments from the data posted and available on the websites. In this direction, authors collected data from Twitter by analyzing tweets with timestamps and hashtags related to mental disorders [15]. Similarity calculations and stochastic gradient descent were exploited for analyzing sentiment distribution to predict the user suffering from severe depression disorders in real-time. Safa et al.

Table 4.2 ML-based techniques for predicting various mental disorders

References	Dataset description	Model utilized/proposed	Mental illness type	Performance measures	Summary
Chao et al. [23]	• Patient data from Tisanshui Third Hospital, China • Participants: Total (32), mental disorder (16), healthy (16)	MNN, FNN, CFNN, RNN	Depression	Mean, AUC, STD, and slope	• Utilized fNIRS to monitor mental state and diagnose psychiatric behavior for depressive disorders • Achieved highest ACC (99.94%) for fear emotion by combining multiple features
Alghowinem et al. [36]	• BDI: depressed (30), Healthy (30) • Pitt: low depression (19), severe depression (19) • AVEC: 340 video clips	SVM	Depression	ACC, and feature-based classification results	• Selected multiple datasets to prove the generalizability of the model • Performed various levels of aggregation to enhance model accuracy • Selected features from different levels of aggregation prove the effectiveness of the model on unseen and different datasets
Safa et al. [13]	Self-generated dataset from Twitter using Twitter API	BoW, NLP, Lexicon analysis	Depression	ROC, ACC, AUC, F1-Score, PR, and RE	• Automatically collected data from user's profile and tweets • For model computational efficiency, reduced feature size using correlation analysis • Highest overall ACC (91%) using the Catboost/GB model
Hao et al. [42]	1425 participants from 18 Chinese universities	CRS method	Depression, and anxiety	NRC, SE, SRC, t-statistics, Goodness of fit, F-statistics, and p-value	• Predicted depression in undergraduates • Multiple regression for correlating depression, anxiety, and social support • Suggested the need for enhanced social support to reduce the risk of depression in students

(continued)

Table 4.2 (continued)

References	Dataset description	Model utilized/proposed	Mental illness type	Performance measures	Summary
Sofia et al. [37]	Questionnaire-based self-generated dataset	DT, KNN, NB	Depression	ACC, and confusion matrix	• Included positive as well as negative questions in the questionnaire dataset • Trained model using three ML techniques and selected the optimal • KNN achieved the highest accuracy among other models
Seuklic et al. [14]	Reddit dataset	SVM, LR, RF	Depression, and bipolar disorder	F1-score, MCC, and ACC	• Exploited three features namely, psycholinguistic, lexical, and Reddit user • Achieved ACC and F1-score more than 86% in comparison to other baselines
Zhou et al. [15]	Twitter hashtag	Cluster analysis	Depression, anxiety, OCD, bipolar and panic disorder	PR, RE, and F1-Measure	• Multipolarity analysis on short texts gathered from social media • Analyzed and refined the 11-dimensional distribution of mental illnesses • Real-time monitoring of mental health but a small sampling dataset restricted accuracy
Wang et al. [18]	• Blog data from APC: 125,559 blogs • Sentiment140 dataset: 1,048,575 tweets • Mental disorder tweet data: 167,072 tweets • Occupation-based tweets: 164,625 tweets	Sequential emotion analysis	Depression, anxiety, OCD, bipolar and panic disorder	PR, RE, and F1-Measure	• Categorized state of mental disorder into various classes such as severe, moderate, mild, confirmed, and tendency • Analyzed occupational-based mental disorders to identify the stressful occupation • Evident that the tendency of mental disorders is more in the waiter profession • Highest ACC (>90%) in various mental disorder types

(continued)

Table 4.2 (continued)

References	Dataset description	Model utilized/proposed	Mental illness type	Performance measures	Summary
Tadesse et al. [17]	• Reddit dataset • Depression posts: 1293 • Non-depression posts: 548	NLP, MLP, and SVM	Depression	ACC, F1-score, PR, and RE	• Extracted multiple features namely, LIWC, LDA, and bigram • NLP and text classification techniques to identify the lexicon of words • MLP with the highest ACC (91%) and F1-score (0.93) integrating all three features
AlSagri and Ykhlef [12]	• Twitter dataset • Total tweets: 300 K	NB, DT, and SVM	Depression	ACC, PR, RE, confusion matrix, AUC and F1-measure	• Extracted LIWC, sentiment, synonyms, and social activity as features • SVM performed better in comparison to other ML techniques • Overall ACC (82.5%) and RE (0.85)
Chiong et al. [16]	• Total instances: Twitter (22,191), Facebook (9178), Reddit (50,000), and Victoria Diary (62) • Depression: Twitter (8807), Facebook (9178), Reddit (50,000), and Victoria diary (62) • Non-depression: Twitter (13,384), Facebook (0), Reddit (0), and Victoria diary (0)	LR, Linear SVM, DT, and MLP	Depression	ACC, PR, RE, and F1-measure	• Detected depression in social media texts without searching for direct words • Used multiple classifiers and algorithms to prove the efficiency of the model • RF achieved the highest accuracy on all datasets in comparison to other ensemble techniques

(continued)

Table 4.2 (continued)

References	Dataset description	Model utilized/proposed	Mental illness type	Performance measures	Summary
Mousavian et al. [24]	• NKI-E: Total (279), Depression (38), CTL (241) • MPI-LMBD: Total (187), Depression (22), CTL (165) • Closed-eyes: Total (72), Depression (51), CTL (21)	KNN, DT, RF, MLP, AdaBoost, Bernoulli NB, Gaussian NB, LDA, SGD, Bagging, Boosting	ROC, AUC, ACC, SEN, SPE, PR, rank-sum test	Depression	• Exploited similarity of spatial cubes for creating features to detect depression in healthy individuals • Utilized random oversampling to address the imbalance in datasets • Selected top features for improved performance using statistical tests • The small dataset size restricted the generalizability of the model

ML: Machine learning, DL: Deep learning, NLP: Natural language processing, ROC: Receiver operating characteristics, ACC: Accuracy, AUC: Area under the curve, PR: Precision, SEN: Sensitivity, SPE: Specificity, MNN: Multilayer neural network, FNN: Feedforward neural network, CFNN: cascade forward neural network, RNN: Recurrent neural network, STD: Standard deviation, fNIRS: Functional near-infrared spectroscopy, BDI: Black dog institute, Pitt: University of Pittsburgh depression dataset, AVEC: Audio/Visual emotion challenge depression dataset, SVM: Support vector machine, CRS: Cluster random sampling, NRC: Nonstandard regression coefficient, SE: Standard error, SRC: Standard regression coefficient, DT: Decision tree, KNN: K-nearest neighbor, SVM: Support vector machine, LR: Logistic regression, NB: Naïve Bayes, MCC: Majority class classifier, OCD: Obsessive–compulsive disorder, MLP: Multilayer perceptron, APC: Admissive psychology community, NKI-E: Nathan Kline Institute-Enhanced, MPI-LMDB: MPI-Leipzig-Mind-Brain-Body, CTL: Psychologically and neurologically typical control, LDA: Linear discriminant analysis, SGD: Stochastic gradient descent

[13] collected Twitter data and hashtags to detect depression automatically. The collected data was evaluated using a multi-modal approach such as n-gram language models, LIWC dictionaries, image tagging, and bag-of-visual words (BoW). Nine ML-based different classifiers were utilized to measure model effectiveness. Similarly, authors evaluated Twitter users' accounts and classified them into depressed and non-depressed accounts by analyzing activities and content features. Multiple ML-based algorithms utilized to predict mental and psychological issues. On the other hand, Reddit user data was examined by various authors to categorize the users into depression and other bipolar disorders [14, 17]. In [14], authors extracted three features namely, psycholinguistic lexical and Reddit user features. These features were processed using three ML algorithms namely, SVM, RF, and LR to predict the bipolar disorders in users. However, authors extracted three features namely, LIWC, LDA, and bigram [17]. These features were analyzed individually and combined using various algorithms such as LR, SVM, Adaboost, RF, and MLP to predict the presence of depression in social media users.

To explore the robustness of the methods, authors investigated Twitter data along with other social networking data such as Facebook, Reddit, and Victoria Diary to predict depression effectively. For this, authors utilized Twitter data for training the model, and model performance was tested on other three public datasets to prove the model's generalizability [16]. Textual features were extracted and processed using various ML-based classifiers to detect depression in the users. On the other hand, authors identified depression in various occupations by gathering individual data from different social media sites. Sequential emotion patterns were examined using sixteen parameters to predict the severity of depression.

To summarize, ML-based algorithms examined clinical data and social networking profiles to identify persons suffering from mental illness. These models extracted multiple features and different ML algorithms to classify the depressed users from the healthy. Model performance was reviewed on multiple datasets to determine the generalizability and predict signs of depression in human beings.

4.3.2 Deep Learning Based Models for Predicting Mental Disorders

Table 4.3 tabulates the salient features of the representation work predicting depression by processing data using DL-based algorithms. DL-based models are exploited to predict severe mental illness and level of mental disorders by utilizing high-level networks such as CNN, RNN, LSTM, and transformers. In this direction, authors integrated bidirectional LSTM and CNN and evaluated the performance on multi-modal datasets containing voice and text information for predicting clinical depression [7]. 1-D and 2-D audio signals were extracted and integrated with mapped numeric values obtained from text information using multiple DL models. The obtained softmax values from DL models were ensembled to predict depression.

However, the performance of the model was evaluated on limited data size due to a smaller number of participants. To address this, authors proposed an Audio-assisted BERT model based on DL architecture to predict depression by integrating multiple modalities such as audio and text information [8]. Pre-trained text and audio models composed of multiple components are processed using the dual self-attentional model to improve classification accuracy and enhance model performance. However, the authors addressed the limitation of smaller datasets by expanding training labels and feature transfer learning [28]. Low- and high-level audio features were extracted using RNN and LSTM. The first and second-degree of audio features were trained using Mel frequency cepstral coefficients to diagnose mental state automatically. In [26], authors analyzed the time–frequency representation of audio signals using DL-based EmoAudioNet. High and low-level audio features were aggregated and processed using the CNN network to obtain classification scores for the identification of major depression disorders.

Clinical prediction of depression is performed using medical data such as EHR, and EER [4, 9–11, 25]. EEG signals record brain waves effectively which can be used for classifying depressed individuals from healthy. In this direction, authors extracted brain signals using generalized partial directed coherence and direct directed transfer function methods [9]. Each individual image was constructed from EEG signals and processed using five different DL models to automatically learn patterns from EEG. These models captured spatial and temporal features from the brain to diagnose major depressive disorder. Similarly, Sharma et al. [10] diagnosed depression by analyzing EEG signals using CNN and LSTM. CNN was used for preprocessing by windowing the EEG signals over time series and LSTM determines sequence learning by extracting local features. In [11], authors exploited CNN and LSTM to identify neurological disorders using EEG signals. Local features were extracted using CNN signals and LSTM to learn local characteristics and patterns in the EEG. LSTM used memory cells to remember important features and update feature weights during training to identify right and left hemispheres EEG signals for diagnosing clinical depression. On the other hand, clinical depression can also be predicted using high-dimensional data from EHR clinical notes [4]. Bidirectional representation learning model with transformer architecture was exploited to model the five features namely, diagnoses, procedure codes, medications, demographics, and clinical data extracted from EHR. The model was pre-trained and fine-tuned to process temporal data for predicting the future possibilities of depression. Similarly, Bertl et al. [25] evaluated the real-time medical claim data and analyzed temporal properties using DL models with GRU decay. Explainability was introduced in the model by incorporating self-attentional model for depression screening for better quality of life.

Early depression signs can also be traced by analyzing social media using DL-based algorithms [2, 27, 35, 38]. In this direction, authors analyzed textual social media data using LSTM and RNN [2]. To manage the model efficiency, two hidden layers in LSTM and two dense layers in RNN were used for predicting early signs of depression and suicidal tendencies. Also, one hot encoding and principal component analysis methods represented the depression symptoms and sentiments in social media data. In [27], authors utilized DL models for tracing the early signs of mental

Table 4.3 DL-based techniques for predicting various mental disorders

References	Dataset description	Model utilized/ proposed	Mental illness type	Performance measures	Summary
Zhang et al. [6]	• Training: 14,915 (mental illness instances), 4538 (healthy instances) • Test: 290 (mental illness instances), 310 (healthy instances)	DL-MIL	Severe mental illness including depression, Schizophrenia, anxiety, and other disorders	AUC, ACC, SEN, and SPE	• Utilized MRI scans to predict severe mental illness • Socio-demographics such as gender and age for better generalizability of the model • Achieved the highest ACC (82%) on both primary and unseen datasets
Jo and Kwak [7]	• EDAIC-WOZ dataset • Total instances (275), depression (66), non-depression (209) • Depression: Training (37), Validation (12), Test (17) • Non-depression: Training (126), Validation (44), Test (39)	Bidirectional-LSTM and CNN	Depression	ACC, F1-score, PR, RE, and confusion matrix	• Combined features extracted from voice and data to predict depression • Better performance on voice data in comparison to text data • Applied data augmentation and tokenization as preprocessing steps • Achieved highest ACC (96.67%) using four stream DL model

(continued)

Table 4.3 (continued)

References	Dataset description	Model utilized/proposed	Mental illness type	Performance measures	Summary
Ramírez-Cifuentes et al. [27]	• Total instances dataset 1: suicide (7075), depression (3015), alcoholism (250), and eating disorders (784) • Total instances dataset 2: mental disorders (11,124), control (20,057)	DL-based predictive model	Depression, eating disorders, suicidal ideation, and other mental disorders	ACC, RE, and F1-score	• Exploited lexicon-based features to characterize mental disorders in Reddit posts • Analyzed social media posts to predict mental conditions considering the psychological risk factors • Achieved highest ACC (98.40%) on baseline model with BoW approach
Ghosh et al. [35]	• Multiple datasets • Total instances: 10,659 • Depression: 5899 • Non-depression: 4749	BiGRU and CNN	Depression, emotion recognition	ACC, F1-score, and PCC	• Analyzed Twitter user data and profiles for predicting depression • Improved model performance using multi-modal features • Achieved overall ACC (70%)

(continued)

Table 4.3 (continued)

References	Dataset description	Model utilized/ proposed	Mental illness type	Performance measures	Summary
Bertl et al. [25]	EHR data of 812,853 patients	Attention-based GRU decay model	Depression	AUC, AUPRC, SEN and SPE	• Integrated attentional module to ensure model explainability • Achieved overall scores as AUC (0.99), AUPRC (0.97), SEN (0.94), and SPE (0.99)
Amanat et al. [2]	• Tweets-Scraped dataset from Kaggle • Total: 4 K tweets	LSTM and RNN	Depression	ACC, PR, RE, F1-Measure, Confusion matrix, and support	• Examined robust features using the one-hot methodology • Stemming and lemmatization along with PCA and one-hot for dataset cleaning and feature extraction • Achieved overall ACC (99%)
Sharma et al. [10]	EEG signal data from the University of Arizona, USA	CNN + LSTM	Depression	ACC, and MAE	• Utilized CNN for temporal learning and LSTM for sequence learning • Automatic learning of model without any feature extraction • Achieved ACC (99.1%) in 45 subjects

(continued)

Table 4.3 (continued)

References	Dataset description	Model utilized/ proposed	Mental illness type	Performance measures	Summary
Thoduparambil et al. [11]	EEG recordings from the University of New Mexico	CNN, and LSTM	Depression	ACC, SEN, and SPE	• Automated system to predict depression utilizing clinical data • Extracted feature using CNN and model learning using LSTM • Achieved ACC right hemisphere (99.07%) and left hemisphere (98.84%)
Othmani et al. [26]	• RECOLA dataset: 25 audio recordings (Avg. duration: 5 min) • DAIC-WOZ depression dataset: Utilized 182 audio recordings (Avg. duration: 15 min)	DL-based EmoAudioNet	Depression	ACC, PCC, RMSE, confusion matrix, F1-score	• Analyzed short-time spectral analysis to predict emotional dimensions • Exploited time–frequency representation and spectrum of audio signal • Achieved ACC (73.25%) and F1-score (82%)

(continued)

Table 4.3 (continued)

References	Dataset description	Model utilized/proposed	Mental illness type	Performance measures	Summary
Rejaibi et al. [28]	DAIC-WOZ depression dataset: Utilized 182 audio recordings (Avg. duration: 15 min) RAVDESS: 1440 audio recordings AVi-D: 150 audio recordings	MFCC-based RNN MFCC-based RNN using TL	Depression	ACC, RMSE, PR, RE, F1-Score, and Confusion matrix	• Weighting strategy to deal with imbalanced datasets • Gender-based assessment to address the gender bias in the evaluation • Achieved avg. ACC (76.27%), RMSE (0.4), and F1-score (46%) for depression
Toto et al. [8]	• 15 thematic datasets extracted from DAIC-WOZ • 189 audio recordings (Avg. duration: 7–33 min)	AudiBERT framework	Depression	F1-score	• Multi-modal utilizing text and audio for depression prediction • Data augmentation using dual self-attention mechanism • Achieved F1-score (0.92)

(continued)

Table 4.3 (continued)

References	Dataset description	Model utilized/proposed	Mental illness type	Performance measures	Summary
Saeedi et al. [9]	EEG dataset: depression (33), non-depression (30)	1D-CNN, 2D-CNN, LSTM, 1D-CNN + LSTM, 2d-CNN + LSTM	Depression	SPE, ACC, F1-score, PR	• Analyzed EEG signal connectivity to the brain using GPDC and dDTF • 16 connectivity methods to construct images from EEG signals for prediction • Among all, 1D-CNN + LSTM achieved the best performance ACC (99.24%)
Meng et al. [4]	Patient's EHR dataset suffering from critical diseases	Bidirectional DL transformer architecture	Depression	ROCAUC, PRAUC	• Pre-training and finetuning for temporal representation of multimodal EHR data • Achieved model interpretability using self-attention weights in EHR sequences • Model prediction can support clinical outcomes for depression risk prediction

(continued)

Table 4.3 (continued)

References	Dataset description	Model utilized/proposed	Mental illness type	Performance measures	Summary
Zhang et al. [38]	Self-created Twitter dataset: depression (2575)	Transformer-based DL model using BERT, XLNet, and RoBERTa	Depression	ACC, F1-score, AUC, PR, RE	• Analyzed depression levels in different people on Twitter • Integrated psychological text features and demographic details for identifying depression trends • Achieved ACC (78.9%) on the collected dataset

CNN: Convolutional neural network, LSTM: Long-short term memory, RNN: Recurrent neural network, ACC: Accuracy, PR: Precision, RE: Recall, RMSE: Root mean square error, EDAIC-WOZ: Extended Distress Analysis Interview Corpus-Wizard of Oz, BoW: Bag-of-words, EHRs: Electronic health record, PCC: Pearson's coefficient correlation, DL: Deep learning, MIL: Multiple instance learning, AUPRC: Area under precision-recall curve, EEG: Electroencephalography, MAE: Mean absolute error, PCA: Principal component analysis, MFCC: Mel Frequency Cepstral Coefficients, TL: Transfer learning, GPDC: Generalized Partial Directed Coherence, dDTF: Direct directed transfer function disorders in social media posts. Cosine similarity and visual approach were defined to identify the variations in the social media data for multi-class classification. Zhang et al. [38] created large datasets containing depressed users and their past tweets. This dataset was processed using three DL-based transformers investigating psychological text features and sociodemographics to monitor group-level and population-level depression trends.

To summarize, DL-based have reported enhanced accuracy and effectiveness in detecting mental disorders by monitoring clinical data and social media data. Integration of speech and text information in the form of clinical notes has shown better efficiency. However, the model's generalizability in predicting future depression could not be assured due to limited training data availability. Also, EER signals have enough potential to represent brain waves and determine early signs of depression. The complex nature of these signals demands pre-processing steps which impacts the computational efficiency of the model. In order to reduce the model complications, social media data is also examined using a recent transformer-based DL model for predicting users suffering from mental illness. The prediction accuracy of these models is highly dependent on the truthfulness of the data shared by users on social media.

4.3.3 Hybrid Models for Predicting Mental Disorders

To address the limitations of ML and DL-based models, hybrid models are investigated for predicting mental disorders [1, 3, 32]. These models integrate the benefits of both ML and DL models along with NLP and supervised and unsupervised algorithms to predict various mental disorders [30, 33, 34]. These models also examined clinical data and social media data for depicting mental illness. Table 4.4 tabulates the salient features of the representative work utilizing hybrid predictive models for various mental disorders.

Authors investigated the signs of depression in EHR data collected from Mercy Care by integrating NLP, LSTM, RF, and GBT [5]. EHR data was annotated using NLP services such as MedCAT and BioYODIE. Feature importance was employed to identify the sensitive features for improving the computational complexity of the system. On the other hand, Sarkar et al. [34] analyzed EEG clinical data using DL models and supervised ML models for tracking depression. In DL-model LSTM with RNN outperformed the other algorithms but in ML techniques SVM and LR were superior in detecting depression in EEG brain waves. In [29], authors investigated the speech samples collected globally by performing sentiment analysis to identify the traces of depression, anxiety, and loneliness. The patterns in speech were extracted by DL-based neural network and clustering technique to predict the patient's behavior and influence of depression.

To predict depression, anxiety, and suicidal ideation using social media, Fatima et al. [3] integrated message-level sentiment analysis with three DL models for robust feature extraction from social media data. Four classifiers were trained to detect the

Table 4.4 Hybrid techniques for predicting various mental disorders

Reference	Dataset description	Model utilized/proposed	Mental illness type	Performance measures	Summary
Graham et al. [3]	28 studies depicting mental illness	ML, DL, and NLP	Depression, Schizophrenia, and Suicidal ideation	ROC, ACC, AUC, PR, SEN, SPE and F1-score	• Analyzed studies using EHRs to predict various neurological disorders • Most of the studies fail to consider age as a feature indicator Utilized NLP before applying supervised ML techniques to predict various mental illnesses
Msosa et al. [5]	• EHRs dataset from Mersey care • Total instance: 94,605 • Depression: 41,967 Non-depression: 52,638	NLP, RF, GBT, and LSTM	Depression	AUC, PR, RE, F1-score, SEN, SPE, ACC	• Utilized textual data of EHRs to analyze the patient's mental health • Annotated unstructured EHR data using NLP services such as MedCAT and BioYODIE • Best performance on RF-trained LSTM with mean AUC (0.901)
Fatima et al. [31]	• Training: Total (887), depression (135), non-depression (752) • Test: Total (820), depression (79), non-depression (741)	NB, CNN, LSTM, and CNN-BiLSTM with attention	Depression	ACC, PR, RE, F1-score, and confusion matrix	• The model can predict the early stages of depression by analyzing social media posts • Training and testing on different datasets proved model generalizability • Attention-based CNN-BiLSTM achieved the best results as ACC (0.97), PR (0.84), RE (0.95), and F1-score (0.92)
Haque et al. [32]	• Self-collected Reddit dataset: 1895 posts • C-SSRS dataset: 500 posts • NLP dataset: 50 K reviews	DL-based SDCNL and unsupervised label correction	Depression and suicidal ideation	ACC, ROC, PR, RE, F1-score, and AUC	• Method can be used as a preliminary approach to screen the depression in a person • Multiple combinations of models to identify the best • GUSE-dense achieved the highest ACC (72.24%), AUC (77.76), and F1-score (73.61)

(continued)

Table 4.4 (continued)

Reference	Dataset description	Model utilized/proposed	Mental illness type	Performance measures	Summary
Wani et al. [1]	• Total instances: FC (5700), YC (14,029), TC (11,590) • Depression instances: FC (2700), YC (7520), TC (6080) • Non-depression instance: FC (3000), YC (6509), TC (5510)	Advanced NLP, CNN, LSTM, and CNN + LSTM	Depression	ACC, PR, RE, and F1-score	• Utilized Word2Vec-based features and TF-IDF-based representation to identify depression symptoms analyzing three popular social media sites • Analyzed depression in comments written in English language only • Achieved the highest ACC (99.02%) with the Word2Vec feature on the YC dataset
Nadeem et al. [33]	• Kaggle dataset • Total tweets: 31 K	SSCL, SVM, LR, LSTM, CNN, BiLSTM, and self-attention mechanism	Depression	ACC, F1-score, Confusion matrix	• Utilized N-gram and TF-IDF techniques to extract features for ML models • For DL models, Word2Vec, FastText, and GloVe were used as feature extraction methods • BiLSTM and CNN with FastText achieved the highest ACC and F1 score on binary data at 96.6%
Sarkar et al. [34]	• EEG brain wave dataset from Kaggle	DL: CNN, RNN, MLP, RNN + LSTM ML: SVM and LR	Depression	ACC and recognition rate	• Among DL-models RNN and RNN + LSTM achieved superior results • ML approach SVM and LR predicted depression efficiently • RNN Model ACC: Training (97.05%) and Testing (96.50%) • ML model ACC: Training (100%) and Testing (97.25%)

(continued)

Table 4.4 (continued)

Reference	Dataset description	Model utilized/proposed	Mental illness type	Performance measures	Summary
Ansari et al. [30]	• CLPsych 2015 Shared task: Total (1146), depression (327) • Reddit: Total (1841), depression (1200) • eRisk: Total (4498), depression (770)	Ensemble model using LSTM and LR	Depression	PR, RE, F1-score, and ACC	• Examined the impact of single and combined lexicon features on model performance • Parallel training of classifiers to emphasize different features • The ensemble model achieved ACC (75%) and F1-score (0.77) in Reddit social media data
Ahmed et al. [29]	• 2173 trained global speech samples	Neural network and clustering technique	Depression, anxiety, loneliness	ACC, PR, RE, detection-score	• Constructed a pool of clusters and patterns for learning speech signals • Segmented samples to extract attributes set • Overall ACC (90%) achieved across various users

ACC: Accuracy, PR: Precision, RE: Recall, BoW: Bag-of-words, GBT: Gradient boosting tree, ROC: Receiver operating characteristics, AUC: Area under the curve, SEN: Sensitivity, SPE: Specificity, EHR: Electronic health record, NLP: Natural language processing, LSTM: Long-short term memory, RF: Random Forest, FC: Facebook corpus, YC: Youtube corpus, TC: Twitter corpus, SSCL: Sequence, semantic, context learning, TF-IDF: Term frequency-inverse document frequency, MLP: Multilayer perceptron, SVM: Support vector machine, LR: Logistic regression, EEG: Electroencephalography early signs of depression from an individual's posts. However, authors attempted to differentiate between depression and suicidal ideations by training multiple models using the social media data and label correction method to reduce the impact of noise in online content [32]. Label correction methods utilized unsupervised clustering to efficiently use large-scale web datasets for better performance. Authors exploited CNN and LSTM to predict the behavior of an individual by gathering data from different social media posts and profiles [1]. Hybrid feature extraction techniques such as Word2Vec and TF-IDF (term frequency-inverse document frequency) used for selecting optimal features for detecting depression. However, authors extracted robust features using the GloVe method to capture the semantics of tweets from Twitter data [33]. Self-attentional DL model analyzed and classified the context to predict the mental health of the individual. The performance of the model was tested on a randomly selected small set of unseen datasets. On the other hand, Ansari et al. [30] compared ensembled and hybrid approaches using text classifiers for the automatic detection of depression. Features were extracted using symbolic and subsymbolic models. These results were passed to a LR module to classify the text. Simultaneously, word embeddings were processed using Attentional LSTM and a linear classifier. The results of LR and linear classifier were averaged to obtain the final results for predicting the major depressive disorders on public depression datasets.

To summarize, hybrid approaches integrated NLP, ML, and DL models to build a robust model for predicting depressive disorder symptoms. Initially, hybrid techniques performed sentiment analysis on the word embedding gathered either from clinical data or social media data. After this, the data was processed using a combination of ML and DL algorithms to extract the outcomes efficiently. The hybrid models also tested the performance of unseen datasets selected randomly which ensures the generalizability and realistic applicability of the model.

4.4 Artificial Intelligence-Based Solutions for Students with Special Needs

Mental disorders are not only prevalent in adults but also its dominance can also be visualized in middle-aged infants and teenage students [43]. Students suffering from mental disability and other psychological disorders have special needs that need to be addressed to improve their teaching–learning experience. Students suffering from neurological disorders not only affect their health but also their families and society. It is essential to recognize the early signs of mental illness in students and treat them as early as possible. For this, AI has provided many tools and applications that not only accommodate the needs of these special students but also generate interest by enhancing their cognitive learning [44]. In addition, AI has incorporated imparting personalized learning and education to students who are under stress and depression due to their bad academic performance and family pressure. The next section will discuss various AI-based optimization methods and tools that not only helpful in predicting the early signs of neurological disorders but also provide solutions to address them.

4.4.1 Artificial Intelligence for Optimizing Mental Disorders in Students

Students are suffering from mental health problems that include stress, anxiety, suicidal thoughts, and depression disorders. These problems are prevalent in students as they are under pressure to perform well in their academics. In order to address neurological disorders in students many AI-based tools and suggestions are recommended by various researchers [45].

AI has provided many tools to analyze the activities of students to improve their attention and focus on their work. These tools help reduce the stress and depression in students that may occur due to fear of non-performance in their academics. In this direction, authors proposed child activity sensing and training tool comprised of 42 unique features to analyze the physical and physiological activities of the students in real time [46]. This tool improves the student's focus by assisting them in real-time with their various academic activities. AI-based wearable vibrating watch known as WatchMinder was proposed by [47]. This device collected the activity and behavior of the wearer and periodically sends reminders to refocus on work. In [48], authors developed an AI-based anxiety scale to measure the anxiety levels in students to motivate them to learn.

To reduce stress, anxiety, and depression in students, a personalized learning experience is recommended. For this, authors explored learning areas such as reading, writing, spelling, and computing [45]. To inculcate reading comprehension, automatic storytelling apps, visual perception, and touching letters were provided. Supportive classroom environment to improve the physical structure of the learning. Daily teaching and learning schedules were defined innovatively to accommodate the needs of the students for better response. The balance between educational and extracurricular activities must be managed to improve the learning outcomes. AI therapies and supportive education for individualized learning and improving social skills in students suffering from mental disorders.

4.4.2 Artificial Intelligence-Based Robots for Special Students

AI-based robots such as chatbots, voicebots, and many others are recommended to address the needs the special students for their mental health [45]. These tools simplify the task and generate interest so that students would not feel depressed and consider learning as a burden.

In [49], authors suggested chatbot therapy to provide self-help for depressive students. The text contents were trained in the chatbot which is approved by professional therapists. The prewritten templates and questionaries were fed into the chatbots which get activated based on individual responses. During the therapy session, the chatbot differentiated among the emotions, thoughts, reactions, and behavior of the user. Based on the outcome, chatbot not only determined the level of negative feelings but also told practical ways to reduce negative feelings and overcome stress.

Arshad et al. [50] reviewed a robot for understanding the mathematics place value system named Mindstorms EV3 using LEGO to improve classroom learning in autistic students. This system interacted with students to enhance the interest, attention, focus, and personalized engagement of the students with special needs. Feedback on teaching and learning was also gathered by interviewing the teachers teaching the special needs students. It has been observed that this has improved the cognitive learning abilities of the students by generating interest and encouraging them to participate more in classroom activities.

In [51], authors discussed audio-based robots that can understand the emotions of autistic students and can support them in creating an interactive teaching and learning environment. These robots can adapt themselves in accordance with the change in emotion of the student such as crying, laughing, and other mental states. Speech data of the students were analyzed and processed based on the captured emotion to improve their engagement in various classroom activities and improve the quality of education.

4.5 Analysis of Mental Health During COVID-19 Pandemic

During the pandemic situation of COVID-19, people were known to suffer from depression, anxiety, and loneliness due to various preventive measures such as lockdowns and isolation. The pandemic restriction prevented travel, and people were forced to work from home. This pandemic has recorded a change in the psychological behavior of the people and led to severe mental disorders and mental health destabilization [29].

In order to analyze the traces of mental depression in Twitter social media users during the COVID-19 pandemic, authors gathered tweets from Twitter using an expression-based search method [38]. A fusion classifier was created to integrate the text features with demographic information to determine the depression features. The model had investigated the group-level and population-level depression trends in the generation during the pandemic. This non-invasive, and non-clinical data-based method can not only predict depression levels in different age groups but also generate awareness to prevent its propagation.

On the other hand, authors proposed a questionnaire set based on Hamilton tools designed in consultation with the psychiatrist to investigate depression in people during the COVID-19 pandemic [37]. The responses from humans were recorded and examined through various ML-based approaches to identify the depression trends in humans. The created questionnaire set was diversified consisting of both positive and negative questions to identify depression at an early stage.

In [29], authors validated speech signals and performed sentimental analysis to identify depression in people during the global lockdown in the whole world. The mental health of the people is severely affected due to isolation situations from the rest of the world. The behavioral patterns in depression-influenced people are analyzed in recorded speech data using the DL-based method. Cluster validation and signal segmentation techniques are utilized to evaluate the speech signal for the presence of depression and other mental disorders.

One of the potential measures recommended to prevent the spread of COVID-19 was social distancing. Social distancing has a potential negative impact on the mental health of children as well as adolescents [52]. The risk of child abuse and exploitation increased as families were isolated at home and parents were under stress either due to the loss of their jobs or work-from-home pressure. The socio-emotional development of both parents and children was impacted and raised concerns about mental health and behavioral disorders.

To determine the impact of COVID-19 on people's mental health, authors reviewed the various psychiatric symptoms to determine the direct and indirect impact of the pandemic on people [53]. It has been analyzed that the pandemic has worsened the situation in people who already have pre-existing depression symptoms such as anxiety, sleeping irregularities, and psychological disorders. There are many other sociodemographic factors such as living status, education, job-related factors, and gender are reported which indirectly have contributed to people's mental disorders.

4.6 Summary

In this chapter, we have reviewed clinical as well as social information for the diagnosis of mental health. Clinical diagnosis of mental disorders is based on the unstructured as well-structured information gathered from the patients. Unstructured information includes textual information such as EHR data and clinical notes containing the patient's medical history. Structured data utilizes imaging data such as MRI and signaling data such as EER for predicting disorders related to mental health. Clinical diagnosis of depression, anxiety, and suicidal ideation requires medical data as well as discussion with the psychiatrist. On the other hand, Diagnosis using social media is based on textual information captured from various social networking sites such as Facebook, Reedit, and Twitter. This information is critical and useful in providing accurate prediction based on the data collected from user's profiles, posts, and tweets. Algorithms based on ML and DL techniques can process this data to predict mental disorders automatically. These predictive models can analyze the psychological disorder symptoms efficiently so that treatment can be started at an early stage.

Students suffering from depression affect not only their personal life but also their studies. These children are not able to concentrate, focus, and be involved with other normal students. Their need is different and demands personalized teaching to improve their classroom learning. AI has provided many tools in terms of chatbots, robots, and apps so that classroom teaching can be improved for such students. AI has also provided monitoring tools that can record the daily activities of these students to provide them better teaching–learning experience. Also, COVID-19 has impacted people of all age groups including children, adults, and senior citizens. It has been analyzed that people are suffering from major depressive orders during COVID-19. The various pandemic restrictions such as isolation, social distancing, and lockdowns worsen the conditions in the people who already were suffering from any other depressive disorder. The depression trends are monitored in the patients using recent algorithms to predict mental health.

References

1. Wani MA, ELAffendi MA, Shakil KA, Imran AS, Abd El-Latif AA (2022) Depression screening in humans with AI and deep learning techniques. IEEE transactions on computational social systems
2. Amanat A, Rizwan M, Javed AR, Abdelhaq M, Alsaqour R, Pandya S, Uddin M (2022) Deep learning for depression detection from textual data. Electronics 11(5):676
3. Graham S, Depp C, Lee EE, Nebeker C, Tu X, Kim H-C, Jeste DV (2019) Artificial intelligence for mental health and mental illnesses: an overview. Curr Psychiatry Rep 21:1–18
4. Meng Y, Speier W, Ong MK, Arnold CW (2021) Bidirectional representation learning from transformers using multimodal electronic health record data to predict depression. IEEE J Biomed Health Inform 25(8):3121–3129
5. Msosa YJ, Grauslys A, Zhou Y, Wang T, Buchan I, Langan P, Folarin A, et al (2023) Trustworthy data and AI environments for clinical prediction: application to crisis-risk in people with depression. IEEE J Biomed Health Inf
6. Zhang W, Yang C, Cao Z, Li Z, Zhuo L, Tan Y, Gong Q (2023) Detecting individuals with severe mental illness using artificial intelligence applied to magnetic resonance imaging. EBioMedicine 90
7. Jo A-H, Kwak K-C (2022) Diagnosis of depression based on four-stream model of Bi-LSTM and CNN from audio and text information. IEEE Access 10:134113–134135
8. Toto E, Tlachac M, Rundensteiner EA (2021) Audibert: a deep transfer learning multimodal classification framework for depression screening. In: Proceedings of the 30th ACM international conference on information & knowledge management
9. Saeedi A, Saeedi M, Maghsoudi A, Shalbaf A (2021) Major depressive disorder diagnosis based on effective connectivity in EEG signals: a convolutional neural network and long short-term memory approach. Cogn Neurodyn 15:239–252
10. Sharma G, Parashar A, Joshi AM (2021) DepHNN: a novel hybrid neural network for electroencephalogram (EEG)-based screening of depression. Biomed Signal Process Control 66:102393
11. Thoduparambil PP, Dominic A, Varghese SM (2020) EEG-based deep learning model for the automatic detection of clinical depression. Phys Eng Sci Med 43:1349–1360

12. AlSagri HS, Ykhlef M (2020) Machine learning-based approach for depression detection in twitter using content and activity features. IEICE Trans Inf Syst 103(8):1825–1832
13. Safa R, Bayat P, Moghtader L (2022) Automatic detection of depression symptoms in twitter using multimodal analysis. J Supercomput 78(4):4709–4744
14. Sekulić I, Gjurković M, Šnajder J (2018) Not just depressed: bipolar disorder prediction on reddit. arXiv preprint arXiv:1811.04655
15. Zhou TH, Hu GL, Wang L (2019) Psychological disorder identifying method based on emotion perception over social networks. Int J Environ Res Public Health 16(6):953
16. Chiong R, Budhi GS, Dhakal S, Chiong F (2021) A textual-based featuring approach for depression detection using machine learning classifiers and social media texts. Comput Biol Med 135:104499
17. Tadesse MM, Lin H, Xu B, Yang L (2019) Detection of depression-related posts in reddit social media forum. IEEE Access 7:44883–44893
18. Wang L, Liu H, Zhou T (2020) A sequential emotion approach for diagnosing mental disorder on social media. Appl Sci 10(5):1647
19. Kumar A, Singh D (Eds.) (2025) Revolutionizing Medical Systems Using Artificial Intelligence: A Breakthrough in Healthcare. Elsevier
20. Bhalla S, Kumar A, Kushwaha R (2023) Comparative analysis on brain stroke prediction using machine learning. In: 2023 14th International conference on computing communication and networking technologies (ICCCNT). IEEE
21. Kumar A (2023) Visual object tracking using deep learning. CRC Press
22. Kumar A, Jain R, Gupta M, Islam SM (2023) 6G-enabled IoT and AI for smart healthcare: challenges, impact, and analysis. CRC Press
23. Chao J, Zheng S, Wu H, Wang D, Zhang X, Peng H, Hu B (2021) FNIRS evidence for distinguishing patients with major depression and healthy controls. IEEE Trans Neural Syst Rehabil Eng 29:2211–2221
24. Mousavian M, Chen J, Traylor Z, Greening S (2021) Depression detection from sMRI and rs-fMRI images using machine learning. J Intell Inf Syst 57:395–418
25. Bertl M, Bignoumba N, Ross P, Yahia SB, Draheim D (2024) Evaluation of deep learning-based depression detection using medical claims data. Artif Intell Med 147:102745
26. Othmani A, Kadoch D, Bentounes K, Rejaibi E, Alfred R, Hadid A (2021) Towards robust deep neural networks for affect and depression recognition from speech. in pattern recognition. ICPR international workshops and challenges: virtual event, January 10–15, 2021, Proceedings, Part II. Springer
27. Ramírez-Cifuentes D, Largeron C, Tissier J, Baeza-Yates R, Freire A (2021) Enhanced word embedding variations for the detection of substance abuse and mental health issues on social media writings. IEEE Access 9:130449–130471
28. Rejaibi E, Komaty A, Meriaudeau F, Agrebi S, Othmani A (2022) MFCC-based recurrent neural network for automatic clinical depression recognition and assessment from speech. Biomed Signal Process Control 71:103107
29. Ahmed ST, Singh DK, Basha SM, Abouel Nasr E, Kamrani AK, Aboudaif MK (2021) Neural network based mental depression identification and sentiments classification technique from speech signals: a COVID-19 focused pandemic study. Front Public Health 9:781827
30. Ansari L, Ji S, Chen Q, Cambria E (2022) Ensemble hybrid learning methods for automated depression detection. IEEE Trans Computat Soc Syst 10(1):211–219
31. Fatima B, Amina M, Nachida R, Hamza H (2020) A mixed deep learning based model to early detection of depression. J Web Eng 19(3–4):429–455
32. Haque A, Reddi V, Giallanza T (2021) Deep learning for suicide and depression identification with unsupervised label correction. In: Artificial neural networks and machine learning–ICANN 2021: 30th international conference on artificial neural networks, Bratislava, Slovakia, September 14–17, 2021, Proceedings, Part V 30. Springer
33. Nadeem A, Naveed M, Islam Satti M, Afzal H, Ahmad T, Kim K-I (2022) Depression detection based on hybrid deep learning SSCL framework using self-attention mechanism: an application to social networking data. Sensors 22(24):9775

34. Sarkar A, Singh A, Chakraborty R (2022) A deep learning-based comparative study to track mental depression from EEG data. Neurosci Inf 2(4):100039
35. Ghosh S, Ekbal A, Bhattacharyya P (2021) What does your bio say? inferring twitter users' depression status from multimodal profile information using deep learning. IEEE Trans Computat Soc Syst 9(5):1484–1494
36. Alghowinem S, Gedeon T, Goecke R, Cohn JF, Parker G (2020) Interpretation of depression detection models via feature selection methods. IEEE Trans Affect Comput 14(1):133–152
37. Malik A, Shabaz M, Asenso E (2023) Machine learning based model for detecting depression during Covid-19 crisis. Scientific African 20:e01716
38. Zhang Y, Lyu H, Liu Y, Zhang X, Wang Y, Luo J (2021) Monitoring depression trends on Twitter during the COVID-19 pandemic: observational study. JMIR infodemiology 1(1):e26769
39. Singh D, Singh D, Manju, Gupta U (2023) Smart healthcare: a breakthrough in the growth of technologies. In: Artificial intelligence-based healthcare systems. Springer, pp 73–85
40. Tripathy SK, Singh D, Jaiswal A (2023) Multi-Layer feature fusion-based deep multi-layer depth separable convolution neural network for Alzheimer's disease detection. In: 2023 International conference on IoT, communication and automation technology (ICICAT). IEEE
41. Tripathy SK, Singh D, Srivastava S, Srivastava R (2023) Multiscale low-level feature fused multilayer convolution neural network for Alzheimer's disease detection
42. Hao F, Pang G, Wu Y, Pi Z, Xia L, Min G (2019) Providing appropriate social support to prevention of depression for highly anxious sufferers. IEEE Trans Computat Soc Syst 6(5):879–887
43. Kumar A, Nayyar A, Sachan RK, Jain R (2023) AI-assisted special education for students with exceptional needs. IGI Global
44. Kumar A, Singh D, Vohra R (2023) Improving learning abilities using AI-based education systems. In: AI-Assisted special education for students with exceptional needs. IGI Global, pp 137–155
45. Barua PD, Vicnesh J, Gururajan R, Oh SL, Palmer E, Azizan MM, Acharya UR, et al (2022) Artificial intelligence enabled personalised assistive tools to enhance education of children with neurodevelopmental disorders—a review. Int J Environ Res Public Health 19(3):1192
46. Sonne T, Obel C, Grønbæk K (2015) Designing real time assistive technologies: a study of children with ADHD. In: Proceedings of the annual meeting of the Australian special interest group for computer human interaction
47. Somma F, Rega A, Gigliotta O (2019) Artificial intelligence-powered cognitive training applications for children with attention deficit hyperactivity disorder: a brief review. Child. Worldw 2(4)
48. Wang Y-Y, Wang Y-S (2022) Development and validation of an artificial intelligence anxiety scale: an initial application in predicting motivated learning behavior. Interact Learn Environ 30(4):619–634
49. Liu H, Peng H, Song X, Xu C, Zhang M (2022) Using AI chatbots to provide self-help depression interventions for university students: a randomized trial of effectiveness. Internet Interv 27:100495
50. Arshad NI, Hashim AS, Ariffin MM, Aszemi NM, Low HM, Norman AA (2020) Robots as assistive technology tools to enhance cognitive abilities and foster valuable learning experiences among young children with autism spectrum disorder. IEEE Access 8:116279–116291
51. Kim JC, Azzi P, Jeon M, Howard AM, Park CH (2017) Audio-based emotion estimation for interactive robotic therapy for children with autism spectrum disorder. In: 2017 14th international conference on ubiquitous robots and ambient intelligence (URAI). IEEE
52. Imran N, Zeshan M, Pervaiz Z (2020) Mental health considerations for children & adolescents in COVID-19 Pandemic. Pakistan J Med Sci 36(COVID19-S4):S67
53. Vindegaard N, Benros ME (2020) COVID-19 pandemic and mental health consequences: systematic review of the current evidence. Brain Behav Immun 89:531–542

Chapter 5
Diagnosis and Prediction of Cardiovascular Disorder Using Artificial Intelligence

Abstract Cardiovascular disorder (CVD) is one of the leading diseases which has a high mortality rate worldwide. Atherosclerosis is a condition which is a major cause of CVD and occurs due to the accumulation of plaque and calcium in the coronary arteries vessels. Intervascular ultrasound (IVUS) is a diagnostic technique that provides artery vessel imaging. To identify the severing of calcification and plaque in artery vessels, it is necessary to segment the IVUS imaging into lumen and media which demands specialized skills. For accurate diagnosis and prognosis of CVD in a patient, AI has provided many predictive algorithms which segment the IVUS imaging effectively. In this chapter, we will review the various deep-learning techniques exploited for IVUS imaging segmentation. We will also highlight the limitation of IVUS imaging that reduces the accuracy and effectiveness of CVD prediction.

Keywords Artificial intelligence (AI) · Intervascular Ultrasound · Percutaneous coronary intervention · Coronary angiography computed tomography · Angiography · Risk assessment

5.1 Introduction

Cardiovascular disorder (CVD) causes coronary artery diseases that occur due to an insufficient supply of oxygen to the human heart. Atheromatous plaques accumulated in the coronary arteries narrow their lumen area and hence, reduce the blood flow to the heart [1, 2]. Plaque ruptures the blood vessels and impacts the cardiac muscles which leads to heart attack/heart failure. The components of atheromatous plaques are examined to determine the severity using a technique known as percutaneous coronary intervention (PCI) [3]. In this, intervascular ultrasound (IVUS) is used as a pre-intervention invasive technique to analyze coronary arteries, vessel regions, and lesions to assess the impact of plaque and extend appropriate treatment to prevent casualty.

A. Kumar and D. Singh, *Artificial Intelligence in Modern Healthcare System*, Transactions on Computer Systems and Networks,
https://doi.org/10.1007/978-981-96-6703-1_5

Early detection of CVD in a patient not only enhances the treatment success rate but also reduces the risk of heart failure in patients. IVUS-guided PCI is considered to be the superior technique in the detection of atheromatous plaque, and stenosis in lesion vessels of the patients [4]. Apart from IVUS, other invasive techniques such as OCT (Optical coherence tomography), and NIRS (near-infrared spectroscopic) techniques can be used for examining the ruptured coronary arteries. Alternatively, non-invasive techniques such as stress ECG (Echocardiography), cardiac MRI (Magnetic resonance imaging), and CCTA (coronary angiography computed tomography) are also popular among cardiologists for prediction of CVD risk. Among these non-invasive techniques, CCTA is a gold standard and widely preferred by cardiologists for the identification of calcified, non-calcified, and mixed attenuation plaque in patients suffering from CVD [5, 6]. CCTA images provide geometrical information on coronary arteries which requires further interpretation and analysis from the expert radiologist to estimate the stenosis severity. This process is time-consuming and non-very reliable as the outcomes from different experts vary in terms of diagnosis and interpretation [7]. On the other hand, IVUS imaging is more accurate in terms of the prediction of plaque from complex lesions of coronary arteries.

IVUS imaging is quite complex and requires specialized skills and knowledge for interpretation and analysis. With the advent of technology, artificial intelligence (AI) has shown significant improvement and efficiency in the analysis of image processing [8, 9], education system [10, 11], tourism [12], and medical imaging [13–15]. AI with machine learning (ML) [4, 16, 17] and deep learning (DL) [18–20] based prediction algorithms have simplified the IVUS imaging segmentation containing complicated lesions into various categories of calcification. Also, the lumen and media segmentation are highly desirable in the complex IVUS images for evaluation of the degree of calcification. The evaluation of calcification in the target lumen area and media in IVUS images is crucial to optimize the stent implementation for accurate outcomes during PCI procedures. AI with its feature extraction and selection capabilities can segment the lumen-intima (LI) and media-adventitia (MA) border from the coronary arteries wall automatically.

Generally, the non-invasive technique of CCTA provides elaborated imaging of the coronary arteries, and CTA images can be processed for predicting the risk of stenosis in coronary vessel boundaries [21]. CTA images can be manually segmented by radiologists and are time-consuming. However, CTA images are quite complex and require specialized skills for their interpretation for predicting CVD risk. Segmenting CTA images for visualization of stenosis in coronary vessels is highly dependent on the expert's skills and may introduce selection and evaluation errors. To resolve this, DL-based methods are extensively explored for segmenting CTA images and have shown superior accuracy [22, 23]. These methods qualify for providing automatic, accurate, and faster segmentation of coronary arteries. However, prediction of CVD risk using IVUS imaging is highly preferred by cardiologists as it provides a clear view of coronary arteries for complex lesions along with identification of vulnerable plaque.

Recently, several techniques and methods have been proposed for the automatic segmentation of the IVUS images for the identification of various categories of plaque. These techniques have extracted handcrafted features such as texture [4], artificial neural networks (ANN) [16], and convolutional networks (CNN) [18] are quite popular. Particularly UNet-based DL algorithms have shown more accurate outcomes and prevent errors in the prediction of calcified plaque from IVUS images [1, 3, 24, 25]. These prediction algorithms can typically segment calcified, non-calcified, and mixed attenuation plaques from the IVUS images. Fibrous and fibro-fatty tissue can also be segmented from IVUS images to analyze the correlation between calcium and atherosclerosis [26].

The key objectives of this chapter are as follows:

- The IVUS imaging procedure is elaborated along with its methods, types, and limitations to gather clinical imaging data for further investigation.
- Various ML and DL-based prediction algorithms are reviewed, and salient features are gathered which ensure the automatic prediction of calcified, non-calcified, attenuated, and mixed attenuated plaque from the IVUS images.
- Apart from IVUS imaging, details about other imaging techniques such as CTCA are also highlighted to compare for accuracy and efficiency.

The rest of the chapter is organized as follows. Section 5.2 elaborates on the key procedure technique that involves gathering clinical data using the IVUS technique. The details about AI-based techniques for automatic segmentation of IVUS imaging to analyze the CVD risk are categorized into ML-based and DL-based techniques and are highlighted in Sect. 5.3. Section 5.4 details the benefits and limitations of the CTCA technique for CVD risk prediction in comparison to IVUS imaging. Lastly, the concluding remarks and future directions are sketched in Sect. 5.5.

5.2 IVUS Imaging Data Acquisition

IVUS images are 360-degree cross-sectional visualizations of coronary arteries. IVUS images can be utilized to analyze lumen and vessel morphology such as shapes, borders, and areas. These images can also be examined for various types of plaque and their composition. Images can also be useful in decision-making for CVD diseases in atherosclerosis and post-surgery examination such as stent underexpansion and stenosis. Figure 5.1 illustrates (a) the IVUS image in which (b) media-adventitia (MA) border, (c) lumen area, and (d) calcified plaque are manually annotated and highlighted.

Basically, IVUS generates HD resolution gray-scale images from the ultrasound (USd) signals reflected by the coronary arteries structure. Coronary arteries can be divided into three parts namely, the innermost part as intima, the middle layer as media, and the outermost layer as adventitia. Atherosclerotic plaque is less echogenic and is accumulated in the intima layer of coronary arteries. In comparison to the inner layer, the adventitia layer (outer layer) is highly echo-reflective. The media layer contains smooth tissues and hence, does not reflect USd signal and appears dark in the IVUS images. Also, the atherosclerotic plaque on the inner layer is moderately

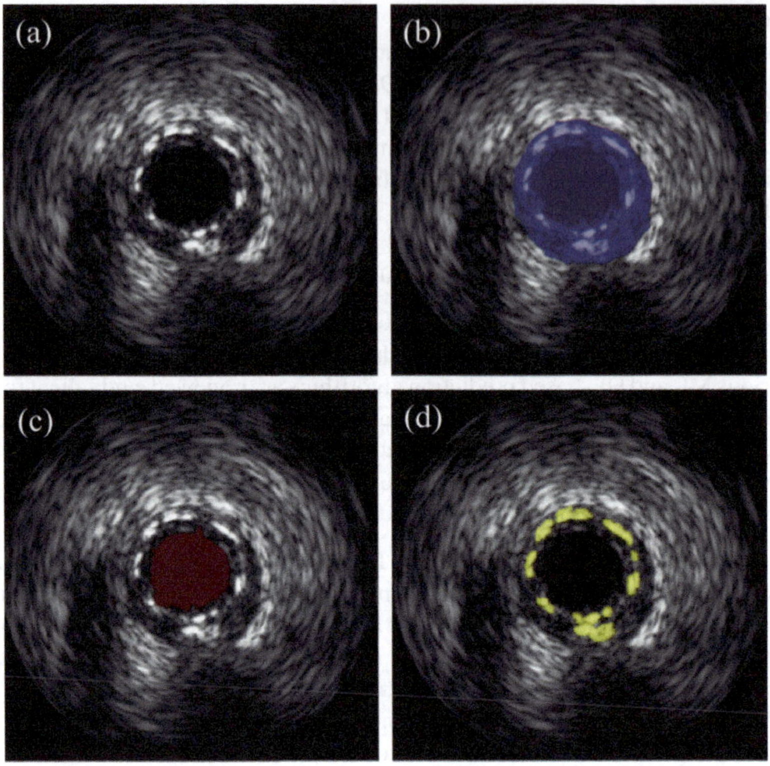

Fig. 5.1 IVUS images with manual annotations as **a** original gray-scale image **b** MA layer region **c** lumen region **d** calcified tissue localization [18]

echogenic in nature. This difference in properties of the three layers of coronary arteries and the plaque enables identification and segmentation easily.

5.2.1 Description of Catheter for Capturing IVUS Images

IVUS catheter is a device used for collecting images from coronary arteries. It is a thin and flexible tube-like structure with a small transducer mounted on it. Like other USd imaging, one end of the catheter is connected to a device that converts the reflected USd signals and displays real-time images of the coronary vessels on the screen. Initially, the IVUS catheter is fed over a guidewire guided by angiography to the region of interest (ROI). IVUS transducer collects the images either distal to ROI and then pulled back through the stenosis region or directed placed at ROI. IVUS images are captured with an automated pullback speed. 60 images/mm at 30 fps are acquired by the IVUS device with the probe withdrawn at a fixed speed of 0.5 mm/sec. The pullback operation can either be automated or manual. Manual pullback is

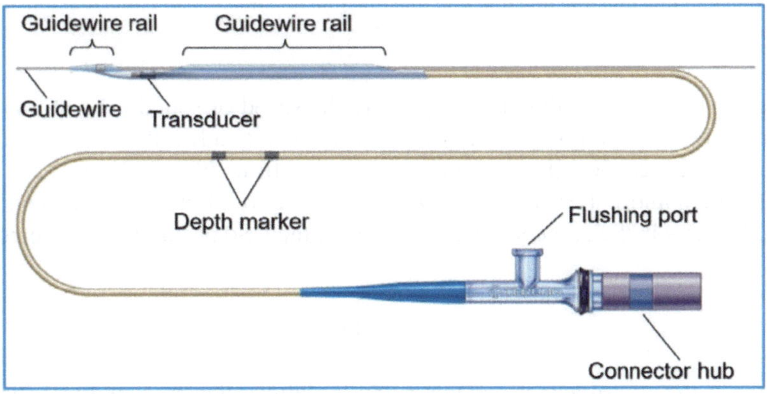

Fig. 5.2 Description of IVUS catheter fed over a guidewire [27]

more popular as it allows the observer to concentrate more on the target regions of the vessels. Figure 5.2 represents the catheter fed over a guidewire.

There exist many commercial catheter manufacturers for capturing IVUS images [27]. IVUS imaging data can be acquired by various equipment at multiple frequencies namely, 20, 50, and 40 MHz. Boston Scientific Corporation, Philips Volcano, and Terumo are a few popular manufacturers providing a wide range of catheters for IVUS imaging. OptiCross catheter with the rotational transducer at a variable frequency of 15, 30, 40, and 60 MHz is manufactured by Boston Scientific. EagleEye, Revolution, and Refinity catheters are manufactured by Philips Volcano. These devices work on transducer frequencies of 20 and 45 MHz with a phased array and rotational transducer. Terumo manufactures catheters with rotational transducers which work at 40 and 60 MHz frequencies known as View IT, AltaView, AnteOwl WR, Navifocus WR, and Intrafocus WR.

5.2.2 *Quality Assessment for IVUS Images*

For accurate and efficient outcomes from the IVUS images, it is necessary to assess the quality of the gray-scale images captured by the IVUS device. IVUS image quality can be evaluated by considering spatial resolution, imaging sensitivity, and contrast [28].

The spatial resolution of an USd image is computed as the minimum distance between two neighboring and differentiated features. The spatial resolution is inversely proportional to the computed distance. For an image with a lesser distance, spatial resolution is high. Spatial resolution is categorized as axial resolution and lateral resolution for an USd image. Axial resolution (ν^{ax}) measures the depth resolution and is defined as the capacity to differentiate close neighboring features along the USd beam axis. ν^{ax} can be computed using Eq. (5.1).

$$\nu^{ax} = \frac{\lambda}{2\beta} \tag{5.1}$$

where λ is the wavelength computing using the speed of sound (c) and transducer frequency (τ) as $\frac{c}{\tau}$ and β is the transducer fractional bandwidth (−6 dB). For USd transducer with 20–50 MHz frequency, ν^{ax} ranges from 70 to 200 μm.

Lateral resolution (υ^L) is the capability measure to differentiate between neighboring features in the perpendicular direction of the USd beam. It is computed using Eq. (5.2).

$$\upsilon^L = \lambda\dot{\phi} \tag{5.2}$$

where λ is the transducer wavelength and $\dot{\phi}$ defines the ratio of focal length to the aperture size of the USd image. Natural focal length $\left(\dot{\phi}_n\right)$ is computed for an unfocused transducer using Eq. (5.3).

$$\dot{\phi}_n = \frac{D_t^2}{4\lambda} \tag{5.3}$$

where D_t is the USd transducer diameter. For USd transducers with 20–50 MHz frequency, lateral resolution (υ^L) ranges from 200 to 250 μm.

Temporal resolution (υ^{tp}) in an USd image represents the capability to differentiate between instantaneous events of rapidly moving structures. During a cardiac cycle, it is defined as the time from the starting of one frame to the next. υ^{tp} is improved by reducing penetration depth, focal points and scan lines/frame. This will allow the USd signal to travel a small distance, prevent scan line duplication, and narrow the size of frames. If ϕ^N is the number of focal points, D_p is penetration depth and S_L is the number of scan lines, then, υ^{tp} is computed using Eq. (5.4).

$$\upsilon^{tp} = \frac{154000}{2 \times \phi^N \times D_p \times S_L} \tag{5.4}$$

The contrast of an IVUS image ($\mathcal{I}_c$) determines the distinguishing capability of the target feature from the nearby tissues. In IVUS images, it is computed from the difference in impedances between the target (T) and the background (B) regions. $\mathcal{I}_c$ is a vital parameter for features that are prominent against background areas. It is calculated using Eq. (5.5).

$$\mathcal{I}_c = \frac{|\delta_T - \delta_B|}{\sqrt{\mu_T^2 - -\mu_B^2}} \tag{5.5}$$

where δ_T and μ_T are the acoustic signal magnitude and S.D. (Standard deviation) in the target region. δ_B and μ_B are acoustic signal magnitudes and S.D. for the background.

The sensitivity parameter is used to detect the USd echo in the presence of background noise. The acoustic attenuation of the USd signal increases due to scattering and abortion which decreases the signal-to-noise ratio (η_{SNR}). η_{SNR} is computed using Eq. (5.6).

$$\eta_{SNR}(dB) = 20\log_{10}\frac{V_{Tis}}{V_{noi}} \tag{5.6}$$

where V_{Tis} and V_{noi} are acoustic signals from the target echogenic region and background region, respectively.

5.3 Artificial Intelligence for Predicting Risk of Cardiovascular Disorder

IVUS images provide lumen information, plaque categorization, and damaged vessels effectively. This information assists in the early diagnosis of cardiological disorder and prevents stroke/heart failure in patients suffering from coronary heart diseases [17, 29]. However, IVUS images are quite complex and require specialized skills to detect impacted lumen border, extent of calcification, and degree of atherosclerosis. In addition, the number of frames in an IVUS sequence for a single patient is huge and hence, requires plenty of time for accurate analysis and diagnosis so that proper treatment and advice can be extended to the patients [2, 30]. For accurate, and faster detection of the severity of coronary artery disease in IVUS images, AI has provided a lot of ML and DL-based prediction algorithms. These algorithms not only provide automatic and faster segmentation of IVUS images but also visualize the morphological features of plaque which help in the identification of their adverse impact on coronary arteries. In this section, we have categorized the various AI-based CVD risk prediction algorithms into two categories namely, ML-based prediction algorithms [31–34] and DL-based prediction algorithms [3, 35–37] for IVUS image segmentation. Tables 5.1 and 5.2 highlight the salient features and compare the ML and DL-based prediction methods for IVUS segmentation, respectively.

Table 5.1 Salient features of ML and DL-based IVUS segmentation algorithms

SN	Attributes	ML-based algorithms	DL-based algorithms
1	Training data availability	Low	Limited
2	Feature extraction	Manual	Automatic or manually
3	Feature selection	✗	✓
4	Computational complexity	Relatively less	High
5	Performance	Moderate	High
6	Statistical measure	Limited	Large
7	Dataset augmentation	✗	✓

Table 5.2 Comparison of ML and DL-based IVUS segmentation algorithms

SN	Attributes	ML-based algorithms	DL-based algorithms
1	Generalizability	✗	✗
2	Model explainability	✗	✗
3	Risk-of-Bias	✓	✓
4	Data classification	✓	✗
5	Unseen datasets	✗	✗
6	Dataset processing	✗	✓
7	Loss function	✗	✓
8	Limitations	Trained on limited datasets	Trained on variable datasets

5.3.1 Machine Learning-Based Algorithms for IVUS Image Segmentation

A lot of ML-based algorithms such as Random forest (RF) [38–40], k-nearest neighbour (KNN) [30], Feedforward neural network (FNN) [38], Support vector machine (SVM) [39], Decision tree (DT) [41], Bagged tree (BT) [41] and many other [32, 34, 42] are proposed for feature-based classification and segmentation of IVUS images for CVD risk prediction. Table 5.3 tabulates the salient features of ML-based IVUS image segmentation techniques.

Generally, multiple features are extracted and selected from grayscale IVUS images to classify the vulnerable plaque tissues into calcified, attenuated, mixed attenuated, fatty, and fibrous fatty along with lumen and media borders. In this direction, Yang et al. [29] proposed a regression network based on coupled contour to resample the lumen and EEL in IVUS images. The anatomical relationship between the lumen and EEL was reconstructed to reduce false prediction. The processing speed of the method was quite fast due to the elimination of post processing step and the reduction in parameters in the decoder. On the other hand, authors proposed a graph-based segmentation technique to segment the lumen and EEL from the coronary vessel surface in IVUS images [32]. The method combined automated segmentation and computer-aided refinement to deal with calcification, shadow, and imaging artifacts. In [39], authors evaluated six ML-based algorithms for the classification of lesions with fivefold cross-validation. The algorithm performance was measured by defining thresholds for maximal accuracies and quality by Matthew's correlation coefficients.

To segment arterial walls for morphological structures such as bifurcations, shadows, echogenic plaques, and normal, features are extracted from the various IVUS images. For this, authors generated a 22-D feature vector for each column of the IVUS image to deal with the multi-classification of various morphological structures [40]. The classification accuracy was enhanced by assessing the feature importance using the RF algorithm. Similarly, authors extracted multiple features from the IVUS images and employed four ML classifiers to segment the calcified

Table 5.3 ML-based algorithms for IVUS image segmentation

Reference	Dataset description	Model utilized/ proposed	Number of participants	CVD prediction outcome	Summary
Matsumura et al. [31]	• Total Images:8076 • Training/Validation/ Test: 6703/1373/437 • IVUS HD imaging using 60 MHz OptiCross catheter paired with iLab Polaris Multi-modality guidance system (Boston Scientific Corporation)	ML-based Unet architecture	234 patient datasets from the USA, Asia, or Europe for training and Validation. Testing on the unseen dataset of 92 patients from the USA	Prediction of appropriate balloon size, lumen, and vessel area to determine agreement rate in the lumen and acute stent area	• Compared ML model outcomes with the manually extracted segmentation results by multiple human radiologist experts • Strong correlation between ML-segmentation results and manually segmented IVUS images (90%) • Ineffective in segmenting the stent expansion, calcified and non-calcified regions, and dissection detection

(continued)

Table 5.3 (continued)

Reference	Dataset description	Model utilized/proposed	Number of participants	CVD prediction outcome	Summary
Jun et al. [38]	• Total Images:12,325 • Training/Test: 9860/2465 • IVUS imaging using 40 MHz motorized transducer (Boston Scientific/SCIMED Corporation)	FNN, KNN, RF, CNN classifiers, ReLU activation, RMSprop optimizer, multiple K values, Euclidean distance metric, KS = 3X3, ELU activation	100 approved patients from Asan Medical Center	Classified thin-cap fibroatheroma caused by plaque rupture	• Augmented images using random rotation • AUC for CNN (0.933), KNN (0.890), FNN (0.884) and RF (0.878) • Pixel range-based feature extraction for better classification accuracy • Performed feature ranking evaluation using the Chi-square test

(continued)

Table 5.3 (continued)

Reference	Dataset description	Model utilized/ proposed	Number of participants	CVD prediction outcome	Summary
Yang et al. [29]	• Total Images:3557 (40 MHz), 3647 (60 MHz) • Training/Validation/ Test: 2739/383/435 (40 MHz), 2945/395/ 307 (60 MHz) • IVUS imaging using 40 MHz (Atlantis, Boston Scientific, MA), and 60 MHz OptiCross catheter (Boston Scientific Corporation, MA)	Coupled contour regressionNet, Adam optimizer	Med-X Research Institute, Shanghai Jiao Tong University, Shanghai	Segmented lumen and EEL contours from the IVUS images	• Augmented dataset by flipping rotation and scaling • Achieved DSC of 0.940 and 0.958 for lumen and EEL, respectively • Employed CNN for predicting coupled contours and object contours using the linear decoder • Fastest method in comparison to other mask-based methods with a 10 ms processing time
Sun et al. [32]	• Total Images:6467 • IVUS imaging using 20 MHz solid state catheters (Volcano IVG3 imaging system)	LOGISMOS dual surface graph-based segmentation	41 datasets	Automatic segmentation of lumen and EEL from IVUS images	• Utilized case-specific morphology to deal with the shadow artifacts • Pre-segmentation of lumen area to define the lumen centerline • Correct local and regional inaccuracies by approximating the desired surface locations

(continued)

Table 5.3 (continued)

Reference	Dataset description	Model utilized/ proposed	Number of participants	CVD prediction outcome	Summary
Lee et al. [39]	• Total Images:1382 • Training/Test: 1063/265 • IVUS imaging using 40 MHz motorized transducer (Boston Scientific/SCIMED Corporation, MA)	Supervised FCN VGG-16 model, LR, ANN, RF, AdaBoost, CatBoost, SVM	1382 patients from Asan Medical Center, Seoul, Korea	Identified ischemia-producing lesions to reduce the need for pressure wires	• Computed 99 IVUS features and 6 clinical variables for of training model with fivefold cross-validation • Evaluated 6 ML classifiers for lesion separation • Utilized GridSearch CV for selecting hyperparameters and better AUC • Manual corrections of all IVUS frames impacted the algorithm speed

(continued)

Table 5.3 (continued)

Reference	Dataset description	Model utilized/ proposed	Number of participants	CVD prediction outcome	Summary
Vercio et al. [40]	• Total Images:435 • Training/Test: 305/130 • IVUS imaging using 20 MHz Eagle Eye catheter Si5 imaging system (Volcano Corporation, CA, USA)	RF, K-fold cross-validation	–	Classified IVUS images into bifurcation, shadow, echogenic plaque, and normal	• Classifiers automatically segmented the images in the classes by obtaining thresholds • Handled dataset imbalance using undersampling and oversampling techniques • Can define RF sensitivity to detect morphological structures
Ziemer et al. [34]	• Total Images:23,970 • Training/Validation/ Test:20,586/1692/1692	Multi-frame CNN, GP regressor	Dataset from 52 patients	Automatic lumen segmentation from IVUS imaging	• Automatic gating and segmentation for assessing the IVUS dataset • Utilized GP regressor to reduce high-frequency noise and contour continuity of lumen boundary • Model generalizability not discussed

(continued)

Table 5.3 (continued)

Reference	Dataset description	Model utilized/ proposed	Number of participants	CVD prediction outcome	Summary
Cui et al. [42]	• Training: 77 groups of 5 consecutive frames	GBF, RTB, CKL, KS = 19X19	Three public datasets	Lumen segmentation with LAD, LCX, and RCA	• Lumen segmentation JS = 96.8 • Validation of real IVUS patient data
Archana and Vanithamani [41]	• Total Images:99 • Training/Test:79/20	KNN, SVM, DT, and BT	Public dataset	Classified plaque from IVUS images of carotid artery	• Extracted multiple features namely, mean, SD, energy, histogram, entropy, and contrast • Accuracy = 91.1% (SVM), 96.2 (KNN), 88.6% (DT), 96.2% (BT) • BT showed superior classification performance among other ML algorithms

(continued)

Table 5.3 (continued)

Reference	Dataset description	Model utilized/ proposed	Number of participants	CVD prediction outcome	Summary
Rezaei et al. [30]	• Total Images:599 • Training/ Validation:440/159 • IVUS imaging using 20 MHz electronic probe (In-Vision Gold, Volcano Therapeutics, Inc.)	Hybrid FCM and KNN	10 patients dataset	Classified different types of plaque from the lumen area	• Extracted color-based features to show non-zero pixels for better classification • Robust feature extraction for accurate and automatic plaque localization • To enhance performance, applied ensemble classification of SVM, ANN, and kNN
Tong et al. [2]	• Total Images:435 Training/Test:109/326	Sparse coding, kernel dictionary, texture feature	Public dataset	Evaluated for general performance, bifurcations, sidevessels, shadow, and none-structures	• Automatically segment the lumen to extract information about various types of plaque • Achieved overall JS = 0.87 • Data pre-processing along with kernel methods minimized the impact of artifacts and shadow

KS = Kernel Size, DoR = Dropout rate, BN = Batch normalization, JS = Jaccard Similarity, AUC = Area under the curve, MA = Media-adventitia, LI = Lumen-intima, ms = milliseconds, KNN = k-Nearest neighbor, SVM = Support vector machine, DT = Decision tree, BT = Bagged tree, RF = Random forest, LR = Linear regression, ANN = Artificial neural network, FNN = Feed forward neural network, DSC = Dice similarity coefficient, EEL = External elastic lamina, GBF = Gradient boosting framework, RTB = Regression tree building, CKL = Convolutional kernels learning, LAD = Left anterior descending artery, LCX = Left circumflex artery, RCA = Right coronary artery, FCM = Fuzzy clustering mean plaque and the normal [41]. An efficient feature selection strategy was deployed to minimize misclassification and enhance accuracy. Rezaei et al. [30] proposed a hybrid model by combining KNN and Fuzzy C-means (FCM) for automatic and accurate segmentation of VH-IVUS (Virtual histology IVUS) images. The color feature was extracted along with pixel clustering, cluster labeling, and outlier removal. Multiple algorithms were also exploited for robust feature extraction such as closed luminal border tracing, open lumen border tracing, confluent components, NC layering, and plaque burden assessment for classifying plaque from luminal border efficiently.

In another line of research, ML algorithms such as Gradient boosting framework (GBF), Regression tree building (RTB), Convolutional kernel learning (CKL), sparse coding, and dictionary learning were also proposed for automatic segmentation of lumen border. In [42], authors utilized a gradient boosting framework along with its quadratic approximation to generate discriminative boundaries for lumen segmentation. Tong et al. [2] proposed dictionary-based IVUS image segmentation utilizing sparse coding and kernel dictionary. The dictionary consisted of positive and negative tissue samples to reduce the impact of artifacts and shadowed for better segmentation results. Also, linear discrimination of pixels in ROI was done using kernel cluster algorithms to improve detection quality during morphological operations. On the other hand, authors exploited multi-frame CNN to segment lumen boundary automatically in IVUS images. Initially, the minimum lumen area and stenosis area percentage were used for making decisions for lumen boundary segmentation. After this, it was subjected to the Gaussian process regression stage for further refinement. The automated gating and regression stage improved the effectiveness and accuracy of the method.

To summarize, ML-based segmentation algorithms utilized methods along with features-based classifiers to improve the accuracy and effectiveness of the method. Multiple features were extracted to segment the border areas such as MA and LI from the lumen border. The trained classifiers were used for the classification of lumen segmentation into multiple categories. Most of the methods were not generalizable and were trained on limited datasets. The clinical deployment of most of the ML-based methods was yet to be explored.

5.3.2 Deep Learning Based Algorithms for IVUS Image Segmentation

DL-based algorithms have recently been used for the automatic and fast segmentation of IVUS images. These methods help the clinical practitioner for accurate diagnosis of atherosclerosis and stenosis to prevent CVD risk in heart patients. DL-based algorithms either utilize deep networks for segmentation or integrate hand-crafted features for better accuracy. The details about various DL-based IVUS image segmentation algorithms are tabulated in Table 5.4.

Table 5.4 DL-based algorithms for IVUS image segmentation

Reference	Dataset specification	Model utilized/ proposed	Number of participants	CVD prediction outcome	Summary
Shinohara et al. [1]	• Total Images:3738 • Training/Test: 3415/323 • IVUS imaging using 60 MHz HD IVUS system (AltaView, Terumo, Tokyo, Japan)	U-Net/ 2-D CNN, DoR = 0.1, KS = 3X3, stride = 1	24 patients from the University of Tokyo Hospital	Classified damaged blood vessels and vessels for treatment	• Manually segment the IVUS imaging into lumen area, media including plaque, calcification, and stent • Classification accuracies: 0.97 and 0.98 • Cannot classify stent in complex lesions of coronary arteries
Dong et al. [25]	• Total Images:675 • Training/Test: 567/108 • IVUS imaging using 40 MHz OptiCross catheter (iLab IVUS, Boston Scientific Corporation)	8-layer fully CNN U-Net, D^S: 3X3 conv, stride = 2X2 U^S: 5X5 deconv, stride = 2	30 patients from The Second Affiliated Hospital of Zhejiang University School of Medicine	Segmented coronary artery lumen and cross-sectional area of EEM	• Augmented the dataset using MeshGrip, Flip, and Rotate before processing • Mean IoU: 0.937 and 0.804 for lumen and EEM, respectively • Accuracy more than manual labeling but no analysis on unseen datasets

(continued)

Table 5.4 (continued)

Reference	Dataset specification	Model utilized/ proposed	Number of participants	CVD prediction outcome	Summary
Li et al. [18]	• Total Images:713 • Cross-validation using 18 sets of images as 17 training and 1 test set for each iteration • IVUS imaging using 20 MHz Solid state IVUS transducer (Philips Volcano, Inc.)	End-to-end DL model with multiple layers, U^S, D^S: KS = 2X2, stride = 2, ReLu activation and BN	18 patients from Show Chwan Memorial Hospital, Taiwan	Automatically segment the IVUS images for media-adventitia, luminal borders, and calcified plaque in coronary arteries tissue images	• Analyzed model performance using three loss functions namely, Dice, Tversky, and focal loss • Average precision: 0.90 and 0.73 for the media-adventitia layer and calcified tissue, respectively • Superior performance in the presence of shadow artifacts or side vessels on the target tissues in IVUS images
Balakrishna et al. [3]	• Total Images:435 • Training/Test: 109/326 • IVUS imaging using 20 MHz Eagle Eye monorail catheter (Volcano Corp., SanDiego, US)	VGG16-UNet, PreLU and Sigmoid activation, BN	Public dataset	Automatically extracted two components from coronary arteries viz. lumen and media	• Augmented dataset using horizontal and vertical flips, width, and height shift, and random rotation • Segmentation accuracy: 90% (Approx.) • Ineffective to handle noise in test data including shadow artifacts and catheter images on the target vessels

(continued)

Table 5.4 (continued)

Reference	Dataset specification	Model utilized/ proposed	Number of participants	CVD prediction outcome	Summary
Szarski and Chauhan [19]	• Total Images:435 • Training/Test: 109/326 IVUS imaging using 20 MHz Eagle Eye monorail catheter (Volcano Corp., SanDiego, US)	UNet with padding at each layer, ReLU activation, BN	Public dataset	Automatically extracted two components from coronary arteries viz. lumen and media	• Augmented dataset using polar transformation, in-line sampling, horizontal flip, and pixel-wise additive Gaussian noise • Faster model with 3 ms multi-class segmentation speed and JS = 0.90 and 0.91 for lumen and media, respectively • Efficiently detect calcification areas in complex lesions with shadow artifacts
Xia et al. [20]	• Total Images: 77 (DB1), 425 (DB2) • Training/Test: 95/58 (DB1), 545/ 326 (DB2) • IVUS imaging using 40 MHz Atlantis SR 40 Pro catheter from iLab IVUS (Boston Scientific, Fremont) (DB1), and 20 MHz Eagle Eye monorail catheter from Si5 (Volcano Corp., SanDiego, US) (DB2)	U-net with multi-scale feature aggregation,	Two public datasets with 20 MHz and 40 MHz	Automatically detects membrane borders for LI and MA by segmenting coronary arteries images from IVUS	• Augmented dataset random rotation, shearing, and horizontal flipping • Faster with a segmentation speed of 0.14 s and evaluated model performance at 20 MHz (JS = 0.90 for MA and LI) and 40 MHz (JS = 0.85, 0.84 for MA and LI, respectively) • Multi-scale inputs and feature fusion in skip connections improved the performance of the proposed network

(continued)

Table 5.4 (continued)

Reference	Dataset specification	Model utilized/ proposed	Number of participants	CVD prediction outcome	Summary
Arora et al. [24]	• Total Images:1097 • Training/Test: 877/220 • IVUS imaging using 40 MHz OptiCross catheter (Boston Scientific Polaris)	U-net with Atrous spatial pyramid pooling, ReLU activation, Adam optimizer, 1X1 conv, BN	12 patients from the Department of Cardiology, PGIMER, Chandigarh	Automatically detect lumen area, severe calcification plaque, and shadow areas behind calcification from complex lesion area	• Utilized two augmentation strategies in real-time namely, noise augmentation (Gaussian or Salt & Pepper or Speckle noise) and spatial augmentation (rotation and flipping) • Obtained mean 0.88, 0.63, and 0.71 for lumen, calcification, and shadows, respectively • Effectively segment lumen borders in the presence of severe calcification and artifacts shadow

(continued)

Table 5.4 (continued)

Reference	Dataset specification	Model utilized/ proposed	Number of participants	CVD prediction outcome	Summary
Masuda et al. [44]	• Total Images:191 • Training/Test: 153/38 • IVUS imaging using 40 MHz IVUS catheter (Atlantis, Boston Scientific, Natick, MA)	Pre-trained GoogleNet Incept v3 model, adam optimizer, DoR = 0.5, ReLu activation	178 patients	Segmented the various types of plaque such as fatty and fibro-fatty	• Augmented dataset using horizontal and vertical flipping, random transformation and rotation, and shifting width and height by shearing and zooming • Insufficient data size and quality for the training model leads to the overfitting problem • Limited discussion on the plaque category. Non-calcified lesions and moderate stenoses were not segmented
Bargsten et al. [46]	• Total Images:620 • Training/Test: 153 + 295/38 + 115 • IVUS imaging using 20 MHz Eagle Eye Platinum catheter (Philips Healthcare, San Diego, USA)	Two DL models namely, DeepLabV3 and UNet,	COCO dataset has a calcium dataset and a wall-lumen dataset	Segment calcification, vessel wall, and lumen	• Augmented dataset using random flip and rotation • Out of two training models, DeepLabV3 showed superior performance in comparison to Unet for calcification, vessel, and lumen segmentation • Due to the small dataset size less accurate in calcification segmentation

(continued)

Table 5.4 (continued)

Reference	Dataset specification	Model utilized/ proposed	Number of participants	CVD prediction outcome	Summary
Zhu et al. [43]	• Total Images:1746 • Training/Test: 1572/174 • IVUS imaging using EagleEye, Volcano Corporation, Cordova, CA, USA	Unet + + with feature pyramid network, RMSprop optimizer	18 patients from 7th People's Hospital Zhengzhou	Extracted lumen and MA border from the coronary arteries IVUS images	• JS of 0.942 and 0.9509 for the lumen and MA border, respectively • Better and faster segmentation accuracy with 0.1203 s • Multi-scale features retain global contour for consistent segmentation accuracy
Olender et al. [37]	• Total Images:553 • Training/Test: 353/200 • IVUS imaging using 20 MHz EagleEye Gold catheter, Philips Healthcare, Andover, MA	CNN, momentum optimizer, BN, ReLU activation	8 patients	Domain-enriched method for classifying tissues into various classes namely, dense calcium, necrotic core, fibrous and fibro-fatty tissue, and media	• Augmented datasets using rotation and reflection • Overall accuracy of 85.65% for classifying plaque on grayscale IVUS images • Slow method with lower feasibility to deploy for classification in real-time

(continued)

Table 5.4 (continued)

Reference	Dataset specification	Model utilized/proposed	Number of participants	CVD prediction outcome	Summary
Li et al. [47]	• Total Images: 435 (DB2) • Training/Test: 109/326 (DB2) IVUS imaging using 40 MHz Atlantis SR 40 Pro catheter from iLab IVUS (Boston Scientific, Fremont) (DB1), and 20 MHz Eagle Eye monorail catheter from Si5 (Volcano Corp., SanDiego, US) (DB2)	U-net, LBP, and HOG features, Shearlet transform, shadow, and relative shadow	Two public datasets	Detecting lumen border, side vessels, bifurcation, shadow, and none-artifacts	• Extracted 193 handcrafted features and 64 high-level features for accurate detection • Employed feature selection based on extended binary cuckoo search • Kernel sparse coding for image classification from 36-D hybrid feature subset and achieved JS (0.88) for lumen border detection
Bajaj et al. [35]	• Total Images:824,750 • Training/Test: 740,013/84,737 • IVUS imaging using 50 MHz Dualpro system (Infraredx, Burlington, MA)	ResNet CNN, Pix2pix conditional GANs,	65 patients from Barts Heart Centre	Automatic and accurate segmentation of EEM and lumen borders in end-diastolic IVUS frames	• Dataset augmentation using random rotations, crossing, rescaling, and brightness adjustments • Methodology applicability was limited to 50 MHz IVUS sequences only • Methodology was ineffective in segmenting stent areas

(continued)

Table 5.4 (continued)

Reference	Dataset specification	Model utilized/proposed	Number of participants	CVD prediction outcome	Summary
Nishi et al. [36]	• Total Images:45,449 • Training/Validation/Test: 31,814/6817/6817 • IVUS imaging using 40-MHz (Boston Scientific Corporation), Terumo Corporation) and 45-MHz IVUS (Volcano Corporation)	Fully CNN DeepLabv3 model, Adam optimizer	Training datasets from Stanford University, and testing on 11 patients' data from AltaView, Terumo, Tokyo, Japan	Segmented vessel, lumen, and stent area along with LI and MA borders for CVD risk assessment	• Superior segmentation results in IVUS images without stents in comparison to those with stents • Model applicability is limited to 40 and 45 MHz IVUS images • Model not able to segment stent expansion, calcified regions, dissection, and thrombotic material

(continued)

Table 5.4 (continued)

Reference	Dataset specification	Model utilized/ proposed	Number of participants	CVD prediction outcome	Summary
Cho et al. [45]	• Total Images:137,989 • Training/Test: 1,13,746/24,243 • IVUS imaging using 40 MHz motorized transducer (Boston Scientific/SCIMED, Minneapolis, MN)	EfficientNet-B3, fivefold cross-validation	778 patients from Asan Medical Center, Seoul, Korea	Segmented plaque into three categories such as calcified, attenuated, and without it in IVUS imaging	• Overall accuracy of the model is 96%. Predicted attenuation plaque and calcified plaque with 93% and 96% accuracies, respectively • Data imbalance due to differences in frames in each category impacted the performance • Model generalizability not quantified
Yang et al. [33]	• Total Images: 78 (DB1), 435 (DB2) • Training/Test: 19/59 (DB1), 109/ 326 (DB2)	DPU-Net, UNet, SegNet, KS = 3X3, stride 2, PreLU, Adam optimizer, BN	Two datasets namely 20 MHz (10 patients) and 40 MHz	Segmented lumen and media for bifurcation, side vessel, shadow, and none from the IVUS images	• Real-time augmenter for augmenting datasets in real-time to improve performance • Training over multiple datasets ensured the model's generalizability • Achieved average JS for 40 MHz and 20 MHz for lumen and media as 0.87, 0.86, 0.90 and 0.92, respectively

KS = Kernel Size, DoR = Dropout rate, BN = Batch normalization, U^S = Up sampling, D^S = Down sampling, JS = Jaccard Similarity, IoU = Intersection Over Union, MA = Media-adventitia, LI = Lumen-intima, ms = milliseconds, ReLU = Rectifier linear unit, EEM = External elastic membrane, HOG = Histogram of gradient, LBP = Linear binary pattern, GAN = generative adversarial network, PreLU = Parametric rectifier linear unit

Mostly, DL-based algorithms utilize U-net and its variants for effective and efficient segmentation of IVUS images. In this direction, Shinohara et al. [1] exploited U-net for automatic segmentation of complex lesions in IVUS images to identify normal, calcified, and stent areas. However, authors utilized deep 8-layer UNet architecture to mask the lumen and EEM in the IVUS images [25]. Feature maps were extracted from the encoding layer whereas feature maps were concatenated in the decoding layer using skip connection. The final predictions were recovered from the class probability maps generated by the softmax in the decoding layer. The method eliminated the pre- and post-processing steps to speed up the segmentation process. In [19], learned translation dependence was integrated with UNet to separate the vessel components in polar coordinated IVUS images. Interior and external vessels were separated using context awareness in multi-class segmentation using spatial content information. Lumen and media vessels were masked by thresholding the softmax probabilities in the post-processing step. The method segmented the vessels in real-time on a 1080 NVIDIA GPU. Balakrishna et al. [3] exploited VGG-16 to design UNet architecture. The proposed method segmented the lumen and media by generating a pixel map rather than considering the whole image. Thresholding-based post-processing step was applied to improve the boundary smoothness for the segmented area.

To improve the efficiency of the feature map for accurate IVUS image segmentation, certain extensions are incorporated into the basic architecture of the UNet framework. In this direction, Zhu et al. [43] proposed UNet++ by integrating feature pyramid maps to utilize feature maps at multiple scales. Multi-scale features were fused, and upscale operators were utilized for scaling the heterogeneous feature scale. Feature pyramid maps generated the final probability feature map using voting mechanism. Self-adapting threshold was used in the post-processing step to obtain the final target area. In [20], authors integrated a feature aggregation module with the UNet to deal with the multi-scale features. Feature aggregation module extracted global semantic information and high-resolution local information from the up-convolutional layer and the encoding layer of the network and fused them efficiently. Authors integrated convolutional block attention modules and atrous spatial pyramid pooling in UNet architecture to preserve the spatial features for detecting vessel components from the crucial channels. These additional modules determined feature importance and give high weights to the target vessel components for better segmentation accuracy in the presence of artifacts and shadows. The method was effective in segmenting the vessel lesions in the presence of severe calcification and shadowing areas behind the accumulated calcified plaque.

Apart from the UNet model, other DL models such as CNN, GoogleNet, InceptionV3, and DeepLabV3 are also explored for segmentation in IVUS images. In [44], authors modified the pre-trained GoogleNet Inception v3 model using transfer learning for the classification of coronary plaque in the lesion area of IVUS images into fibrous, fatty, or fibro-fatty plaques. On the other hand, authors utilized the EfficientNet-B3 neural network for categorizing plaque into three categories [45]. The model's performance was enhanced by reducing memory occupancy in the network architecture. The efficient and optimal hyperparameters were searched using compound scaling to optimize the network for efficient results. Olender et al. [37] exploited CNN architecture for classifying the tissues into pathological and non-pathological utilizing spatial and geometric constraints. The ROI pixels classified plaque into four types. Stochastic gradient descent was used to reduce the training time for faster classification. In [18], authors utilized deep CNN along with a cascaded network to detect the MA borders, lumen, and calcium. The prediction was performed in two stages namely, the segmentation stage and the location stage. In the first stage, plaque regions were identified in CNN and probability feature maps were generated. In the final stage, convolutional operations were applied to locate the calcified tissues in the plaque. The efficiency of the method was proved by testing the performance on multiple metrics and loss functions. Bajaj et al. [35] segmented lumen borders and EEM in IVUS images in real time using ResNet deep neural network architecture. Hyperparameters were selected empirically and GAN structure with a fixed learning rate. The proposed method was accurate, automatic, and faster in comparison to other methods. The segmentation was performed in 60 s for a 30 mm coronary segment. The method was limited to 50 MHz IVUS images only and had not analyzed stent area images in the training dataset. Authors segmented IVUS images for lumen and vessel borders along with stent area using the DeepLabv3 network and ResNet encoder [36]. The segmentation performance of the proposed model was relatively higher in the images without stent area than the one with stent.

To investigate the IVUS image segmentation capabilities, performance from multiple deep neural networks was compared and evaluated. In this direction, Bargsten et al. [46] compared the performance from DeepLabV3 and UNet for vessel wall and lumen segmentation for calcified plaque. For comparable performance, the training datasets and network capabilities were kept identical. Both networks had 40 M parameters and three downsamplings in the network encoder. In [33], the authors proposed Dual path UNet (DPU-Net) and trained two other deep networks namely, SegNet and UNet to generate a prediction map for segmenting the IVUS images. The transposed convolutional layer was added in the last upsampling stage for fair comparison. A real-time augmenter was also integrated into the DPUNet to improve the processing speed and generalizability of the model on small training datasets. Apart from this, authors integrated 193-D handcrafted features with 64-D high-level features extracted from UNet [47]. The hybrid feature vector of 257-D was exploited to improve the discriminability of the network for the lumen boundaries. The dictionary was loaded with lumen and non-lumen images to improve segmenting accuracy in complex lesions. Pre-processing steps were employed to improve the feature selection and post-processing was to enhance the model accuracy to detect vessel boundaries efficiently.

To summarize, due to the high volume of gray-scale images generated by the IVUS device, it became important to analyze those images not only accurately but also in real time. Since IVUS is an invasive technique, radiologists can't analyze such a huge number of images quickly, and give predictions for the patients who are waiting for the angiography in the operation theater. DL-based prediction algorithms are quite faster to examine the IVUS images and provide outcomes in real time.

5.4 Other Imaging for Predicting Risk of Cardiovascular Disorder

IVUS is an invasive technique for CVD risk prediction by analyzing coronary arteries in real time. Apart from invasive techniques, non-invasive techniques such as the CCTA technique are also quite popular among cardiologists for the prediction of atherosclerosis [5]. CCTA images can be reviewed to determine the narrowing of coronary arteries due to the accumulation of calcified plaque. DL-based methods are preferred over conventional methods as these methods provide better results by handling the attenuation caused by the accumulated plaque in the coronary vessels.

Basically, DL-based UNet architecture is explored to segment CCTA images for predicting coronary disorders [21, 22]. In [22], authors proposed 3D multi-channel UNet to segment the CTA (Computed tomography angiography) image for identification of vessel stenosis automatically. The method applied the preprocessing steps to remove the irrelevant tissues from the CTA images. Activation unit ReLU, max pooling for downsampling, and DCE as a loss function were used in the DL architecture. The dataset was augmented by flipping and rotating the background of the vessel regions. On the other hand, Pan et al. [21] exploited 3D dense UNet to segment the coronary arteries CTA images. Focal loss function was adopted to address the class imbalance between the background area and the coronary arteries region. The preprocessing step was employed to prepare the training dataset by transforming the HD images to low resolution maintaining global and correlation information. The method was faster taking an average running time of 10–15 secs for segmentation. The method was trained on a relatively larger dataset hence, it was expected that the method performs well on unseen data as well.

To address the limitation of poor quality and contrast of CTA images, authors proposed a region-based DL method based on supervised attention UNet [48]. The model utilized a hybrid loss function combining logistic and Dice functions to measure the relationship between the predicted and training data. The method utilized five-fold validation for segmenting the left ventricular myocardial contours from the coronary vessels. In [23], authors exploited 3D deep attention UNet for segmenting the epicardial adipose tissues from the coronary arteries automatically to examine the deposited fat. To enlarge the training dataset, various augmentation techniques such as flipping, rotating, and scaling were applied. Five-fold validation was used to demonstrate the segmentation results in a better way. The method had achieved high accuracy and precision, but validation of the method on multiple datasets from different vendors was substantial before its clinical deployment. Figure 5.3 represents the CTA image with the right coronary artery and aorta.

To identify the narrowing of coronary vessels and aorta, authors proposed a 2D UNet architecture to segment these components in the CTCA images [5]. Fast and multiple preprocessing techniques were applied to adjust brightness, pixel intensity, and scaling to convert the input image into 8-bit PNG (Portable network graphics) format. The proposed model was fully automatic by including the sigmoid function, batch normalization, and dropout layer to reduce overfitting and improve the

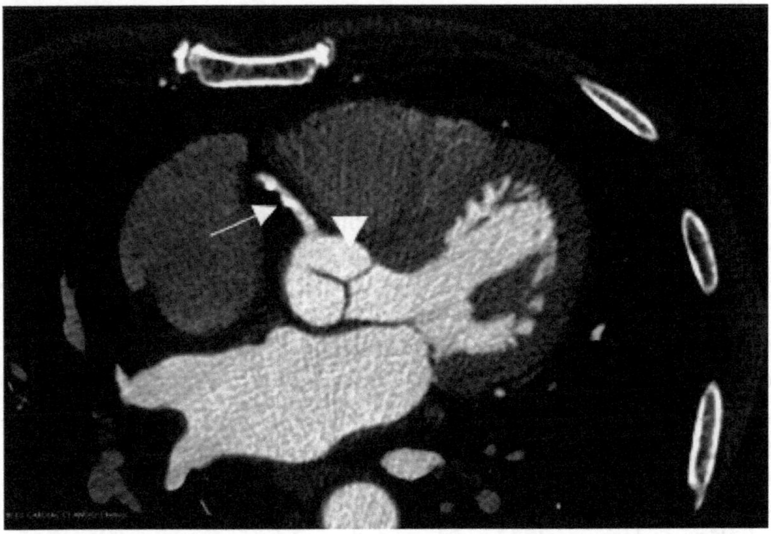

Fig. 5.3 IVUS image representing coronary arteries (arrow) and aorta (arrowhead) [5]

model's stability and performance. However, authors embedded a global feature network using semantic information for segmenting the boundaries in CTA images accurately [49]. Noisy activation function and improved active contour loss were integrated to suppress noise effect and optimize the network predictions for accurate and automatic segmentation of vessel boundaries. The model was equipped with multi-level semantic information to obtain refined vessel boundaries with high-quality score maps.

To summarize, DL-based methods are highly effective and efficient in segmenting the coronary vessels from the CTA images. These methods are not only accurate but also fast in comparison to manual segmentation techniques. DL-based methods help in the early identification of CVD to prevent the risk of heart failure by predicting vessel narrowing by plaque accumulation. However, there are certain limitations in terms of dataset availability and clinical deployment of the method. Various augmentation techniques such as flipping, rotation, scaling, and many more are adopted to increase the training data which do not ensure that the method performs well on the unseen datasets. Methods also suffered from the selection bias and the evaluation bias. The outcomes of DL methods are compared with the manually annotated data. This introduces the evaluation bias in the accuracy as it is highly dependent on the radiologist's capabilities and experience.

5.5 Summary

In this chapter, we have reviewed the AI-based methods for segmentation of IVUS imaging for the prediction of CVD risk at an early stage. AI-based IVUS image segmentation methods are categorized into ML-based techniques and DL-based techniques. The salient features, similarities, differences, and limitations are explored to determine the achievement and improvement in the area. ML-based methods are limitedly explored for segmenting the vulnerable calcified plaque from the coronary vessels due to complex and tedious regions in the IVUS images. On the other hand, DL-based methods including CNN architecture and different variants of UNet are widely popular among researchers due to better accuracy and precision in the segmented results. DL-based methods have shown successful results in segmenting critical components such as MA border, LI borders, fibrous and non-fibrous plaque from the complex lesion of coronary arteries. But DL-based methods are data hungry and to satisfy the requirements of training data, various data augmentation techniques are widely explored. Data augmentation techniques increase the training data size, but these techniques do not guarantee the model accuracy on unseen real-time datasets. In addition, attentional networks, feature aggregation, and other parameters are introduced to improve accuracy, but they impact the speed of the model. Also, the DL-model outcomes are compared with the manual annotated datasets which introduce the selection and evaluation bias in the final results.

In the future, DL-based models can be explored for generalizability by selecting the datasets from various sources. Also, the emphasis should be given public availability of the code so that the exhaustive performance evaluation can be performed before the clinical deployment of the DL model.

References

1. Shinohara H et al (2021) Automatic detection of vessel structure by deep learning using intravascular ultrasound images of the coronary arteries. PLoS ONE 16(8):e0255577
2. Tong J et al (2021) Automatic lumen border detection in IVUS images using dictionary learning and kernel sparse representation. Biomed Signal Process Control 66:102489
3. Balakrishna C, Dadashzadeh S, Soltaninejad S (2018) Automatic detection of lumen and media in the IVUS images using U-Net with VGG16 Encoder. arXiv preprint arXiv:1806.07554
4. Taki A et al (2013) An IVUS image-based approach for improvement of coronary plaque characterization. Comput Biol Med 43(4):268–280
5. Cheung WK, et al. (202) A computationally efficient approach to segmentation of the aorta and coronary arteries using deep learning. IEEE Access 9:108873–108888
6. Song A et al (2022) Automatic coronary artery segmentation of CCTA images with an efficient feature-fusion-and-rectification 3D-UNet. IEEE J Biomed Health Inform 26(8):4044–4055
7. Zhou Z et al (2019) Unet++: redesigning skip connections to exploit multiscale features in image segmentation. IEEE Trans Med Imaging 39(6):1856–1867
8. Kumar A (2023) Visual object tracking using deep learning. CRC Press
9. Kumar A, Sarren P, Raja (2023) Deep learning-based multi-object tracking, in object tracking technology: trends, challenges and applications. Springer, pp 183–199

10. Kumar A, et al (2023) AI-Assisted special education for students with exceptional needs. IGI Global
11. Kumar A, Singh D, Vohra R (2023) Improving learning abilities using AI-based education systems. In: AI-Assisted special education for students with exceptional needs. IGI Global, pp 137–155
12. Kumar A, Vohra R (2023) Impact of deep learning models for technology sustainability in tourism using big data analytics. Deep learning technologies for the sustainable development goals: issues and solutions in the post-COVID era. Springer, pp 83–96
13. Kumar A, Singh D (Eds.) (2025) Revolutionizing Medical Systems Using Artificial Intelligence: A Breakthrough in Healthcare. Elsevier
14. Maindarkar M, Kumar A (2025) Artificial intelligence based disease diagnosis using ultrasound imaging. In Revolutionizing Medical Systems using Artificial Intelligence, pp 147–161. Academic Press
15. Kumar A, Sareen P, Arora A (2023) Healthcare engineering using AI and distributed technologies. Smart distributed embedded systems for healthcare applications. CRC Press, pp 1–14
16. Su S et al (2017) An artificial neural network method for lumen and media-adventitia border detection in IVUS. Comput Med Imaging Graph 57:29–39
17. Yang J, et al (2018) IVUS-Net: an intravascular ultrasound segmentation network. In: Smart multimedia: first international conference, ICSM 2018, Toulon, France, August 24–26, 2018, Revised Selected Papers 1. Springer
18. Li Y-C et al (2021) Automatic detection of atherosclerotic plaque and calcification from intravascular ultrasound images by using deep convolutional neural networks. IEEE Trans Ultrason Ferroelectr Freq Control 68(5):1762–1772
19. Szarski M, Chauhan S (2021) Improved real-time segmentation of intravascular ultrasound images using coordinate-aware fully convolutional networks. Comput Med Imaging Graph 91:101955
20. Xia M, et al (2020) Extracting membrane borders in IVUS images using a multi-scale feature aggregated U-Net. In: 2020 42nd annual international conference of the IEEE engineering in medicine & biology society (EMBC). IEEE
21. Pan L-S et al (2021) Coronary artery segmentation under class imbalance using a U-Net based architecture on computed tomography angiography images. Sci Rep 11(1):14493
22. Chen Y-C, et al (2019) Coronary artery segmentation in cardiac CT angiography using 3D multi-channel U-net. arXiv preprint arXiv:1907.12246
23. He X et al (2020) Automatic segmentation and quantification of epicardial adipose tissue from coronary computed tomography angiography. Phys Med Biol 65(9):095012
24. Arora P, et al (2023) CADNet: an advanced architecture for automatic detection of coronary artery calcification and shadow border in intravascular ultrasound (IVUS) images. Phys Eng Sci Med 1–14
25. Dong L et al (2021) Automatic segmentation of coronary lumen and external elastic membrane in intravascular ultrasound images using 8-layer U-Net. Biomed Eng Online 20(1):1–9
26. Athanasiou LS, Fotiadis DI, Michalis LK (2017) Atherosclerotic plaque characterization methods based on coronary imaging. Academic Press
27. Peng C et al (2021) Recent advances in transducers for intravascular ultrasound (IVUS) imaging. Sensors 21(10):3540
28. Ng A, Swanevelder J (2011) Resolution in ultrasound imaging. Continuing Education in Anaesthesia, Critical Care & Pain 11(5):186–192
29. Yang Y, et al (2023) Coupled contour regression for efficient delineation of lumen and external elastic lamina in intravascular ultrasound images. IEEE J Biomed Health Inf
30. Rezaei Z et al (2017) Automatic plaque segmentation based on hybrid fuzzy clustering and k nearest neighborhood using virtual histology intravascular ultrasound images. Appl Soft Comput 53:380–395
31. Matsumura M, et al (2023) Accuracy of IVUS-based machine learning segmentation assessment of coronary artery dimensions and balloon sizing. JACC: Advances 2(7):100564

32. Sun S, Sonka M, Beichel RR (2013) Graph-based IVUS segmentation with efficient computer-aided refinement. IEEE Trans Med Imaging 32(8):1536–1549
33. Yang J, Faraji M, Basu A (2019) Robust segmentation of arterial walls in intravascular ultrasound images using Dual Path U-Net. Ultrasonics 96:24–33
34. Ziemer PG et al (2020) Automated lumen segmentation using multi-frame convolutional neural networks in intravascular ultrasound datasets. European Heart J-Digital Health 1(1):75–82
35. Bajaj R et al (2021) Advanced deep learning methodology for accurate, real-time segmentation of high-resolution intravascular ultrasound images. Int J Cardiol 339:185–191
36. Nishi T et al (2021) Deep learning-based intravascular ultrasound segmentation for the assessment of coronary artery disease. Int J Cardiol 333:55–59
37. Olender ML et al (2020) A domain enriched deep learning approach to classify atherosclerosis using intravascular ultrasound imaging. IEEE J Selected Topics Sig Process 14(6):1210–1220
38. Jun TJ et al (2019) Automated detection of vulnerable plaque in intravascular ultrasound images. Med Biol Eng Compu 57:863–876
39. Lee J-G et al (2020) Intravascular ultrasound-based machine learning for predicting fractional flow reserve in intermediate coronary artery lesions. Atherosclerosis 292:171–177
40. Vercio LL, Del Fresno M, Larrabide I (2017) Detection of morphological structures for vessel wall segmentation in IVUS using random forests. In: 12th International symposium on medical information processing and analysis. SPIE
41. Archana K, Vanithamani R (2021) Classification of plaque in carotid artery using intravascular ultrasound images (IVUS) by machine learning techniques. Ann Roman Soc Cell Biol 4751–4760
42. Cui H, Xia Y, Zhang Y (2020) Supervised machine learning for coronary artery lumen segmentation in intravascular ultrasound images. Int J Numer Methods Biomed Eng 36(7):e3348
43. Zhu F et al (2022) A deep learning-based method to extract lumen and media-adventitia in intravascular ultrasound images. Ultrason Imaging 44(5–6):191–203
44. Masuda T et al (2022) Deep learning with convolutional neural network for estimation of the characterisation of coronary plaques: validation using IB-IVUS. Radiography 28(1):61–67
45. Cho H et al (2021) Intravascular ultrasound-based deep learning for plaque characterization in coronary artery disease. Atherosclerosis 324:69–75
46. Bargsten L, et al (2021) Deep learning for calcium segmentation in intravascular ultrasound images. In: Current directions in biomedical engineering. De Gruyter
47. Li K et al (2021) Automatic lumen border detection in IVUS images using deep learning model and handcrafted features. Ultrason Imaging 43(2):59–73
48. Jun Guo B et al (2020) Automated left ventricular myocardium segmentation using 3D deeply supervised attention U-net for coronary computed tomography angiography; CT myocardium segmentation. Med Phys 47(4):1775–1785
49. Gu J et al (2020) Segmentation of coronary arteries images using global feature embedded network with active contour loss. Comput Med Imaging Graph 86:101799

Chapter 6
Diagnosis and Prediction of Cardiovascular Risk in Retinal Imaging Using Artificial Intelligence

Abstract Cardiovascular illness is a primary cause of death and disease globally, highlighting the need for innovative methods in the early assessment of cardiovascular risk and detection. This article examines the latest developments in the domain of artificial intelligence, including deep learning and machine learning, to identify cardiovascular risk. The paper focuses on the use of AI (Artificial intelligence) to analyze various data modalities, specifically highlighting retinal fundus photos as a possible non-invasive risk assessment tool. The paper examines the current state of AI-based cardiovascular disorder detection, emphasizing the difficulties, developments, and potential paths forward in applying these technologies to enhance cardiovascular prediction and facilitate early intervention.

Keywords Artificial intelligence (AI) · Non-invasive · Retinal imaging · Feature selection · Risk assessment

6.1 Introduction

Cardiovascular disorder (CVD) is a broad category of medical illnesses that impact the flow of blood to the heart. It is one of the foremost reasons for illness and death affecting individuals across all age groups from infancy to old age and a significant contributor to disability and reduced productivity in adults. It is driven due to excessive consumption of fat or genetic factors. CVD leads to stroke and heart diseases, which are the primary reasons for mortality worldwide. One in three adults is impacted by one or more types of CVD. The likelihood of this happening grows with age and varies across different groups with varying ethnicities, races, and geographies [1, 2]. In many countries, the prevalence of CVD is substantial and is on the rise. In some countries, the onset of the first heart attack occurs about 10 years earlier than in other countries. The key risk factors including smoking, diabetes, lipids, hypertension, diet, alcohol consumption, physical activity, obesity, and psychosocial factors accounted for 86% of CVD [3]. Timely assessment of CVD-related risk factors is extremely important to lower the frequency of cardiac events and consequently the

A. Kumar and D. Singh, *Artificial Intelligence in Modern Healthcare System*,
Transactions on Computer Systems and Networks,
https://doi.org/10.1007/978-981-96-6703-1_6

death rate. Although detection of CVD risk can be done through traditional methods recently AI-based CVD detection is outperforming traditional methods owing to its automated nature and reduced manual intervention. Most recently, researchers have proposed methodologies utilizing RFI (Retinal fundal imaging) for CVD detection leveraging DL-based techniques [4]. This approach can certainly seem very promising owing to the non-invasive nature of retinal examination. Moreover, it is also accessible to low-income group people.

CVD refers to a condition impacting the blood vessels or heart, resulting in damage in arteries located in the brain, heart, kidneys, and eyes. It generally involves the presence of fatty deposits within arteries, posing a risk of blood clot formation. The main types of CVD are as follows [5]:

- *Coronary Heart Disease (CHD)*: Impacts the arteries supplying blood to the muscles of the heart.
- *Cerebrovascular Disease (CeVD)*: Impacts the blood vessels that provide supply to the cerebral region.
- *Peripheral Artery Disease (PAD)*: Impacts the arteries that provide blood to the arms and legs.
- *Rheumatic heart disease (RHD)*: Streptococcal bacteria result in harm to both the cardiac muscle and valve structures.
- *Congenital heart disease*: Congenital anomalies that interfere with the regular growth and operation of the heart, stemming from irregularities in the heart's structure.

Most CVDs could be averted through effective intervention addressing cardiovascular risk factors. Prompt diagnosis and treatment are essential in this context. The primary cardiovascular condition is ischemic heart disease more commonly observed in men. Following this, there are instances of stroke, heart failure, and irregular heart rhythms [6]. The risk factors responsible for CVD can be non-modifiable and modifiable. Non-modifiable risk factors are fixed and can't be changed or varied with time. These factors include age, sex, ethnicity, family history, and socioeconomic level. On the other hand, modifiable risk factors are time-varying and include lipid abnormalities, excessive blood pressure, Diabetes mellitus, obesity, sedentary lifestyle, smoking, and tobacco usage [6, 7]. Figure 6.1 depicts the CVD risk factors that increase of possibility of heart attacks in human beings.

Assessing cardiovascular risk is a critical aspect of determining the most effective treatment for a patient. This process utilizes tools to compute the chances of experiencing a cardiovascular activity in a specified timeframe, typically within the next 10 years, aiding in informed decision-making regarding the most suitable treatment. Several widely recognized CVD risk assessment tools are utilized globally to guide preventive strategies and personalized treatment plans. Framingham CVD assesses 10-year risk of CVD events including CHD, PAD, heart failure, and stroke [8]. Pooled Cohort Equation (PCE) derived from the American cohort is utilized for estimating the 10-year risk for arteriosclerotic CVD including non-fatal myocardial infraction, CHD death, and fatal or non-fatal stroke [9]. Systematic Coronary Risk Evaluation (SCORE) derived from the European cohort is utilized for estimating the 10-year risk

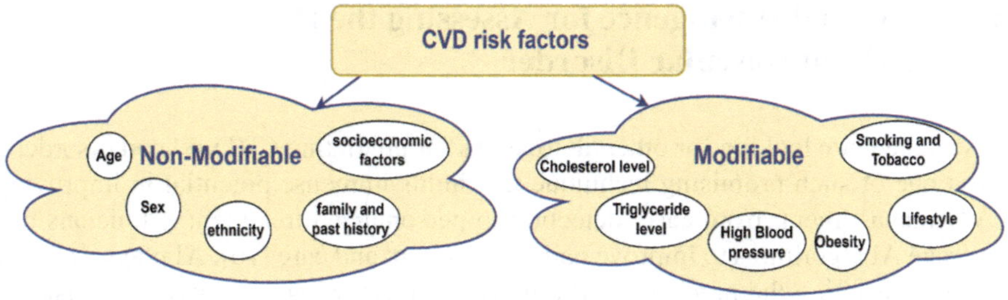

Fig. 6.1 Description of various CVD risk factors

of fatal CVD events [10]. QRISK3 derived from the United Kingdom cohort is used for assessing the 10-year risk of CVD events including CHD, ischemic stroke, or transient ischemic stroke. American College of Cardiology/American Heart Association (ACC/AHA) risk prediction tool provides a 10-year risk estimate of atherosclerotic CVD [9]. Modified Framingham risk score (FRS) serves as an official risk assessment tool employed in Singapore [11]. Aortic stenosis and ventricular dysfunction are the categories of CVD that are frequently identified by Cardiac Magnetic Resonance (CMR) and Electrocardiograms (ECG) [12–15].

The purpose of this chapter is to provide insights into the identification and evaluation of CVD risk through the application of AI techniques, including ML and DL utilizing RFI. The key objectives of this chapter are as follows:

- The salient features of traditional techniques and AI-based techniques for CVD risk predictions are elaborated to determine the importance of ML and DL techniques in CVD detection.
- The chapter reviews the various state-of-the-art (SOTA) techniques for CVD detection, with a specific focus on AI-based techniques.
- Details about CVD detection using RFI within the context of DL are gathered and reviewed to highlight the potential benefits and limitations of using RFI for CVD risk prediction.

The chapter is organized as follows: Sect. 6.2 discusses about AI as tool for CVD risk assessment. It further discusses traditional ways of detecting CVD risk and the limitations pertaining to traditional methods. Section 6.3 discusses about automated detection of CVD using ML-based techniques. The intersection of retinal imaging and cardiovascular health is elaborated in Sect. 6.4. Section 6.5 discusses CVD risk prediction techniques through DL-based techniques. Lastly, the conclusion and future directions of CVD detection through RFI are detailed in Sect. 6.6.

6.2 Artificial Intelligence for Assessing the Risk of Cardiovascular Disorder

Researchers are looking for other alternatives for predicting CVD-related disorders. AI is one of such promising technique that holds immense potential in improving CVD management, from early detection to personalized treatment. Clinicians can leverage AI algorithms to improve patient treatment and care [16]. AI plays a crucial role in enhancing the precision and efficiency of assessing an individual's risk for developing CVD. According to [17], AI is set to revolutionize medicine, particularly in cardiovascular and medical imaging, enhancing efficiency and empowering physicians with advanced computational tools for improved patient care. AI can make use of various ML (Machine learning) and DL (Deep learning) models to analyze larger datasets for identifying patterns and relationships that may contribute to CVD risk. It has the capability to assess medical images, like CT scans (Computed tomography) and MRI (Magnetic imaging resonance), identifying initial signs of cardiovascular diseases such as arterial plaques, stenosis, and other irregularities linked to CVD. It can scrutinize data from electronic health records [16] to detect patterns and associations that may suggest an elevated risk of cardiovascular diseases, encompassing details like blood pressure, cholesterol levels, and medical history. Compared to conventional techniques, AI holds the potential to automate the assessment of CVD risk in a faster and more efficient way. Moreover, some patients detected with CVD risk may need to opt for the surgery to reduce the risk of death. In response, AI techniques including DL can be utilized by researchers to perform tasks such as visual tracking [18, 19] during laparoscopic surgery which can lead to an increase in the success of the surgical procedure involved.

6.2.1 Limitations of Traditional Methods for Diagnosing Cardiovascular Disorder

Although standard risk assessment tools offer a proactive approach to patient treatment but these tools such as Framingham CVD, FRS, PCE, QRISK3 and SCORE rely on cohorts predominantly comprising individuals of Western descent [4, 20]. Given the emerging understanding of diverse ethnicities having distinct risk factor profiles, these tools may not provide the highest accuracy for all populations [4]. According to [21], prediction equations for assessing the risk of CVD might not be effective in today's scenario, leading to insufficient or excessive treatment of risk factors for CVD. Authors investigated that conventional cardiovascular risk may result in over- or underestimation of CVD risks, providing limited benefits to clinical outcomes [22]. Currently, determining the optimal risk assessment models remains challenging due to variations in categories of risk, the presence of comparable cohorts, and diversity among populations prone to risk.

Furthermore, ECG and CMR, diagnostic tools are utilized for patients showing symptoms rather than asymptomatic individuals [14]. Challenges in the form of high costs and the need for specialized technical expertise limit their application as screening tools for the general population, hindering early CVD diagnosis [13, 23]. As a result, numerous patients remain undiagnosed until the disease reaches its advanced stages, leading to inferior outcomes for those cases [24].

6.2.2 Comparison of Traditional and Artificial Intelligence-Based Approach for Predicting Cardiovascular Disorder Risk

CVD risk assessment can be done either using a traditional approach or a modern approach. The traditional approach follows conventional techniques for CVD prediction, but modern approaches are based on AI-based methods. The salient comparison of traditional versus AI approaches for detecting CVD risk is depicted in Table 6.1.

Table 6.1 Traditional versus AI-based approach CVD risk assessment

Attributes	Traditional approach	AI-based approach
Data processing	Manual collection and input of data	Automated data integration and analysis
Data types	Limited to basic clinical parameters	Diverse data types, including genetics, medical imaging and lifestyle factors
Risk assessment models	Relies on predefined risk equations	Utilizes advanced predictive modeling techniques adapted to evolving data
Feature selection	Manual or limited feature selection	Utilizes sophisticated algorithms for comprehensive feature selection and prioritization
Computational efficiency	May involve time-consuming manual processes	Enhances efficiency through automated processes, reducing computation time
Accuracy	Relies on predefined equations and clinical judgment	Utilizes advanced algorithms for improving accuracy, especially in complex and dynamic scenarios
Integration with healthcare systems	May have limitations in seamless integration	Can be integrated into electronic health records and existing healthcare systems

6.2.3 Challenges in Utilizing Artificial Intelligence to Assess Cardiovascular Risk

Detecting CVD through AI techniques poses several challenges. A few of the challenges are described and enumerated below:

- *Quality of the dataset*: The effectiveness of AI models is strongly impacted by the standard of training data. Poor performance and incorrect outcomes could result from incomplete, inaccurate, or biased data [13].
- *Lack of Interpretability*: The difficulty of understanding AI models' decision-making processes often leads to their being perceived as "black boxes." This lack of interpretability poses difficulties in trusting the outcomes of AI-driven CVD detection [25].
- *Regulatory issues*: Before using AI-based systems for CVD diagnosis in practical scenarios, regulatory approval from an appropriate authority must be obtained which can incur significant expenses and time investment [16].
- *Integration with the current framework*: Healthcare systems are complex and protecting data security and privacy is crucial. These factors make it difficult to integrate AI-driven CVD detection systems into the current healthcare framework [26].
- *Expense of deployment*: Adoption of AI-driven CVD detection systems may be hampered by the high cost of development and implementation, particularly in environments with restricted funding [27].

6.3 Machine Learning-Based Models for Predicting Risk of Cardiovascular Disorder

ML is becoming increasingly crucial in the early identification and diagnosis of critical diseases [28], emphasizing the ability to draw inferences based on emerging information through the detection of concealed patterns within observations [29]. Various SOTA ML-based techniques are adept at uncovering meaningful patterns in large datasets to address clinical queries, demonstrating significant potential in stratifying risk across various populations. ML contributes to the identification of predictors and their relationship, revealing potential risk factors that traditional models might fail to recognize [30]. Researchers are widely utilizing ML-based techniques for the detection of various life-threatening ailments including cervical cancer [31], brain stroke [28], and breast cancer [32]. Advances in ML have sparked a renewed enthusiasm for assessing the likelihood of a patient being diagnosed with heart disease. Many SOTA algorithms utilizing ML techniques have been proposed by various researchers in order to predict CVD-related disorders. Table 6.2 depicts various SOTA approaches for predicting CVD risk utilizing ML-based techniques.

In [30], authors have proposed a prospective study for predicting CVD events in hospitalized diabetic patients using an ML-based approach. The methodology has

Table 6.2 ML-based approach for predicting CVD risk

Reference	Dataset description	Model utilized/ proposed	Patient's features	Number of participants	Summary
Nabrdalik et al. [30]	A prospective cohort of diabetic patients from Zabrze, Poland, diabetology ward	RUSBoost and MLR for fitting	Demographic, clinical, laboratory and pharmacotherapy-related details	1735	• Predicted CVD using ML approach for diabetic patients • Recognized diabetes people who have a high chance of having a new cardiovascular incident • Predicted CVD in hospitalized diabetic patients using clinical, laboratory, and demographic data
Ozcan et al. [29]	Comprehensive heart disease dataset	Supervised ML-based CART algorithm	Medical and non-medical features	1190	• Proposed an ML algorithm for predicting heart disease • Provided a graphical depiction of the models and analyzed rule inferences • Features related to ECG tests including ST Slope and Oldpeak features to detect heart disease
Chandrasekhar et al. [33]	Cleveland and IEEE Dataport	Ensemble classifier combining random forest, K-NN, LR, NB, gradient boosting, and AdaBoost	Various medical factors considered according to the dataset	Cleveland = 303 IEEE Dataport = 1190	• Enhanced accuracy of CVD prediction due to soft voting ensemble method • Optimized hyperparameters using GridSearchCV with five-fold cross-validation • Identified the optimal model parameters and evaluated performance using accuracy and negative log loss

(continued)

Table 6.2 (continued)

Reference	Dataset description	Model utilized/ proposed	Patient's features	Number of participants	Summary
van Dalen et al. [25]	Retrospective data	XGBoost	Clinical data, medication, and imaging data, including CACS, Rb-82, and PET imaging	1007	• Utilized ML model to predict oCAD based on clinical data, CACS, and PET imaging • Identified patients with oCAD who have never had coronary artery disease (CAD) before
Trigka et al. [34]	Coronary prediction dataset	NB, LR, Multilayer Perceptron, K-NN, RF, Rotation Forest, J48, Stacking, Voting, Bagging	–	No. of instances—3655 Participants with CAD—556 (15.2%), non-CAD participants—3099 (84.8%)	• Long-term risk prediction of coronary artery disease using ML models with tenfold cross-validation with use or non-use of SMOTE
Bhatt et al. [35]	CVD dataset from Kaggle	RF, decision tree classifier, multilayer perceptron, XGBoost, K-mode clustering and GridSearchCV	–	No. of instances-70000 No. of features-11	• Increase generalization of CVD prediction results by utilizing a larger and more diverse dataset • Utilized K-modes for improving convergence, XGBoost for developing predictive models, and GridSearchCV for hyperparameter tuning
Dalal et al. [36]	Heart failure prediction dataset sourced from UCI ML repository	Ensemble model based on QUEST, Neural Network, RF, Bayesian Network, C5.0	–	No. of instances-70000 No. of features-11	• Predicted heart failure using ensemble learning models • Achieved high accuracy and efficiency in CVD prediction

MLR: Multiple logistic regression, *CART:* Classification and regression tree algorithm, *ECG:* Electrocardiogram, *LR:* Logistic regression, *NB:* Naïve Bayes, *K-NN:* K-nearest neighbor, *RF:* Random forest, *CACS:* CT-based Coronary Artery Calcium Scoring, *Rb:* Rubidium, *PET:* Positron emission tomography, *oCAD:* obstructive coronary artery disease, *SMOTE:* Synthetic minority oversampling technique integrated neighborhood component analysis with a hybrid sampling/boosting classification algorithm, offering options such as unsupervised hierarchical clustering or multiple logistic regression (MLR). With just 12 easily obtainable predictors, the model demonstrated strong generalization and outperformed traditional treatment strategies in clinical utility. However, single-center study limits interoperability to broader populations. Observational constraints including limited natriuretic peptide access, may impact heart failure diagnosis, particularly with preserved ejection fraction. Ejection fraction categorization was not conducted, a notable gap in heart failure phenotyping. The methodology proposed by [29] forecasted cardiac disease and created decision rules that make the correlations between input and output variables using the Classification and regression tree (CART) algorithm, a supervised machine learning technique. Additionally, the study prioritized the characteristics that affect heart disease based on their importance. However, the proposed model lacked in considering certain patient information like socioeconomic status and smoking which are essential for a fair and unbiased AI model.

Authors have assessed six ML algorithms: RF (Random forest), KNN (K-nearest neighbor), LR (Logistic regression), NB (Naïve Bayes), Gradient Boosting, and AdaBoost [33]. To adjust hyperparameters, evaluate accuracy, and calculate negative loss metrics, the model utilized GridSearchCV with five-fold cross-validation. All classifiers were combined using a soft voting ensemble method to improve the overall accuracy of the model. Nevertheless, the model had limited processing capabilities since it trained on a small sample of patient data—between 303 and 1190 individuals in the dataset. van Dalen et al. [25] have conducted a retrospective analysis and created the XGBoost model to diagnose obstructive coronary artery disease (oCAD) with the use of clinical information, PET scan data, and the coronary artery calcium score (CACS). The model has improved risk classification and supported decisions for patients with low to intermediate risk by acting as a post-test likelihood estimation for oCAD. In addition, it has utilized feature importance learning to identify important predictors. However, using data from a single hospital limits generalizability. Individuals were identified as having oCAD using invasive coronary angiography follow-up, which may have underestimated the positive cases. Additionally, CACS calculation using the Agatston method has its own limitations in maintaining high accuracy and precision in prediction outcomes.

In [34], authors have assessed several ML models, including multilayer perceptron, LR, NB, KNN, RF, rotation forest, J48, stacking, and bagging. Following SMOTE preprocessing, the authors used a stacking ensemble model with tenfold cross-validation to predict long-term coronary artery disease (CAD) risk. Authors have investigated various ML algorithms including RF, decision tree classifier, multilayer perceptron, and XGBoost for the prediction of CVD [35]. The authors employed an automated approach namely GridSearchCV method for hyperparameter tuning and K-modes clustering algorithm to improve the convergence of the model. However, the study was trained and tested on a single dataset limiting its applicability to a diverse population. Also, the model focused on a restricted set of demographic and clinical variables overlooking lifestyle and genetic factors. Moreover, study did not assess model performance on new data and lacked evaluation of result interpretability. Dalal et al. [36] used a variety of machine learning methods, including as QUEST, RF, neural networks, Bayesian networks, and C5.0 to study the prediction of heart failure. The single model might be vulnerable to issues such as variance and bias, therefore various models were combined into a single ensemble to reduce error and improve predictions.

To summarize, ML-based approach for CVD risk prediction can provide an automated way of analyzing the process, saving time and resources for healthcare professionals, and potentially improving the efficiency of CVD risk assessment. Numerous patient features and attributes, such as clinical, genetic, and lifestyle characteristics are efficiently integrated by ML algorithms. These algorithms can also indicate early detection of CVD risk, facilitating prompt intervention and preventive actions. However, ML algorithms discussed in this section mainly relied on datasets containing clinical factors that strongly impact CVD risk prediction. Capturing such clinical data at times may involve invasive procedures and be difficult to access. Therefore, other approaches for CVD risk prediction which are non-invasive and easily accessible can be further explored.

6.4 Intersection of Retinal Imaging for Predicting Cardiovascular Disorder

RFI can be used to directly inspect the neuro-vasculature using a non-invasive imaging method. Since the retina shares morphological and physiological traits with other organs like the brain and kidneys, the state of retinal vessels serves as an indirect indicator of the overall condition of the systemic microvasculature. Researchers in [37] have reviewed the utility of RFI in detecting various systemic parameters and diseases such as age and gender, smoking and alcohol status, body composition factors, CVD, hematological parameters, neurodegenerative disease, renal diseases, metabolic diseases and hepatobiliary diseases through the use of AI. The next section discusses about the basics of retinal fundus imaging and its utility for predicting critical diseases.

6.4.1 Retinal Fundus Imaging

The process of fundus imaging involves utilizing a single-lens camera to take a two-dimensional picture of the back of the eye. Microaneurysms, minute red-dot-shaped structures, typically arise from an inadequate oxygen supply and the dilation of capillaries. On occasions, when blood supply is completely cut off due to certain arteriolar blockages, the formation of pale soft patches occurs, identified as soft exudates. Hemorrhages, recognized as dark red patches, may occur when there is increased pressure in arterioles, causing the burst of retinal vessels. Hard exudates

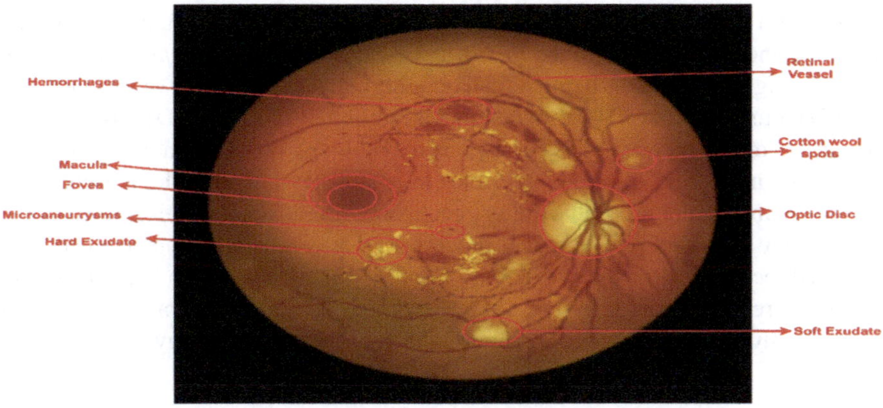

Fig. 6.2 Illustration of the abnormal fundus in the retinal image

are generated when proteins and lipids escape from defective vessel walls, taking on the appearance of solid, yellow, waxy formations as depicted in the diagrammatic representation of the abnormal fundus image in Fig. 6.2.

Analyzing the existence of these abnormalities, in conjunction with assessing retinal indicators such as optic disk, fovea, macula, and blood vessels, offers valuable perspectives on key disorders, contributing significantly to their accurate diagnosis. Advancements in computer vision and DL technologies have exhibited significant progress and potential in the analysis of fundus images. Advanced image processing methods can now extract key features from a given fundus image, highlighting details such as microaneurysms, exudates, hemorrhages, etc. These features despite constituting a small portion of pixels in a fundus image, are instrumental in diagnosing diseases at an early stage [38].

6.4.2 Correlation Between Retinal and Cardiovascular Parameters

The vascular network within the retina is often referred to as a gateway providing insights into the condition of the heart. The narrowing of retinal arterioles has been linked to the existence and extent of coronary artery occlusion as identified through cardiac angiography. Over the last two decades, the constriction of retinal arterioles and the dilation of venules have been correlated with numerous cardiovascular risk factors, both in individuals with established CVD and across the entire age spectrum in the general population.

Recently, there has been growing use of dynamic retinal vessel analysis as a means of diagnosis for assessing cardiovascular risk. New insights suggest that both static retinal vessel analysis and dynamic retinal vessel analysis possess the potential to function as a distinctive screening tool, specifically in evaluating systemic cardiovascular risk and diseases within the microvascular context. Authors have found much evidence suggesting that utilizing retinal imaging can predict the risk of various types of CVD including stroke, CHD, myocardial infarction, PAD, and CVD mortality [39]. Changes in retinal blood vessels correspond to pathology in the coronary circulations such as the narrowing of retinal arterioles in is strongly linked to the presence and severity of angiographic coronary artery occlusion. Moreover, it was examined by the authors that abnormal physiological processes in microvessels are associated with advanced techniques for the prediction of CVD [40]. According to [21], the easily accessible retinal vascular system serves as a unique and non-invasive biological model for studying microvascular abnormalities and CVD pathology.

6.4.3 Benefits of Utilizing Retinal Imaging for Predicting Cardiovascular Disorder

Retinal imaging has the potential to predict the risk of CVD at an early stage and can prevent casualties among people. The key benefits of utilizing RFI in CVD risk prediction are as follows:

- It is based on directly examining the health of retinal blood vessels, which exhibits similarity to microcirculation in parts of the body. As a result, retinal imaging provides comprehensive insights into the existence or absence of observable vascular damage.
- Suitable for telemedicine or preliminary screening, particularly in communities with low income where access to medical services may be limited [37].
- Access to RFI is non-invasive and cost-effective [22, 26].
- Relatively simple Non-radiation procedure as compared to conventional radiation-based CT-scan procedures [27].
- Given the routine capture of retinal photographs in optometric practices, their deployment requires no substantial additional investment in primary care, offering a cost-effective approach. [41].
- There are often visible signs of CVD, such as hypertensive retinopathy and cholesterol emboli in the eye. Moreover, the assessment of several retinal properties, vessel caliber, bifurcation or tortuosity, microvascular changes, and vascular fractal dimensions is made possible with the non-invasive visualization of blood vessels in RFI. These features could function as markers of the cardiovascular system's general health and possible dangers in future [42].
- RFI provides a non-invasive way to visualize atherosclerotic vascular anomalies and offers additional information for assessing the risk of CVD [22].
- DL allows for automated analysis of RFI without the need for manual feature extraction, such as grading. This allows for the analysis of huge image data without sacrificing the ability to identify various parameters of retinal abnormalities, such as vasculature [22].

- When compared to traditional models, the use of RFI and the DL algorithm for Ischemic CVD risk assessment is faster and less expensive [43].
- This alternative holds significance for both the general population and patient care, particularly in regions having limited health care support [43].

Using DL-based approaches, the next section explores the usefulness of RFI in CVD prediction.

6.5 Automatic Prediction of Cardiovascular Events Using Deep Learning-Based Models

The integration of DL techniques in analyzing fundus images for cardiovascular event prediction represents a groundbreaking convergence of ophthalmology and cardiovascular health. By leveraging advanced algorithms on retinal images, this innovative approach aims to detect subtle vascular changes, indicative of CVD risk. Fundus images, commonly used in eye exams, provide a non-invasive means to gain insight into overall health.

The integration of DL has the potential to elevate early detection and risk assessment, promoting a comprehensive approach to cardiovascular well-being [44–46]. This fusion of ophthalmic imaging and AI provides a promising avenue for proactive cardiovascular care. Various researchers have utilized fundus images for predicting CVD risk leveraging DL-based techniques. Table 6.3 depicts various SOTA methodologies proposed by different researchers for predicting CVD risk through the use of RFI.

In [47], authors have linked decreased retinal microvascular parameters to an elevated risk of incident CHD. Despite a large sample size and standardized data collection using AI for retinal vessel analysis, limitations include a predominantly healthy volunteer group lacking fundus image data. Reliance on self-reported diagnoses introduces recall bias, and the UK Biobank's mainly white participants may limit generalizability to other ethnic populations. Authors have sought to improve incident CVD prediction using DL on retinal photos from a diabetic retinopathy program in a large diabetic cohort [48]. However, limitations include using single-entry images, focusing on incident CVD and three risk factors without considering ocular opacity's impact on image gradability. Omitting external validation due to minimal improvements and potential enhancements to DL architecture may boost performance but is unlikely to alter reported ratios.

Vaghefi et al. [41] have introduced CVD-AI, a DL-based algorithm that takes as input a single retinal picture. This method not only determines the exact factors that contribute to an individual's 10-year CVD risk score of suffering a cardiovascular event, but it also assesses that risk score. Authors have designed and validated an innovative system for stratifying cardiovascular risk, leveraging deep learning to predict Coronary Artery Calcium (CAC) from retinal images [27]. It was found that retinal images outperform individual clinical parameters for predicting CAC. The

Table 6.3 DL-based approach for predicting CVD risk through retinal fundus images

Reference	Dataset utilized	Model utilized	Retinal parameter	Cohort details	Number of participants and retinal photographs	Summary
Fu et al. [47]	UK Biobank	–	Retinal microvasculature including fD, number of vS, vSD, and vAD	UK biobank cohort without incident coronary heart disease	• Participants = 57,947 • Mean age 55.6 ± 8.1 years; 56% female	• Reduced retinal vascular network complexity and density could be a sign of a higher incidence risk of coronary heart disease • Analyzed retinal microvasculature complexity (Df and vS) as a biomarker to predict CHD occurrences • Reduced Df, NS, and VSD of retinal microvasculature were associated with an increased risk of incident CHD
Mellor et al. [48]	SDRN-NDS	CNN with ResNet-101 as a backbone network	Bilateral gradable RFI	Cohorts with T1DM and T1DM from SDR Screening Program	• Participants = 24,012 and 202,843 with T1DM and T2DM, respectively • Retinal photographs = 11,910 for T1DM and 101,512 for T2DM	• Predicted future risk of CVD in diabetic patients using retinal photos • Exploited serial images to evaluate the outcomes and deployment contribution to clinical prediction of CVD

(continued)

Table 6.3 (continued)

Reference	Dataset utilized	Model utilized	Retinal parameter	Cohort details	Number of participants and retinal photographs	Summary
Vaghefi et al. [41]	UK Biobank and the US-based AREDS 1 dataset	CVD-AI (Ensemble of CNNs + Inception-ResNet-V2)	RFI	UK and US cohort-based	55,118 patients	• Predicted a person's risk score using CVD-AI for CVD events using only a retinal image as input, as well as the 10-year risk that an individual would experience • 10-year CV risk scores provided by CVD-AI were considerably higher than those of patients who did not experience a CVD event in both the UK Biobank testing dataset and the external validation dataset • The median 10-year CVD risk was higher for those who had a CVD than for those who did not in the internal validation UK Biobank sample

(continued)

Table 6.3 (continued)

Reference	Dataset utilized	Model utilized	Retinal parameter	Cohort details	Number of participants and retinal photographs	Summary
Rim et al. [27]	CMERC-HI, SEED, and UK Biobank	CNN with Efficientnet as a backbone network	RFI	South Korean, Singapore, and UK-based cohort	Retinal photographs = 216,152 Participants • HSC1 (South Korea)—2536 • HSC2 (South Korea)—8707 • CMERC-H1—527 • SEED-8551 • UK biobank—47,679	• Derived Coronary Artery Calcium score for CVD risk stratification (RetiCAC)
Barriada et al. [49]	Images from clinical trial (PRECISE study)	CNN with VGG-16, VGG-19 and ResNet	RFI	Cohort with T1DM	• Participants -76 • Retinal photographs-152	• Predicted CVD from RFI utilizing coronary artery calcium score as a biomarker • Predicted CAC as a binary classification problem with CAC > 400 and CAC < 400
Poplin et al. [42]	EyePACS and UK Biobank	Ensembled CNN using Inception-v3 as a backbone	RFI	The majority of patients in EyePACS were Hispanic, whereas the majority of those in the UK Biobank were Caucasian	• Participants—284,335 (48,101 from UK Biobank and 236,234 from EyePACS) • Retinal images—UK Biobank (96,082), and EyePACS (1,682,938)	• Predicted CVD-related risk factors including age, systolic blood pressure, Hemoglobin, diastolic blood pressure, Current smoker, and gender using RFI

(continued)

Table 6.3 (continued)

Reference	Dataset utilized	Model utilized	Retinal parameter	Cohort details	Number of participants and retinal photographs	Summary
Chang et al. [50]	Clinical database of HPC-SNUH	CNN with Xception model as a backbone	RFI	Retrospective single-center cohort study on Koreans	Participants—32,227	• Proposed modified model named DL-FAS • Utilized RFI to predict carotid artery plaque and confirmed clinical consequences using retrospective cohort analysis • The incidence of CVD fatalities was higher in those with DL-FAS > 0.66 than in those with DL-FAS < 0.33
Son et al. [51]	Data from Seoul National University Bundang Hospital HSC	CNN based on ImageNet pre-trained Inceptionv3	RFI	–	• Participants—20,130 • Retinal images—44,184	• Predicted high CAC score from retinal fundus images which indicates the possible risk of CVD

(continued)

Table 6.3 (continued)

Reference	Dataset utilized	Model utilized	Retinal parameter	Cohort details	Number of participants and retinal photographs	Summary
Lee et al. [22]	Data from SMC and UK Biobank	Integrated DL (CNN utilizing DenseNet-169 + DNN)	Multimodal data (clinical risk factors + RFI)	–	• Participants from SMC -development (2,026), and Validation (517) • Participants from UK Biobank-11,091 • Retinal Images from SMC—development (CVD cases-1758, and non-CVD—1760), validation (CVD cases-1421, and non-CVD cases—1533) • Retinal images from UK Biobank—CVD cases—613, and control (10,685)	• Utilized multimodal data, such as retinal fundus pictures and clinical risk variables, to predict CVD • Exploited multimodal DL approach estimated CVD risks from RFP • Employed clinical risk factors like conventional risk assessment instruments

(continued)

Table 6.3 (continued)

Reference	Dataset utilized	Model utilized	Retinal parameter	Cohort details	Number of participants and retinal photographs	Summary
Nusinovici et al. [52]	Korean dataset from the department of Ophthalmology, Yonsei University, and UK Biobank	CNN with VGG16 as a backbone network	RFI	Korean and UK-based cohort	• Participants—UK biobank (56,301) and Korean cohort (40,480) • Retinal photographs (Korean cohort)-129,236	• Proposed a novel approach named RetiAGE to compute biological age from retinal images • Linked RetiAge to CVD diseases and mortality in a manner that is independent of chronological age and phenotypic biomarkers • Utilized RetiAge as a biomarker for risk stratification of CVD morbidity and mortality

(continued)

Table 6.3 (continued)

Reference	Dataset utilized	Model utilized	Retinal parameter	Cohort details	Number of participants and retinal photographs	Summary
Cheung et al. [53]	SEED	SIVA-DLS (based on CNN)	Retinal vessel caliber estimated from RFI	For training: Singapore-based cohort with ethnic groups: Chinese, Indian, and Malay For external validation: Multiethnic and multicountry-based cohort (Singapore, Hongkong, Australia)	Retinal images—70 K	• Evaluated CVD risk using RFI to estimate retinal vascular caliber • Showed a pattern of correlation between SIVA-DLS retinal-vessel caliber and traditional CVD risk that was essentially consistent with human data • Reduced CRAE, as measured by SIVA-DLS, was linked to the occurrence of CVD and all-cause mortality in two prospective cohorts
Ma et al. [43]	Medical checkup dataset	CNN based on Inception-Resnet-v2 as a backbone	RFI	Chinese cohort	Participants—411,518	• Predicted 10 -year Ischemic CVD risk in the Chinese population

(continued)

Table 6.3 (continued)

Reference	Dataset utilized	Model utilized	Retinal parameter	Cohort details	Number of participants and retinal photographs	Summary
Al-Absi et al. [26]	Qatar Biobank	CNN Stem + MLP Stem + Classification head CNN stem based on AlexNet, VGGNet-11, VGGNet-16, ResNet-18, ResNet-34, DenseNet-121, SqueezeNet-0, and SqueezeNet-1	Multi-modal data fusion of (dual-energy X-ray absorptiometry (DXA) and retinal images)	Qatari cohort	• Participants—500 • Retinal images—1839	• Non-invasive early identification of CVD • Accuracy of more than 75% when using retinal pictures to separate the CVD group from the control group
Gerrits et al. [54]	Qatar Biobank	CNN utilizing MobileNet v2 as a backbone	RFI	Qatari cohort	• Participants-3000 • Retinal images-12000	• Forecasting cardiometabolic risk using RFI

fD: fractal dimension, *vS:* vascular segments, *vSD:* vascular skeleton density, *vAD:* vascular area density, *CHD:* Chronic heart disease, *SDRN-NDS:* Scottish diabetes research network dataset, *CNN:* Convolutional neural network, *RFI:* Retinal fundus imaging, *T1DM:* type1 diabetes mellitus, *T2DM:* type2 diabetes mellitus, *HSC:* Health screening center, *CAC:* Coronary artery calcium, *HPC-SNUH:* Health Promotion Center of Seoul National University Hospital, *DL-FAS:* Deep learning-funduscopic atherosclerosis score, *SMC:* Samsung Medical Center study proposes RetiCAC as a comparable, non-radiation system for predicting CVD events using simple retinal photographs for resource-limited settings, leveraging large-scale CT data for DL. However, the proposed model's validation is limited to specific populations, introducing potential bias. The training set obtained from health screening centers may not accurately reflect the characteristics of the overall population. Misclassification errors might occur in survival models that incorporate data related to death and hospitalization. Short follow-up in one cohort led to rare cardiovascular outcomes. Further studies are needed for a direct comparison between CT-measured CAC scores and RetiCAC.

DL-based method on RFI to predict CVD risk in diabetics is proposed by [49]. The study has employed CNN (Convolutional neural network) to train the model to predict CAC scores, showing promising accuracies in preliminary experiments on clinically verified patients. The research highlighted a positive correlation between elementary clinical data and cardiovascular risk, underlining the complementary nature of results from both cues. However, there are some research challenges including data acquisition and model enhancements. Improving results involves expanding clinical data variables and the image dataset with further refinement in DL architecture.

According to a study by [42], the prediction of several cardiovascular risk factors, including age, gender, and systolic blood pressure, is possible when DL is applied to RFI alone. Given that these factors form the core components of multiple CVD risk calculators, the model demonstrated the potential to directly predict CVD risk. Nevertheless, the study did not assess the potential correlation between the risk of CVD and particular retinal-vessel characteristics, such as venular caliber broadening or retinal arteriolar caliber narrowing. Also, the study trained on a small dataset with a 45° field of view, which demands further investigation of model performance for generalizability. Missing input features, like lipid panels and a definitive diabetic diagnosis, could improve cardiovascular risk prediction. Some variables were available in only one dataset, and self-reported variables may introduce bias. In [50], authors have crafted a DL-based model, DL-FAS to predict atherosclerosis utilizing RFI associated with CVD mortality and substantiating its clinical implications through a retrospective cohort study. However, the study suffers from limitations including single-center data with limited generalizability, DL-FAS accuracy concerns at the designated threshold, and a lack of information on incident cardiovascular diseases. The study also lacks access to medical charts for verifying CVD mortality outcomes, introducing potential bias. Authors examined the increased deposition of CAC by employing cost-effective and radiation-free screening through DL technologies on RFI [51]. The study employed specifically utilized inception-v3, to assess the performance in distinguishing high CACS from CACS of 0 at various thresholds. Additionally, vessel-inpainted and fovea-inpainted images were utilized as inputs to explore areas of interest in determining CACS. However, the current system's performance is insufficient for deployment in clinical settings, requiring improvement and rigorous validation in diverse external datasets. Further investigation into directly predicting CACS from retinal fundus images is suggested.

Further, an AI model was proposed by [22] for the identification of CVD by incorporating multimodal data, which includes both clinical risk factors and fundus photographs. The model's predicted scores were indicative of future CVD events. Combining fundus photographs, clinical risk factors, and non-invasive clinical risk factors in the proposed multimodal model improved reclassification, suggesting the potential for predicting and preventing complex diseases like CVD. However, there are certain limitations. CVD cases were defined as individuals diagnosed at a specific medical center, potentially introducing bias as some patients may have been diagnosed elsewhere. Moreover, the model was trained on a relatively small sample, excluding participants with missing retinal fundus images or electronic medical records. DL algorithm namely RetiAGE to forecast biological age through retinal photographs has been proposed by [52]. The authors assessed its performance in stratifying the risk of death and major diseases within varied demographics. The model demonstrated significant associations with mortality from all causes, CVD mortality, cancer mortality as well as CVD and cancer events. Notably, these associations remained independent of chronological age and traditional phenotype biomarkers. Although the results seem to be promising, confirmation in other populations and an evaluation of clinical utility is needed.

Cheung et al. [53] validated the DL-based model, SIVA-DLS for automated retinal vessel caliber measurement, showing comparable or superior performance to expert graders across diverse datasets. The model demonstrated an association with CVD factors and baseline assessments correlated with incident CVD, indicating the potential for clinically applicable DL systems for CVD prediction. However, the model undergoes training and testing solely on gradable retinal images. DL models, despite standardized training, might be impacted by intergrader variability. The quantitative predictions of retinal-vessel caliber, though visually highlighted by SIVA-DLS, may pose challenges for physicians, in the detection of inaccuracies. Regression models indicated low R2 values, suggesting retinal vessels explain only a limited portion of CVD risk factor variability. The authors developed and validated the DL algorithm to forecast 10-year Ischemic CVD via retinal fundus images in the Chinese population [43]. The algorithm demonstrated consistent performance, effectively identifying individuals with borderline, intermediate, or higher Ischemic CVD risk, underscoring its robust and reliable performance. However, Ischemic CVD risk is derived from cross-sectional data, not longitudinal data impacting the reliability of the algorithm. Moreover, the algorithm's applicability in clinical settings requires validation, and additional research is needed to explore the connection between RFI and future Ischemic CVD incidence in prospective cohorts.

In [26], authors employed a multi-modal strategy integrating data from retinal images and dual-energy X-ray absorptiometry (DXA) for CVD diagnosis. The research proposed a DL-based approach aiming to differentiate between Qatari's cohort control and CVD groups. The method proposed makes early and relatively non-invasive detection of CVD possible. However, the study focused on a Qatari dataset, restricting its generalizability to the local population. Improved accuracy of the model could be achieved with more and better-quality RFI. Gerrits et al. [54] investigated the potential of RFI in predicting various cardiometabolic risk factors, encompassing age, sex, blood pressure, smoking status, glycemic status, sex steroid hormones, bioimpedance measurements, and total lipid panel. Researchers found that age and sex played a significant role in predicting cardiometabolic risk factors from retinal fundus images. However, the study might have restricted applicability as the data originates from a Middle Eastern population.

To summarize, identifying CVD risk at an early stage is very pertinent otherwise it may be life-threatening. CVD risk detection through RFI via DL-based techniques seems to be a promising approach. It offers early identification of abnormalities, enhancing accessibility through a non-invasive nature and cost-effective screenings. However, there are many challenges including model generalizability, data quality impact, external validation needs, and ethical considerations. These factors must be addressed for effective implementation in clinical settings.

6.6 Summary

The use of AI, particularly ML and DL-based methods, in cardiovascular risk detection has great potential to advance preventive healthcare. The examined literature underscores AI's capacity to recognize subtle patterns indicative of cardiovascular risk factors, offering a non-invasive and accessible approach for early detection. Particularly, the examination of CVD through RFI leveraging DL-based techniques has been the main focus of the review. Though significant progress has been made, before AI-based CVD risk detection systems can be extensively deployed, concerns about data quality, interpretability, regulatory approval, and system integration still need to be considered.

AI-based studies have utilized strategies to focus on resolving these challenges, enhancing models, and carrying out comprehensive clinical validations to ensure the reliability and effectiveness of algorithms for the identification of cardiovascular risk in practical healthcare scenarios. Furthermore, examining multimodal data integration, or combining RFI with additional clinical data, also appears to be a feasible way to increase predictive accuracy. This multimodal approach suggests an interesting future direction that might strengthen the AI model's robustness and increase its applicability to a wider range of racial and ethnic groups. Healthcare professionals and data scientists will need to collaborate to properly utilize AI. This will improve the identification of CVD and lessen the prevalence of CVD globally.

References

1. Upadhyay RK (2015) Emerging risk biomarkers in cardiovascular diseases and disorders. J Lipids 2015
2. Mensah GA, Brown DW (2007) An overview of cardiovascular disease burden in the United States. Health Aff 26(1):38–48
3. Garg N, Muduli SK, Kapoor A, Tewari S, Kumar S, Khanna R, Goel PK (2017) Comparison of different cardiovascular risk score calculators for cardiovascular risk prediction and guideline recommended statin uses. Indian Heart J 69(4):458–463

4. Yi JK, Rim TH, Park S, Kim SS, Kim HC, Lee CJ, et al (2023) Cardiovascular disease risk assessment using a deep-learning-based retinal biomarker: a comparison with existing risk scores. Eur Heart J Digital Health 4(3):236–244
5. Namasivayam V, Senguttuvan N, Saravanan V, Palaniappan S, Kathiravan MK (2022) Artificial intelligence and its application in cardiovascular disease management. Machine learning and systems biology in genomics and health. Springer, pp 189–236
6. Sacramento-Pacheco J, Duarte-Climents G, Gómez-Salgado J, Romero-Martín M, Sánchez-Gómez MB (2019) Cardiovascular risk assessment tools: a scoping review. Aust Crit Care 32(6):540–559
7. Payne RA (2012) Cardiovascular risk. Br J Clin Pharmacol 74(3):396–410
8. D'Agostino Sr RB, Vasan RS, Pencina MJ, Wolf PA, Cobain M, Massaro JM, Kannel WB (2008) General cardiovascular risk profile for use in primary care: the Framingham Heart Study. Circulation 117(6):743–753
9. Goff Jr DC, Lloyd-Jones DM, Bennett G, Coady S, D'agostino RB, Gibbons R, et al (2014) 2013 ACC/AHA guideline on the assessment of cardiovascular risk: a report of the American College of Cardiology/American Heart Association Task Force on Practice Guidelines. Circulation 129(25_suppl_2):S49–S73
10. Conroy RM, Pyörälä K, Fitzgerald AE, Sans S, Menotti A, De Backer G et al (2003) Estimation of ten-year risk of fatal cardiovascular disease in Europe: the SCORE project. European Heart J 24(11):987–1003
11. Goh L, Chua S, Sum C, Low L, Tan H, Sethi S (2011) Screening for cardiovascular disease and risk factors. MOH Clinical Practice Guidelines 1
12. Lenstrup M, Kjaergaard J, Petersen C, Kjaer A, Hassager C (2006) Evaluation of left ventricular mass measured by 3D echocardiography using magnetic resonance imaging as gold standard. Scand J Clin Lab Invest 66(8):647–658
13. Vaid A, Johnson KW, Badgeley MA, Somani SS, Bicak M, Landi I, et al (2022) Using deep-learning algorithms to simultaneously identify right and left ventricular dysfunction from the electrocardiogram. Cardiovasc Imag 15(3):395–410
14. Saikrishnan N, Kumar G, Sawaya FJ, Lerakis S, Yoganathan AP (2014) Accurate assessment of aortic stenosis: a review of diagnostic modalities and hemodynamics. Circulation 129(2):244–253
15. Kumar A et al (2024, June) Deep Learning-Based Diabetic Retinopathy Detection Using Enhanced ResNet-152. In International Conference on Data Analytics & Management, pp 101–112. Singapore: Springer Nature Singapore
16. Kumar A, Singh D (Eds.) (2025) Revolutionizing Medical Systems Using Artificial Intelligence: A Breakthrough in Healthcare. Elsevier
17. Maurovich-Horvat P (2021) Current trends in the use of machine learning for diagnostics and/or risk stratification in cardiovascular disease. Oxford University Press
18. Kumar A (2023) Visual object tracking using deep learning. CRC Press
19. Kumar A, Walia GS, Sharma K (2020) Recent trends in multicue based visual tracking: a review. Expert Syst Appl 162:113711
20. Zhang Y, Miao H, Chia YC, Buranakitjaroen P, Siddique S, Shin J, et al (2022) Cardiovascular risk assessment tools in Asia. J Clin Hypertens 24(4):369–377
21. Cheung CY-l, Wong PP-y, Wong TY (2019) Retinal biomarkers and cardiovascular disease: a clinical perspective. In: Computational retinal image analysis. Elsevier, pp 299–318
22. Lee YC, Cha J, Shim I, Park W-Y, Kang SW, Lim DH, Won H-H (2023) Multimodal deep learning of fundus abnormalities and traditional risk factors for cardiovascular risk prediction. npj Dig Med 6(1):14
23. Shrivastava S, Cohen-Shelly M, Attia ZI, Rosenbaum AN, Wang L, Giudicessi JR et al (2021) Artificial intelligence-enabled electrocardiography to screen patients with dilated cardiomyopathy. Am J Cardiol 155:121–127
24. Elias P, Poterucha TJ, Rajaram V, Moller LM, Rodriguez V, Bhave S et al (2022) Deep learning electrocardiographic analysis for detection of left-sided valvular heart disease. J Am Coll Cardiol 80(6):613–626

25. van Dalen J, Koenders S, Metselaar R, Vendel B, Slotman D, Mouden M et al (2023) Machine learning based model to diagnose obstructive coronary artery disease using calcium scoring, PET imaging, and clinical data. J Nucl Cardiol 1–10
26. Al-Absi HR, Islam MT, Refaee MA, Chowdhury ME, Alam T (2022) Cardiovascular disease diagnosis from DXA scan and retinal images using deep learning. Sensors 22(12):4310
27. Rim TH, Lee CJ, Tham Y-C, Cheung N, Yu M, Lee G et al (2021) Deep-learning-based cardiovascular risk stratification using coronary artery calcium scores predicted from retinal photographs. Lancet Dig Health 3(5):e306–e316
28. Bhalla S, Kumar A, Kushwaha R (2023) Comparative analysis on brain stroke prediction using machine learning. In: 2023 14th international conference on computing communication and networking technologies (ICCCNT). IEEE
29. Ozcan M, Peker S (2023) A classification and regression tree algorithm for heart disease modeling and prediction. Healthcare Anal 3:100130
30. Nabrdalik K, Kwiendacz H, Drożdż K, Irlik K, Hendel M, Wijata AM et al (2023) Machine learning predicts cardiovascular events in patients with diabetes: the Silesia diabetes-heart project. Curr Probl Cardiol 101694
31. Kumar A, Rai RS, Gheisari M (2021) Prediction of cervical cancer using machine learning. Cancer prediction for industrial IoT 4.0: a machine learning perspective, p 107
32. Kumar A, Ahluwalia R (2021) Breast cancer detection using machine learning and its classification. Chapman and Hall/CRC
33. Chandrasekhar N, Peddakrishna S (2023) Enhancing heart disease prediction accuracy through machine learning techniques and optimization. Processes 11(4):1210
34. Trigka M, Dritsas E (2023) Long-term coronary artery disease risk prediction with machine learning models. Sensors 23(3):1193
35. Bhatt CM, Patel P, Ghetia T, Mazzeo PL (2023) Effective heart disease prediction using machine learning techniques. Algorithms 16(2):88
36. Dalal S, Goel P, Onyema EM, Alharbi A, Mahmoud A, Algarni MA, Awal H (2023) Application of machine learning for cardiovascular disease risk prediction. Computat Intell Neurosci 2023
37. Betzler BK, Rim TH, Sabanayagam C, Cheng C-Y (2022) Artificial intelligence in predicting systemic parameters and diseases from ophthalmic imaging. Front Dig Health 4:889445
38. Goutam B, Hashmi MF, Geem ZW, Bokde ND (2022) A comprehensive review of deep learning strategies in retinal disease diagnosis using fundus images. IEEE Access 10:57796–57823
39. Kawasaki R, Wong TY (2012) Retinal vascular imaging for cardiovascular risk prediction. Digital teleretinal screening: teleophthalmology in practice. Springer, pp 77–89
40. Hanssen H, Streese L, Vilser W (2022) Retinal vessel diameters and function in cardiovascular risk and disease. Prog Retin Eye Res 91:101095
41. Vaghefi E, Squirrell D, Yang S, An S, Marshall J (2022) Use of artificial intelligence on retinal images to accurately predict the risk of cardiovascular event (CVD-AI). medRxiv, p 2022.10. 12.22281017
42. Poplin R, Varadarajan AV, Blumer K, Liu Y, McConnell MV, Corrado GS et al (2018) Prediction of cardiovascular risk factors from retinal fundus photographs via deep learning. Nat Biomed Eng 2(3):158–164
43. Ma Y, Xiong J, Zhu Y, Ge Z, Hua R, Fu M et al (2021) Development and validation of a deep learning algorithm using fundus photographs to predict 10-year risk of ischemic cardiovascular diseases among Chinese population. medRxiv, p 2021.04. 15.21255176
44. Singh D, Singh D, Manju, Gupta U (2023) Smart healthcare: a breakthrough in the growth of technologies. In: Artificial intelligence-based healthcare systems. Springer, pp 73–85
45. Tripathy SK, Singh D, Jaiswal A (2023) Multi-layer feature fusion-based deep multi-layer depth separable convolution neural network for Alzheimer's disease detection. In: 2023 international conference on IoT, communication and automation technology (ICICAT). IEEE
46. Kumar A, Aelgani V, Vohra R, Gupta SK, Bhagawati M, Paul S et al (2024) Artificial intelligence bias in medical system designs: a systematic review. Multimedia Tools and Applications, 83(6):18005–18057

47. Fu Y, Yusufu M, Wang Y, He M, Shi D, Wang R (2023) Association of retinal microvascular density and complexity with incident coronary heart disease. Atherosclerosis 380:117196
48. Mellor J, Jiang W, Fleming A, McGurnaghan SJ, Blackbourn L, Styles C et al (2023) Can deep learning on retinal images augment known risk factors for cardiovascular disease prediction in diabetes? A prospective cohort study from the national screening programme in Scotland. Int J Med Inf 175:105072
49. Barriada RG, Simó-Servat O, Planas A, Hernández C, Simó R, Masip D (2022) Deep learning of retinal imaging: a useful tool for coronary artery calcium score prediction in diabetic patients. Appl Sci 12(3):1401
50. Chang J, Ko A, Park SM, Choi S, Kim K, Kim SM et al (2020) Association of cardiovascular mortality and deep learning-funduscopic atherosclerosis score derived from retinal fundus images. Am J Ophthalmol 217:121–130
51. Son J, Shin JY, Chun EJ, Jung K-H, Park KH, Park SJ (2020) Predicting high coronary artery calcium score from retinal fundus images with deep learning algorithms. Translational Vision Science & Technology 9(2):28–28
52. Nusinovici S, Rim TH, Yu M, Lee G, Tham Y-C, Cheung N et al (2022) Retinal photograph-based deep learning predicts biological age, and stratifies morbidity and mortality risk. Age Ageing 51(4):afac065
53. Cheung CY, Xu D, Cheng C-Y, Sabanayagam C, Tham Y-C, Yu M et al (2021) A deep-learning system for the assessment of cardiovascular disease risk via the measurement of retinal-vessel calibre. Nat Biomed Eng 5(6):498–508
54. Gerrits N, Elen B, Craenendonck TV, Triantafyllidou D, Petropoulos IN, Malik RA, De Boever P (2020) Age and sex affect deep learning prediction of cardiometabolic risk factors from retinal images. Sci Rep 10(1):9432

Chapter 7
Diagnosis and Prediction of Diabetic Foot Ulcer in Modern Healthcare Using Artificial Intelligence

Abstract Diabetic foot ulcer (DFU) is a common problem, especially affecting diabetes patients. Many patients complain of delayed wound healing and ankle wounds that often develop from calluses, and these wounds themselves can lead to more complicated systemic infections. Therefore, early diagnosis of diabetic foot ulcers in medical diagnostic tools is necessary for timely treatment. Medical imaging is constantly evolving over the years for the detection and prediction of DFU. It is seen that advanced machinery and processes are integrated into daily medical practice to achieve accurate results. The topic of diabetic foot ulcer diagnosis through various medical imaging techniques was addressed. Finally, in the conclusion part, a different artificial intelligence model that would contribute a state-of-the-art solution to a small group of people was proposed. Improvements in medical imaging and artificial intelligence are expected to unlock new insights and make better decisions. Early detection can provide timely and preventive intervention to diabetes patients and improve the quality of life of patients, especially those with diabetes. This integration can be employed in a professional care setting to provide triage to higher-risk servers. With advancements in medical imaging capturing light and 3D imaging or spectroscopy technology, research on the development of predictive algorithms for this data seems to hold promise, as it may uncover important new biological and phenotypic features. This area is a ripe space for future work, especially as effective medicine and therapeutics come online through this interaction. Various technologies for early detection of diabetic foot ulcers to assist clinicians have been addressed in different sections.

Keywords Diabetic foot ulcer (DFU) · Diabetes mellitus · Diabetic complications · Wound classification · Chronic wounds · Foot ulcer diagnosis · DFU staging and grading

A. Kumar and D. Singh, *Artificial Intelligence in Modern Healthcare System*, Transactions on Computer Systems and Networks,
https://doi.org/10.1007/978-981-96-6703-1_7

7.1 Introduction

One of the most severe complications of diabetes is the possibility of developing diabetic foot syndrome (DFS). Most DFS features develop due to neuropathy and vasculopathy associated with diabetes. At the stage of foot deformations, ulceration develops in 85% of patients. This syndrome critically affects the prognosis and the quality of life of diabetes sufferers, resulting in a reduction of physical activity, including immobilization [1]. In the worst cases, gangrenous complications often result in amputation or even death in affected patients. To develop a rational way to prevent new cases of diabetic foot and to reduce this syndrome's high incidence among all diabetic sufferers, global awareness and adoption of preventive interventions are necessary. The possibility of identifying and objectively measuring the first signs of diabetic foot in all patients with diabetes predisposed to this syndrome allows physicians to plan prevention paths in an optimized manner [2, 3].

In the last decade, information technology has experienced exponential growth. This powerful tool, particularly through applications of artificial intelligence and machine learning, allows us to build algorithms with the highest precision, often exceeding human skills. A great opportunity is thus expanding: the possibility of developing an automatic screening algorithm that can identify the probability of detecting diabetic foot syndrome based on data available in patient photographic image collections and providing observance to the identified suspect cases when the patient appears for other diabetic follow-up tests from reference and treatment centers for diabetes [4]. This text is thus aimed at evaluating the application of some artificial intelligence technologies, particularly deep learning and the BERT language model, which is particularly effective in identifying complex patterns in the analysis of natural language and other disciplines. The support represents a new frontier in a reinvigorating way of medical activities in the whole universe of diagnostic and monitoring activities, enabling precision, objectivity, and usability, particularly making accessible diagnostic experiences in contexts like developing countries, with scarce reference figures in this regard [5]. The properties of the models' algorithms and the fact that enormous quantities of data required to return precise tactics are the main issues in the effectiveness of the complex model algorithms, which are often weakened and distorted [6].

7.1.1 Overview of Diabetic Foot Ulcers

Foot ulcer is one of the health problems associated with diabetes worldwide that affects up to 15% of patients with diabetes during their lifetime. Moreover, it is also one of the leading causes of hospitalization. Without proper early identification and treatment, the condition can lead to partial or complete removal of the foot or limb. Medical or surgical management still requires a long time to recover and incurs high costs [7]. However, the proportion of patients who develop foot ulcers varies according to their feet, with a minimum of 5% and a maximum of 16%. Initial control of risk factors, including blood sugar, is fundamental to preventing ulcers. Other intervention trials control significant risk factors [8]. Control of foot health is an important prevention measure, reducing the severity of the disease and mortality rates. Currently, medical imaging studies for foot ulcers include radiographs or clinical photographs, physical examination of foot pressure, and sensory tests, after which doctors or other healthcare specialists provide treatments to reduce infection affecting the wound. Respondents revealed that they visited general medicine to ask about their foot ulcers, and only a small proportion contacted podiatry [9]. Unfortunately, podiatry in developing countries is expensive, and care is not covered by healthcare. The use of this expensive and very limited control requires care centers to have personnel with specific knowledge, thus limiting their use. During the long-term management of this disease, affected persons have to work with the disease; they understand the important parts of recognition, management, and prevention when dealing with foot ulcers. These research results can provide useful suggestions for the application of health education technologies, diabetes support, and systems based on mobile control, which are programmed to monitor and automatically warn of the need for a podiatrist revision [10].

7.1.2 Significance of Early Detection

In partial loss of protective sensation (LOPS), the patient may feel touch pressure but not pain. The most significant current contributing factor is the estimated 43% and growing diabetic population. These individuals may have vascular disease, neuropathy, infection, and peripheral arterial disease and are often associated with more severe soft tissue infections. Because of neuropathy in the lower extremities, these patients are often unable to identify the need for appropriate medical treatment prior to infection [11]. This leads to foot wounds that, by the time patients seek care, are often infected with bacterial biofilms sticking to soft tissue and bone.

A bacterium is most commonly seen and identified in lower extremity wounds of many patients with diabetes. This bacteria produces a light-reactive molecule that, when exposed to ultraviolet light, fluoresces red in the presence of nitric oxide. Because of slow tissue repair in diabetic patients, the window for early diagnosis and appropriate treatment after injury is limited. Early detection of such an infection using an imaging or sensing modality could be intelligently used to trigger antibiotic patches [11–14]. The key contributions in this chapter are as follows:

- The risk factors that lead to the chances of Diabetic foot ulcer occurrence are elaborated in detail to provide awareness pointers in order to prevent its spread among humans.
- We have categorized AI-based predictive algorithms either as ML-based techniques or DL-based techniques. The strengths and limitations of each category are reviewed to highlight the salient features.

- Emphasizes the importance of utilizing large, diverse datasets for training AI models and discusses publicly available datasets specific to DFUs.

The rest of the chapter is organized as follows. Section 7.2 elaborates the background and pathophysiology of AI-based predictive algorithms for Diabetic foot ulcers. In addition, Role of Medical Imaging in Diabetic foot ulcer are discussed in Sect. 7.3. AI-based techniques for automatic segmentation of Diabetic foot ulcer to analyze the foot ulcer that are categorized into ML-based and DL-based techniques in the Sect. 7.4. Datasets and Preprocessing methods for Diabetic foot ulcer are discussed in Sect. 7.5. Performance Metrices to check the efficiency of the AI models are discussed in Sect. 7.6. The challenges of AI-based techniques and future directions are in Sect. 7.7. Lastly, the concluding remarks and future directions are sketched in Sect. 7.8.

7.2 Diabetic Foot Ulcer: Background and Pathophysiology

In this section, we present a brief overview of the remarkable complications of diabetes related to its potential enlargement of the number of chronic foot ulcers. Patients with diabetic foot ulcers are classified as patients at high risk of lower extremity amputations, with the majority of individuals who have undergone the procedure being diabetics. Diabetes mellitus is a group of metabolic conditions characterized by hyperglycemia [15]. Diabetic foot ulcers are infected sores as a result of underlying complications with diabetes, such as insensitive or ischemic neuropathy. Currently, there have been numerous techniques for diagnosing diabetic foot ulcers, such as observations made using clinical examination or medical imaging [16].

Once ulcers have developed, they are best treated with the use of advanced modalities. The simplest and most economical way to help avoid complications with the feet in diabetes is to identify individuals who have either type of diabetes or prediabetes at an early stage of the disease and refer them for appropriate care [17]. This will reduce the number of chronic foot ulcers, rule out unnecessary amputation, reduce the cost of care, and help improve the quality of life for diabetic patients. The use of artificial intelligence has been researched and previously applied in other medical image-related scenarios to assist with diagnostics. However, for medical imaging in the field of diabetic foot ulcers, it is still relatively limited. The goal of this study is to highlight recent research that has used artificial intelligence tools, algorithms, and methods in diabetic foot ulcer-related scenarios using medical imaging [18].

7.2.1 Definition and Types of Diabetic Foot Ulcers

Everyone with diabetes is at risk of developing a foot ulcer, which can become infected. Approximately over 135 million people globally and 16 million people in just the United States report a 15–25% risk of developing a diabetic foot ulcer. However, several of the risk factors for developing a diabetic foot ulcer are neuromuscular abnormalities, peripheral vascular disease, trauma, pathological calluses, and local infections in the foot [19]. These altered distal parts of the legs become insensitive to injury, which often leads to new ulceration. Therefore, the most efficient treatment for diabetic foot ulcers is preventing ulcer formation. Regular monitoring and preventive measures for ulcers are vital to avoid complications in diabetic patients [20].

The awareness of diabetic foot ulcers and their complications has increased awareness of diabetes and its dreadful complications among the population. Diabetic foot ulcers develop neuropathy in 30–40% of patients due to the prolonged period since diabetes was diagnosed [21]. This neuropathy causes the patient's skin and underlying tissues to hurt, making the patient susceptible to pressure. However, muscle imbalance causes hypercallusification both at the ulcer's border and at the plantar surfaces. These prominent areas become unexpectedly and randomly exposed to high pressure, and the tissue becomes subsequently ultra-deep sores involving bones and blood vessels [22]. Left untreated, an ulcer may cause serious consequences for patients, such as chronic pain, osteomyelitis, foot amputations, 50% mortality within five years after the first amputation, and significant annual financial expenses related to their foot problems [23].

7.2.2 Pathophysiological Mechanisms

This process is multifactorial and is primarily associated with hyperglycemia, microvascular disease, neuropathy, and immune disturbances that result in reduced skin strength, perspiration, hydration, and compromised wound healing. The inflammatory process is primarily due to high glucose that affects local immunity, blood flow, oxygen, and capillary integrity [24]. IL-1 and IL-6 from the keratinocytes and cytokines secreted by macrophages affect local inflammation. Other elements exacerbate this, creating a pro-inflammatory environment dominated by IL-17, IL-18, and the Th1 and Th17 receptors. Diabetes is also related to systemic inflammation by combinations of abnormalities in immune and pro-inflammatory mediators, including neutrophil counts [25].

The skin surface is the first element of inflammatory and acute-phase responses that protect the body from possible complications and is involved in vital homeostatic functions such as immunity, thermoregulation, transmission, and control of body fluids [26]. It is affected during type 2 diabetes due to obesity, restricted vessel flow, skin phase, and inflammation. Diabetic ulcers originate primarily from reduced peripheral blood flow secondary to microangiopathy and macroangiopathy, resulting in tissue ischemia and hypoxia. The body's regenerative capability is further compromised by most patients having infections in their wounds, such as damaged formation of multiple keratin, melanin metabolism, and immune disorders. Characteristic lipid lesions, such as decreased ceramide levels and structural changes, can also be seen

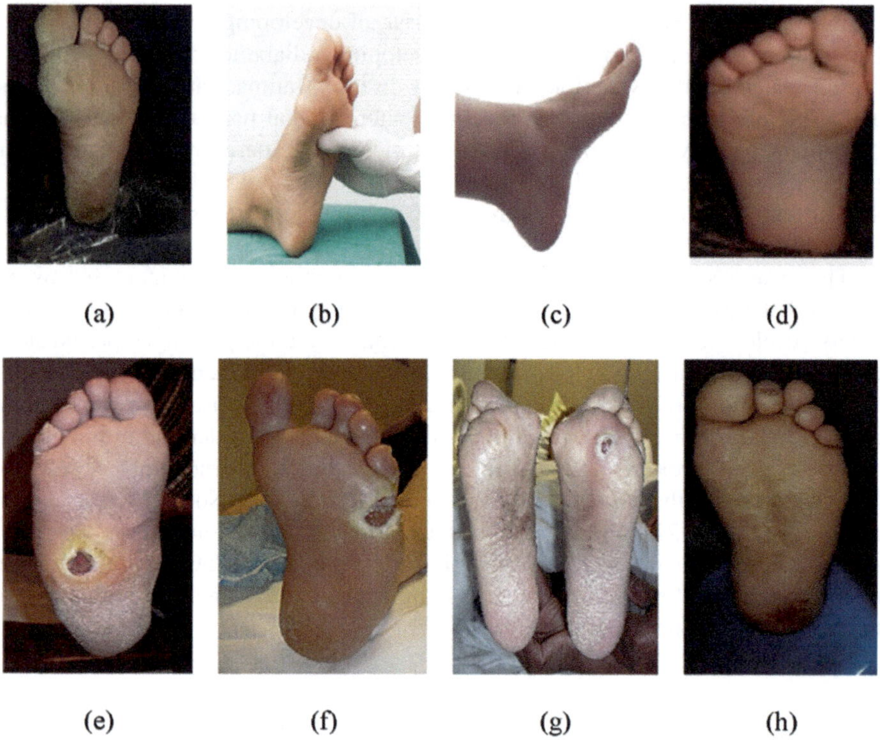

Fig. 7.1 Diabetic foot images Ist Row **a–d** Normal and healthy foot, IInd Row **e–h** Foot affected by a diabetic foot ulcer

on the skin's surface of diabetic foot ulcer patients, reducing the stratified barrier's function [27, 28] (The Diabetic foot images of Ist Row for Normal and healthy foot and IInd Row for Foot affected by a diabetic foot ulcer are shown in Fig. 7.1).

7.3 Role of Medical Imaging in Diabetic Foot Ulcer Detection

Medical imaging procedures obtain the visualization of the physical condition inside complications. Imaging techniques visualize the growth and presence of Diabetic Foot Ulcer (DFU) anatomically and physiologically. The medical imaging techniques used in DFU include the CT scan, X-ray, MRI, and ultrasound. Among these techniques, ultrasound is a trending and reliable method because it does not use ionizing radiation. Ultrasound sends high-frequency sound waves toward the tissues, reflecting echoes and forming images. These images are visualized on the screen that monitors the ulcer in anatomical and physiological aspects. The clinical experts are experienced in rapid analysis of the tissue and determination of the specific medication and treatment. The ultrasound imaging provides significant information and non-uniform visual characteristics. The visual features need to be extracted and provided to the AI model to automate the DNET [28, 29].

The complexity of feature extraction and accuracy increases when the direct images are fed into deep learning models for the feature extraction process without performing the pre-processing steps. Therefore, image pre-processing is essential for hypo-subjects and hyper-subjects. Pre-processing involves equalization, format conversion, and any enhancement model. The equalization process enhances contrast and adjusts the differences in the pixel values in the ulcer region. In this way, the texture and color intensity are preserved in the pre-processed images. Pre-processing treatment is required to build a system that is robust to uncontrollable elements with arbitrary overfitting in conventional pre-processing techniques. Feature selection operates in diverse ways, but the purpose of feature selection is to prohibit irrelevant features. Finally, we focus on extracting the visual characteristics in the ultrasound images of the diabetic foot ulcer. These characteristics can be non-committal to bore, such as irregular shape, obscure margin, and hyperechoic regions that simulate the sinus tract. In a few cases, the texture of the foot that comprises perforation and bone-related features is detected in the ulcer region. Additionally, the hyperechoic regions such as the retina wall or cartilage handling frame data loss are shown in the owner modification and are operating in terms against the malignance. Minor renditions within the DFU region, such as spontaneous dermal flow exudates, require detailed attention. These features will be noticeable and considered by the experienced radiologist to guide and increase the development of a DNET model.

7.3.1 Common Modalities Used

Ultrasonic imaging is an important non-invasive inspection method in medical treatment. It can observe the patient's internal structure in real time and dynamic changes. These advantages are incomparable to other imaging methods. It can detect the patient's diabetic foot ulcer wounds quickly, allowing for rapid diagnosis and treatment. Many doctors and nurses also face difficulties [30]. They often miss the early detection of ulcer wounds and the best treatment time. In recent years, some studies have combined artificial intelligence and medical imaging to assist doctors in the treatment of ulcerous wounds, effectively improving treatment time, accuracy, and quality. This paper uses ultrasonic images as training samples and employs a Faster R-CNN algorithm to detect foot ulcer wounds [31].

In recent years, the detection of diabetic foot ulcers has mainly relied on medical imaging such as color images, magnetic resonance images, etc., combined with artificial intelligence, convolutional neural networks, and Faster Region-based Convolutional Neural Networks to locate and identify the ulcers, assisting clinical medical diagnosis and treatment. Among them, color images are widely used in foot wound detection because they are inexpensive and non-invasively used [32]. Magnetic resonance images are also an important modality of medical imaging; they have rich color information and can be converted to grayscale images for comparison. This is also popular in current wound research. The aforementioned methods depend on the anatomical features of the wound and the basic appearance features such as color, pattern, and scale; while ultrasound scanning, the additional technology used by radiologists usually has higher clinical specificity and sensitivity compared to other modalities [33].

7.3.2 Advantages and Limitations

Medical imaging of chronic diseases relies on multiple medical imaging platforms, associated workflows, and an entire range of imaging personnel from the radiology department. The advantage of using an Artificial Intelligence (AI) platform for medical imaging imparts good quality, enhances workflow, and reduces the usage of medical personnel in the clinical hospital setting [34]. End-to-end deep learning using AI algorithms is performed at the image pixel level, and its architecture performs the subsequent image analysis. It is suitable for many clinical tasks as it has a strong feature learning ability once the training procedure is successful. Deep learning for medical imaging can excellently compete with the evaluations and interpretations of medical personnel, and it can achieve them without a huge cost using robotic tools or instrumentation [35, 36].

Limitations of AI in the healthcare setting include different factors such as the availability of hardware components, the training complexity of medical personnel due to increased virtual methodology, cost factors, legal concerns, registration barriers, and data copyright issues. AI algorithms are effective in finding hidden knowledge from the acquired data about patients. The learning methods of AI algorithms can stimulate the expertise of radiologists and identify unnoticed abnormalities in medical images as well [37]. However, they are presently unable to reach the level of empathy of a radiologist. Echocardiography imaging during the COVID-19 period presents a novel challenge, as any AI algorithm must reproduce results within a short time frame due to exigency. Currently, AI algorithms for radiologists are add-on tools, and clinical diagnoses related to patients need to be attentive to both technical and humanistic medical evaluations [38]. Additionally, the large amount of healthcare data generated by AI-provisioned radiologic imaging raises concerns regarding transparency and accessibility. This creates issues of medical record intimacy, anonymization, and privacy. At this point, the generation of patient data disclosure may also have legal consequences. There is a technical barrier to the application of AI algorithms to medical images, as only centralized data can provide the high-dimensional images useful for training and developing AI models, which may be particularly dependent on institution-based AI models. This phenomenon is often referred to as overfitting [39, 40].

7.4 AI Techniques for Diabetic Foot Ulcer Detection

There is an urgent need for related studies that need to be performed based on efficient AI techniques for the early diagnosis of DFU. Efficient and leading advancements in machine learning by supervised learning, semi-supervised learning, and unsupervised learning, whose methodologies include reinforcement learning and AI enhancements of weakly supervised learning and Mixup machine learning [41]. Different architectures of supervised learning, like convolutional neural networks with transfer learning models, are pre-trained with various layers of supervised and weakly supervised models to solve the problem of training on limited labeled image data because transfer learning has a good record when the labeled training data are limited, and newly formulated images can be created regarding the weakly labeled or unlabeled images with leading performance even in the phase of training iteration and validation [42, 43].

Various kinds of preprocessing techniques, like denoising and data enhancement methods such as image rotation, yield strong and efficient images to solve the issues of handling low-quality medical images and the problem of training on limited labeled medical images. To develop interpretable solutions with the aid of initial visualization by providing an understanding of how features on input images correspond to changes in the predicted outcomes [44]. A feature reduction-based method survey using attribute evaluation is performed on 18 different algorithms for feature selection, leading to the discovery of the most important attributes that can be employed for understanding attribute importance. The usage of bio-inspired AI heuristic algorithms naturally aids in the enormous nature of the problem-solving, leading to advancements in the existence of poor data or complex image data challenges. The important aspect of interpretability in AI is to develop a machine learning model that is carefully considered for the usage of the model [45, 46].

The consideration and understanding of the 18 distinct architectures that pre-train the models help understand the training requirements and the number of layers that allow for the development of machine learning, aiding in providing efficient and strong solutions for the problem of handling limited labeled medical images. The utility of numerous methods is supported through the related results, highlighting the performance gain through the contributions and guiding better stakeholders that are currently handled in subproblems of the practical and prominent sector [47, 48].

7.4.1 Machine Learning Algorithms

Diabetic foot ulcer is a major health burden as their wounds become severe and chronic in a very short interval. The current clinical recognition techniques did not reveal the diabetic foot ulcer at a very early stage, which is a major drawback. Early detection and care are required as they would prevent the occurrence of diabetic foot ulcers [49]. As much attention and care are necessary at the initial stages of diabetic foot ulcers, they are not getting the proper care and attention. The proposed methodology is an AI-based model used to diagnose the diabetic foot ulcer at a very early stage. The AI-based model is used to extract the features. These techniques would help to initially detect the diabetic foot ulcer at a very early stage. The diabetic foot ulcers appear in the wound position on the foot, and these ulcers look similar to other non-diabetic foot ulcers, making it a time-consuming task for a physician to determine them, which may result in data interpretation and identification issues. So, an automatic classification system is required to identify the diabetic foot ulcer at a very early stage [50].

The use of machine learning for the classification of diabetic foot ulcer detection at all stages is found to be effective, and the size of the patches and training methodologies determine the ability to recognize them. The prediction would not be accurate with lower performance and may not satisfy the detection of diabetic foot ulcer features. The role of machine learning classification is developed with higher accuracy in diabetic foot ulcer independent detection, and this method presents the diabetic foot ulcer detection of internal and external characteristics [51]. A diabetic foot ulcer is a very chronic health-related problem. Foot care and self-management are always essential for diabetic patients with foot ulcers. The deep neural systems help find the diabetic foot ulcer with a higher proportion. For the detection of diabetic foot ulcers, it is easy to use hand feature extractors. The previous image classification methods generate a few features [31].

7.4.2 Deep Learning Architectures

The deep learning models can be broadly categorized into Convolutional Neural Networks, Recurrent Neural Networks, and Balanced Decision Trees. The major advancements obtained in computer vision through CNN architectures generated extensive acceptance by research studies. CNN models are primarily skilled at learning innovative feature representations that facilitate local and global accumulations in the input data. The general architecture consists of various stages, such as convolution, activation function and pooling layers for feature extraction, followed by dropout, fully connected, and output layers as the classification phase. The trained model demonstrates the ability to produce predictions with higher accuracy and efficient training time, as a contemporary stride of CNN algorithm functionality. The architectural development in CNN models is a continuous and ongoing research area, but several state-of-the-art models have emerged to address specific computer vision domains. Each model exhibits differential features in terms of the number of layers and the size of fully connected and output layers, with the ability to handle large datasets [49, 50].

The performance of RNNs is extensively exploited for sequential data inputs, such as videos, audio, and Natural Language Processing domains, considering the time factor present in the input data. The unique advantage of the RNN model is the utilization of its internal memory states to capture the significance of the inputs from historical timestamps in sequential data. The formulations of softmax, sparsemax, and automatic solvers are employed at various dimensions of RNN layers, such as time distribution, state-to-state activities, and temporal attention, demonstrating improved performance. The RNN architecture possesses state-of-the-art LSTM and Gated Recurrent Unit modified versions, which are skilled at handling longer input sequences efficiently. The Balanced Decision Trees architecture facilitates the production of effective deep models that can be trained in a balanced manner, leveraging GPU computational power. The training of a BDT architecture can be performed with a smaller input batch size, which in turn reduces the burden on the CPU and memory. The BDT supports efficient training even with larger dimensions of neural networks [52, 53].

7.5 Datasets and Preprocessing

This study uses three medical imaging datasets to build and test the performance of the DFDetector verification system, including Wound, Colormat, and Fluorescent, Choukroun, and PCH4DPod3.0. Each image of this extensive system is reviewed for any visual indications of DFU-related issues. The image texture, features, and morphological characteristics can predict the presence of DFU accurately. We conducted a study to recognize diabetic foot disease using three datasets, which were collected from four different medical devices. AlexNet, which is used to detect the presence of disease, is pre-trained to create self-attention-based models. We devised a novel method for reusing trained models. Reducing mortality by tracking infected wounds is the main advantage of the DF detection model. In a public experiment, the proposed integrated model demonstrated 93% accuracy toward all problems, specifically in terms of infection recognition.

The superpixel over-segmentation technique is widely used as a preprocessing technique to split the image into superpixels of similar size and connection independently, and there is no trivial way to include spatial information when this technique is applied. The ASPP enables the introduction of double-scale spatial and context features. After analyzing the extracted superpixels, we noticed that the quotient image contains not only information necessary for distinguishing one texture from another, but it also contains information about the form of the presented wound. In the next step, our method involves the detection, prediction, processing, and classification of unhealthy images. We first concentrate on checking the separated superpixels and finding the smudge class. In our technique, this is the last step. The flexibility of our solution means that it can be extended by many modules that can be arranged according to the disorders detected in the picture.

Table 7.1 Details of publicly available datasets with of training, testing, and validation values

Datasets	Training	Testing	Validation
Diabetic foot images	2000	500	500
DFU image	1500	300	500
Kaggle DFU detection	2000	200	800
American heart association	Not specified	Not specified	Not specified
Podiatry network image	Not specified	Not specified	Not specified
DFU segmentation	1000	200	300
Gumuchian	1200	300	400

7.5.1 Publicly Available Datasets

In this study, we have used two datasets. One is DFU Dataset [53] The first digital color photographic image dataset has been released to coincide with the publication of a paper that presented one of the first major demonstrations of artificial intelligence's ability to detect infection in diabetic foot ulcers. The study compared six AI systems. Each AI system is accompanied by the human reference standard from which it has been developed. It represents a population of 1 million people with all clinical coded diagnoses.

The second digital color photographic image and depth image dataset has been collected with the aim of assisting with the detection and assessment of diabetic foot ulcers and to evaluate the performance of the developed algorithms. Moreover, the dataset can be used by students and researchers who wish to test their own algorithm's capabilities. The images were acquired with a novel smartphone camera application, incorporating a depth sensor to record depth data of the wound [54] (Table 7.1).

7.5.2 Preprocessing Techniques

Preprocessing plays a crucial role in any machine learning model. The process of learning depends on the data that we provide. Preparing data by cleaning the entire dataset, removing duplicate entries, or filling in missing values is the starting point for a successful model. Preprocessing images by making them the same size is done to make the operations easier. Preprocessing is the first step toward an AI model. Each preprocessing step helps our model learn more easily. Random removal of data and the use of preprocessing techniques like resizing and masking help the model learn accurately [55]. This preprocessing step changes the input data to the model with reduced dimensionality, which in turn gives high throughput for further steps. This step involves cleaning specific data before feature selection. The difference between a regular image and a preprocessed image is that each image is made the same size so that, if desired, we can perform operations on the images. It simplifies the model by making all images the same size and then reduces the decision flow. We used gray color preprocessing for images. Masking of the preprocessed images and resizing the images are used as preprocessing techniques. This is done to make the images and operations easier. Masking helps maintain consistency during training [56].

7.6 Performance Metrics and Evaluation

With the help of artificial intelligence, using CNN, the radiologist can easily diagnose the diabetic foot ulcer at the earliest and thereby reduce treatment costs, time, and resources. The performance has been evaluated using the metrics mentioned below. True positive is the ulcerous gap area that was correctly predicted in the actual image. True negative is the absence of the ulcerous gap area, which was correctly classified as the normal state in the actual image. False positive is the false unobservable image or gap area that was not present in the actual image, but it is the result of predicting the actual ulcerous gap area. High specificity values also help in identifying non-ulcerative activating areas. In comparison, a negative ulcer image based on an increase in the sensitivity value also aids in identifying the active ones, even indicating a positive ability to be diagnosed by non-ulcer images [57].

High accuracy values reduce the erroneous identification of ulcer and non-ulcer images. Depending on the confusion matrix above, recall and precision evaluation metrics (true positive, true negative, false positive, and false negative) determine whether the algorithm correctly detects ulcerous gap areas in an image, the actual appearance of the non-ulcerous gap area or tumor on the images, or non-detection of the non-ulcerous gap or non-tumor area. In terms of diabetes, early diagnosis is also a crucial problem because it has the potential to decrease the rising type 2 diabetic population and control the disease. Dropout alternatives between its layers have also been implemented to avoid severe overfitting. High dropout regularization values place a decent Gaussian dip along the side [58].

7.6.1 Accuracy, Sensitivity, and Specificity

The accuracy, sensitivity, and specificity obtained from the proposed screening system to detect diabetic foot ulcers among diabetes patients are given. The robustness of the model screened a large number of diabetic patients during classification with a high true positive rate. The accuracy, which measures class confusion among non-diabetic and diabetic ulcers, was classified at 92%, which displays that the robustness of the proposed system is good. This high accuracy was attributed to the large database used and the efficient classifier in distinguishing diabetic ulcers from non-diabetic ulcers using combined statistical properties, fractal texture features, and ANN. These results show that the learning-based artificial intelligence approach has clinical relevance in rapidly screening supportive healthcare services to classify the treatment-prone location in the healthcare system.

In this research work, therefore, the ultimate goals were achieved by developing a powerful Candidate Recognition System for diabetic ulcers with high quality. The system used 109 patients when accessing the diabetic ulcer images, collecting isolated areas of the diabetic ulcer images, pre-processing the regions consisting of 31 sets of features defined to describe diabetic ulcers, and building the classification/diagnosis model in an artificial neural network, which also identifies the number of patients with a small number of paired datasets used. If the patients' paired data increases, we can discard the unnecessary feature set in the preprocessing section. We applied an independent testing database to validate the method. The proposed method, then, would prove to be of considerable value, enabling a non-expert operator to detect diabetic foot ulcers non-invasively and automatically.

7.6.2 Receiver Operating Characteristic (ROC) Curve

Based on the concept of sensitivity and specificity, the ROC curves plotting sensitivity and specificity at all significance levels simultaneously are useful for evaluating the effectiveness of diagnostic technology. A characteristic of ROC curves is that the plateau line represents "the absence of a diagnostic effect." In addition, the area under the ROC curve (AUC) is defined to distinguish different technologies, and it is considered that the more the AUC approaches 1, the better the diagnostic effect of the technology becomes; AUC $= 0.5$ represents that "the technology does not have diagnostic effectiveness." When AUC < 0.5, one technology is compared with its negative correlation effect. In other words, AUC > 0.5 proves the reliability of the technology. Receivers operating characteristic curve and area under the curve can be used to provide convincing evidence of the effectiveness of an artificial intelligence system.

7.7 Challenges and Future Directions

To the best of our knowledge, this is the first effort to provide a comprehensive review of the state of the art in diabetic foot ulcer classification, focusing on recent advances in medical imaging technologies and the emerging attention of researchers in the field. Nevertheless, there are several challenges and open research directions that need to be addressed.

Medical imaging technologies and artificial intelligence are potent technologies that need more attention and application in the domain of diabetic foot ulcer detection and classification. There is an emerging increasing attention by researchers on both technologies in the field. This chapter provides detailed motivation and insights about the problem and reviews the current state of the art for enabling medical and biomedical students, researchers, and practitioners to utilize these powerful technologies. Future research directions and challenges are also provided. The aim is to attract more attention and enhance the percentage and quality of research in this special issue and close the chapter.

7.7.1 *Interpretability and Explainability*

Interpretability and explainability have been the primary necessities in the development of machine learning and artificial intelligence models, as we have been relying on them to make judgments in the real world. In the healthcare domain, where the model and AI system's judgment may affect someone's life, explainability and interpretability become even more crucial and directly dictate the adoption of AI. However, as interpretation methods inherit from deep learning models, most models and published works have, until today, exploited validation metrics and visualization. There is only a limited number of comparison algorithms to gauge insights into model decisions, and decision metrics adapted to evaluation are also rare. These may prevent the progress of more precise and effective AI models applied within the healthcare domain.

To support the improvement of our detection model and further advanced research, in this section, we introduce more systematic and multi-view model evaluation methods from the viewpoint of model interpretability, explainability, and performance. We have proposed several interpretability tools such as activation heatmaps, class activation maps, gradient-guided class activation maps, and CAM-enhanced analysis. We have performed simple post-hoc visualizations based on the gray-level gradient calculated from activation maps. With Grad-CAM, we have integrated not only a simple grid visualization but also the heat maps and overlay of the original images. These results have indicated encouragement from compound equipment-dependent models to present much more reliable visual aids for radiologists and clinician decision support. The depression of gradient overfitting from significant statistical decrease in model performance supports evidence of these findings. Although improvements are needed in the literature on model interpretation and evaluation, to facilitate feasible clinical implementation, the assurance of model reliability and credibility is of the utmost importance.

7.7.2 *Generalization to Different Populations*

When building and testing a predictive model using medical images and their reported DFU status from a current population, it is important to consider what population the model should generalize to and ensure the model is not learning a relationship that only applies to a specific subpopulation. We specified that it was not necessarily the intention of our model to predict the probability that an unseen person had a DFU but to detect the location and size of an existing DFU in an image. Future design of studies that explore the relationship between imaging features and the risk of future DFU may not be suitable for derived learning weight methods in which selection of the model will be based on an unseen validation dataset representative of a population for which the model will be used, not the population-specific dataset.

7.8 Conclusion and Implications

Our current study shows the deployment of AI techniques in the field of computer-assisted diagnostics. The development of computer-aided diagnostic methods to detect and locate regions of the Diabetic Foot Ulcer (DFU) is vital to increase the efficiency of the diagnosis; to improve the processing speed of the diagnostic technique and reduce the time delay between diagnosis and treatment. Hence, we developed a newly presented Deep Neural Network model named Modified Hierarchical Monkey Spider Inspired Capsule Networks (MHS-CapsNet) to accurately segment the diabetic foot ulcers from non-foot ulcers in medical images. Also, we improved some of the architecture of the CapsNet which includes more than one level of feature aggregation and channel re-calibration to improve the overall performance of the network. The model also used implementation-based ensembling techniques such as logical operators for further operational fusion that enhances the significance of providing lower FIR filter size to lessen the computational expense and memory load, however, it contains more firepower to conquer the visual domain. To make it further operational, we especially used a novel data augmentation technique that significantly strengthened the categorization capability of the proposed network, especially with the small number of training samples. The conclusions of the investigation presented in this chapter are reliable and relevant in different aspects that include scientific, practical, and methodological.

The modified Capsule Network, MHS-CapsNet, achieved very notable results in the detection of diabetic foot ulcers from non-foot ulcers when compared with other existing convolutional neural network models for the image classification problem. Even for the small dataset, the sensitivity obtained by the proposed model is very high and comparable to the deep learning models available for generalized medical-imaging datasets. The results demonstrate the ability of the proposed model and reveals the capacity to classify the potential areas of high sensitivity for diabetic foot in clinical imaging databases.

The proposed deep learning model's reliable prediction, fast processing, and readily available scanning systems for medical screenings enables us to apply our model to considerably healthy medical images. Moreover, our methods avoid using the point annotation and utilize the bounding box object annotation through our proposed deep learning model for level-sensitive pixel annotations. From our study, it is clear that the most important impact derives from the destined application as follows: In the medical-imaging datasets, the accurate identification of the foot ulcer regions plays a critical role for the lesser experts and the most extended medical services. Due to the advancement of technology, we developed an advanced medical-image model to identify the foot ulceration, thus extending the applicability within the more significant population. Even though our current Deep Learning technique provides an in-depth annotation, the slightly lower size does not affect the robust model architecture. In our case, the end-to-end hierarchy is performed to capture the desired spatial behavior and efficient object recognition tasks. Our Model is not over-reliant on the objectness level, but proves to be a practical method to exhibit the meaningful features of the foot ulcers, hence, enabling a broader audience of computer applications in the medical domain.

7.9 Summary of Key Findings

The efficacy of computer-aided diabetic foot ulcer (DFU) wound detection can be helpful for both the patient and doctor to increase the chances of wound healing. The proposed technique tries to address the drawbacks of conventional techniques such as block processing and to exhibit the importance of the CNN model in automating the task of wound detection. Despite the numerous successes in the field of computer vision, the progression of models for wound detection applications is still in its infancy. Overall, the conducted experiments are performed generally to show the convincing performances of the proposed model, i.e., more accuracy compared to the existing model with less processing time. However, the generalization of depth CNN models for the wound detection task is a promising direction for upcoming research in the field of wound detection.

The dataset utilized in the study is not very large. However, when compared to other existing studies, the dataset is larger and exclusive. An annotated dataset is propagated to motivate the rest of the researchers. With consistent annotation, computer-aided applications for wound healing can possibly be developed. Even though the dataset is created through ethical guidelines, the success of the proposed model could be critically evaluated if and only if the model performance is examined with multiple independent datasets established from multiple sources. Although training the models on a larger dataset possibly mitigates the evidence that is incurred by the CNN models, the architecture selection is another paramount consideration. When inspecting the available literature, it is understood that only a few existing models have been solely developed and tested for the wound detection process. Thus, more specific models are needed for the task of distinguishing the difference between normal tissue and wound surfaces. In addition, the performance of the CNN models depends largely on the selection of hyperparameters, making this process more heuristic for researchers who lack expertise in computer vision tools. Future work should go beyond a small preliminary analysis and conduct a more in-depth investigation of the impact of doctoral factors on detection and segmentation accuracy. Furthermore, researchers should include sufficient details about data expansion and the loss functions used to help the reader decide whether one representation is more plausible than others to depend on the outcome.

7.9.1 Clinical and Research Implications

Diabetic foot ulcers (DFUs) are severe complications of diabetes, which can lead to an increase in healthcare costs and the amputation of lower extremities. In addition to patient care, medical images also provide an understanding of the physiological structure of the human body to support medical practitioners, particularly with patient assessment by foot risk classifications. From the transplantation of the tissue to its removal, good planning and an understanding of the location of the amputated site and the foot structure are required, as this is always necessary before surgery. Even after the surgical process, medical imaging is needed to monitor the healing of the foot and the risk of recurrence. Therefore, medical imaging is very useful in the diagnosis and treatment of DFUs.

Medical imaging also requires a long time for diagnosis if it is done manually by radiologists, especially at a hospital with a large number of patients. Therefore, a computer-aided diagnosis (CAD) with artificial intelligence (AI) approach is proposed to assist radiologists in diagnosing DFUs through medical imaging. From the results of the experiments, we found that the proposed approaches can diagnose several medical imaging cases in terms of the AUC–ROC score with quite high sensitivity and specificity. With radiologists being provided with AI tools that can support them, they may be able to make accurate diagnoses and obtain an optimal treatment plan.

References

1. Primadhi RA, Herman H (2021) Diabetic foot: Which one comes first, the ulcer or the contracture? World J Orthop 12(2):61
2. Bardyugov PS, Parshikov MV (2022) Foot deformities in patients with diabetic foot disease (literature review)
3. Tsaklis PV, Tentolouris N (2024) Biomechanics of the diabetic foot: the road to foot ulceration. The diabetic foot: medical and surgical management. Springer, pp 233–252
4. Nanda R et al (2022) Machine learning algorithm to evaluate risk factors of diabetic foot ulcers and its severity. Med Biol Eng Comput 60(8):2349–2357
5. Mancilla RB et al (2020) Detection of sore-risk regions on the foot sole with digital image processing and passive thermography in diabetic patients. In: 2020 17th international conference on electrical engineering, computing science and automatic control (CCE). IEEE
6. Xiaoling W et al (2024) Enhancing diabetic foot ulcer prediction with machine learning: a focus on localized examinations. Heliyon
7. Kumar A, Singh D (Eds.) (2025) Revolutionizing Medical Systems Using Artificial Intelligence: A Breakthrough in Healthcare. Elsevier
8. Chamberlain RC et al (2022) Foot ulcer and risk of lower limb amputation or death in people with diabetes: a national population-based retrospective cohort study. Diab Care 45(1):83–91
9. Edmonds M, Manu C, Vas P (2021) The current burden of diabetic foot disease. J Clin Orthop Trauma 17:88–93
10. Jalilian M, Ahmadi Sarbarzeh P, Oubari S (2020) Factors related to severity of diabetic foot ulcer: a systematic review. Diabetes, Metabolic Syndrome and Obesity, pp 1835–1842
11. McDermott K et al (2023) Etiology, epidemiology, and disparities in the burden of diabetic foot ulcers. Diab Care 46(1):209–221

12. Alam S et al (2021) Diabetes mellitus: insights from epidemiology, biochemistry, risk factors, diagnosis, complications and comprehensive management. Diabetology 2(2):36–50
13. McGurnaghan SJ et al (2021) Risks of and risk factors for COVID-19 disease in people with diabetes: a cohort study of the total population of Scotland. Lan Diab Endocrinol 9(2):82–93
14. Yosef T, Nureye D, Tekalign E (2021) Poor glycemic control and its contributing factors among type 2 diabetes patients at Adama Hospital Medical College in East Ethiopia. Diabetes, metabolic syndrome and obesity, pp 3273–3280
15. Dörr S et al (2023) Bacterial diversity, antibiotic resistance, and the risk of lower limb amputation in younger and older individuals with diabetic foot infection. Int J Low Extrem Wounds 22(1):63–71
16. Rodríguez-Rodríguez N et al (2022) Wound chronicity, impaired immunity and infection in diabetic patients. MEDICC Rev 24:44–58
17. Wang X et al (2022) Diabetic foot ulcers: classification, risk factors and management. World J Diab 13(12):1049
18. Al-Mohaithef M et al (2022) Screening for identification of patients at high risk for diabetes-related foot ulcers: a cross-sectional study. J Int Med Res 50(3):03000605221087815
19. Monteiro-Soares M et al (2024) Classification of foot ulcers in people with diabetes: a systematic review. Diab Metab Res Rev 40(3):e3645
20. Demirkol D et al (2022) Analysis of risk factors for amputation in patients with diabetic foot ulcers: a cohort study from a tertiary center. Acta Orthop Traumatol Turc 56(5):333
21. Ahmad M et al (2004) The use of artificial intelligence in three-dimensional imaging modalities and diabetic foot disease–a systematic review. JVS-Vasc Insights, p 100057
22. Cassidy B et al (2023) Artificial intelligence for automated detection of diabetic foot ulcers: a real-world proof-of-concept clinical evaluation. Diab Res Clin Pract 205:110951
23. Chun J-W, Kim H-S (2023) The present and future of artificial intelligence-based medical image in diabetes mellitus: focus on analytical methods and limitations of clinical use. J Korean Med Sci 38(31)
24. Kaselimi M et al (2022) A review of non-invasive sensors and artificial intelligence models for diabetic foot monitoring. Front Physiol 13:924546
25. Mishra S et al (2024) An extensive analysis of diseases associated with diabetes. J Pharma Insights Res 2(3):174–187
26. Zakir M et al (2023) Cardiovascular complications of diabetes: from microvascular to macrovascular pathways. Cureus 15(9)
27. Kaur M et al (2023) Understanding the role of hyperglycemia and the molecular mechanism associated with diabetic neuropathy and possible therapeutic strategies. Biochem Pharmacol 215:115723
28. Goldney J, Sargeant JA, Davies MJ (2023) Incretins and microvascular complications of diabetes: neuropathy, nephropathy, retinopathy and microangiopathy. Diabetologia 66(10):1832–1845
29. Kelso MR, Jaros M (2024) Improving wound healing and infection control in long-term care with bacterial fluorescence imaging. Adv Skin Wound Care 37(9):471–479
30. Henshaw FR et al (2020) Point of care ultrasound imaging as a wound assessment tool in diabetic foot ulcers: a case series. J Wound Care 29(Sup8):S28–S34
31. Godavarty A et al (2023) Diabetic foot ulcer imaging: an overview and future directions. J Diabetes Sci Technol 17(6):1662–1675
32. Lucas Y et al (2021) Wound size imaging: ready for smart assessment and monitoring. Adv Wound Care 10(11):641–661
33. Doulamis A et al (2021) A non-invasive photonics-based device for monitoring of diabetic foot ulcers: architectural/sensorial components & technical specifications. Inventions 6(2):27
34. Kumar A, Bhalla S (2025) Exploring brain tumor detection through artificial intelligence. In Revolutionizing Medical Systems using Artificial Intelligence, pp 93–120. Academic Press
35. Zhang Z et al (2023) End-to-end learning-based image compression with a decoupled framework. IEEE Trans Circuits Syst Video Technol

36. Zulfiqar A et al (2021) AI-ForestWatch: semantic segmentation based end-to-end framework for forest estimation and change detection using multi-spectral remote sensing imagery. J Appl Remote Sens 15(2):024518–024518
37. Yang CC (2022) Explainable artificial intelligence for predictive modeling in healthcare. J Healthc Inf Res 6(2):228–239
38. Purohit RM, Verma JP, Jain R, Kumar A (2025) FedBlocks: federated learning and blockchain-based privacy-preserved pioneering framework for IoT healthcare using IPFS in web 3.0 era. Cluster Computing, 28(2):139
39. Hunter B, Hindocha S, Lee RW (2022) The role of artificial intelligence in early cancer diagnosis. Cancers 14(6):1524
40. Mansour RF et al (2021) Artificial intelligence and internet of things enabled disease diagnosis model for smart healthcare systems. IEEE Access 9:45137–45146
41. Panayides AS et al (2020) AI in medical imaging informatics: current challenges and future directions. IEEE J Biomed Health Inform 24(7):1837–1857
42. Hardy M, Harvey H (2020) Artificial intelligence in diagnostic imaging: impact on the radiography profession. Br J Radiol 93(1108):20190840
43. Zhou SK et al (2021) A review of deep learning in medical imaging: Imaging traits, technology trends, case studies with progress highlights, and future promises. Proc IEEE 109(5):820–838
44. Bekbolatova M et al (2024) Transformative potential of AI in Healthcare: definitions, applications, and navigating the ethical landscape and public perspectives. Healthcare, MDPI
45. Tariq M et al, Principles and perspectives in medical diagnostic systems employing artificial intelligence (AI) algorithms. Int Res J Econ Manag Stud IRJEMS 3(1)
46. Thakur GK et al (2024) Deep learning approaches for medical image analysis and diagnosis. Cureus 16(5)
47. Zhang X et al (2022) Deep learning with radiomics for disease diagnosis and treatment: challenges and potential. Front Oncol 12:773840
48. Gill AY et al (2023) Revolutionizing Healthcare: how machine learning is transforming patient diagnoses-a comprehensive review of AI's impact on medical diagnosis. J World Sci 2(10):1638–1652
49. Marini N et al (2022) Unleashing the potential of digital pathology data by training computer-aided diagnosis models without human annotations. npj Digital Med 5(1):102
50. Chauhan AS et al (2024) Unleashing the power of advanced technologies for revolutionary medical imaging: pioneering the healthcare frontier with artificial intelligence. Discov Artif Intell 4(1):58
51. Asif S et al (2024) Advancements and prospects of machine learning in medical diagnostics: unveiling the future of diagnostic precision. Archives of computational methods in engineering
52. ArunKumar KE et al (2021) Forecasting of COVID-19 using deep layer recurrent neural networks (RNNs) with gated recurrent units (GRUs) and long short-term memory (LSTM) cells. Chaos, Solitons Fractals 146:110861
53. (2021) https://www.kaggle.com/datasets/laithjj/diabetic-foot-ulcer-dfu. Available from: https://doi.org/10.1007/s13369-021-05691-8
54. Cassidy B et al (2021) The DFUC 2020 dataset: analysis towards diabetic foot ulcer detection. touchREVIEWS in Endocrinology 17(1):5
55. Salvi M et al (2021) The impact of pre- and post-image processing techniques on deep learning frameworks: a comprehensive review for digital pathology image analysis. Comput Biol Med 128:104129
56. Liu Y et al (2021) Ore image classification based on small deep learning model: evaluation and optimization of model depth, model structure and data size. Miner Eng 172:107020
57. Silva RHLe, Machado AMC (2021) Automatic measurement of pressure ulcers using support vector machines and GrabCut. Comput Methods Programs Biomed 200:105867
58. Jaganathan Y et al (2023) Validating wound severity assessment via region-anchored convolutional neural network model for mobile image-based size and tissue classification. Diagnostics 13(18):2866

Chapter 8
Diagnosis and Prediction of Breast Cancer Using Artificial Intelligence

Abstract Breast cancer is one of the most widespread and prevalent types of cancer with the highest mortality rate in women. It is a disorder that leads to the unconditional growth of malignant cells in the breast of the patient. Early diagnosis of breast cancer not only extends impactful treatment but also prevents the chances of death of the patient. However, early prediction of malignant cells in breast cancer is not an easy task with regular and frequent examination. AI has provided predictive algorithms based on machine learning and deep learning which can classify malignant cells from healthy cells with good accuracy. These algorithms detect cancerous cells at an early stage and hence, enhance the chances of the patient's survival. This chapter compares the features, limitations, and efficiency of various AI-based techniques as machine learning and deep learning predictive algorithms for predicting breast cancer in women.

Keywords Classification · UNet models · Risk factors · Hormonal therapy · Dataset · Generalizability

8.1 Introduction

Breast cancer is a health disorder and a leading cause of death among women. It is one of the most common types of cancer that occurs one in every nine women and its traces can also be seen in males [1]. This is one of the most dangerous types of cancer with a high mortality rate in humans. Breast cancer is a condition that occurs due to superfluous increase in the number of cells in the breast. This intensification of cells leads to a lump-like structure in the breast most commonly termed a tumor. Basically, these tumors can be categorized as either benign or malignant [2].

Generally, benign tumor lumps are non-dangerous and non-cancerous. These tumors do not cause any problems or pain in the breast and its surrounding tissues as well. It is not mandatory to operate these tissues and get them removed from the body. The reason for such types of lumps in the breast could be cysts, hyperplasia, fat necrosis, and many others [3]. On the other hand, malignant tumors are

A. Kumar and D. Singh, *Artificial Intelligence in Modern Healthcare System*,
Transactions on Computer Systems and Networks,
https://doi.org/10.1007/978-981-96-6703-1_8

cancerous and require proper treatment when diagnosed. These tumors can spread and destroy neighboring tissues and lead to metastatic breast cancer. For accurate diagnosis and prognosis of breast cancer, the tumors can further also be classified as benign (adenosis, fibroadenoma, phyllodes, and tubular adenoma) and malignant (ductal carcinoma, lobar carcinoma, mucinous carcinoma, and papillary carcinoma) [4].

Mainly, the female breast constitutes various regions namely, glandular, lobes, and ducts. All these regions are susceptible to breast cancer. The redness, swelling, fluid discharge, and deformation in size are common symptoms that lead to the formation of a lump or tumor in the breast. Breast cancer can be categorized from stage 0 to stage 4 depending on the tumor size and its spread to neighboring tissues [5]. Initial stage 0 is the preliminary level of breast cancer which can be cured with proper treatment. But stage 4 is the most advanced stage which leads to invasive breast cancer. Women above 50 years of age are found to be most infected with advanced-stage cancer with a high mortality rate. However, microscopic examination of breast lesions is essential for accurate and efficient prediction of cancerous cells.

To enhance the survival rate of breast cancer patients, early diagnosis and proper treatment are very substantial [6]. However, societal barriers, socioeconomic status, illiteracy, and lack of knowledge lead to its delayed detection. In addition, the lack of advanced technology and methodologies also prevents its early diagnosis. Breast cancer detection can be either done through conventional techniques that involve manual check-ups or advanced AI-based algorithms. Manual check-up involves the physical examination of the breast for redness, swelling, and irritation and self-assessment by touching the breast for tenderness and variations in breast structure [3]. However, these preliminary assessments are not accurate and require further evaluations using some concrete and trustworthy methodologies. For this, various imaging modalities such as Magnetic resonance imaging (MRI) [7], Ultrasound (USd) [2, 8], and mammography [9, 10] are used for screening for the early diagnosis of cancer. In addition, Electronic health records (EHR) containing patient details such as socio-demographic information and pathological reports are also exploited for the prediction of breast cancer or its chances for reoccurrence [11].

Imaging modalities can examine the various breast tissue components to detect abnormalities in a better way [12–15]. For manual analysis, radiologists study the imaging modalities and interpret the results to characterize the breast lesions for the presence of cancerous cells. However, manual interpretation is not only time-consuming but also, sometimes results from different radiologists are conflicting. To avoid such ambiguity in results, AI-based techniques are considered a better choice for breast cancer prediction by various medical practitioners [5, 16–19]. AI-based algorithms not only study the minute and fine-grained information from imaging data but also reduce the inspection time. These techniques are faster, efficient and can categorize breast cancer into its various classes for better treatment and microscopic examination [4, 20].

In this chapter, we have analyzed various AI-based algorithms for the prediction of breast cancer at an early stage. These algorithms analyze the imaging data for identifying the breast lesions for the presence of cancerous tumors. These algorithms not only addressed the limitations of manual methods for breast cancer detection but also, were fast, accurate, and effective in their outcome. The key contributions in this chapter are as follows:

- The risk factors that lead to the chances of breast cancer occurrence are elaborated in detail to provide awareness pointers in order to prevent its spread among humans.
- We have categorized AI-based predictive algorithms either as ML-based techniques or DL-based techniques. The strengths and limitations of each category are reviewed to highlight the salient features.
- DL-based techniques have utilized either UNet and its variant or non-UNet architecture for extraction of potential features for examination of breast constitutes for the presence of cancer.

The rest of the chapter is organized as follows. Section 8.2 elaborates on the AI-based predictive algorithms for breast cancer prediction. In addition, AI-based techniques for automatic segmentation of breast lesions to analyze breast cancer are categorized into ML-based and DL-based techniques in this section. ML-based techniques and exploited dataset details are highlighted in Sect. 8.3. Section 8.4 details about salient features of the DL-based predictive model into two categories UNet and non-UNet architectures. AI-based techniques are compared to analyze the merits, demerits, and limitations in each category in Sect. 8.5. Lastly, the concluding remarks and future directions are sketched in Sect. 8.6.

8.2 Artificial Intelligence Based Algorithms for Breast Cancer Prediction

Early diagnosis of breast cancer is very critical to extend proper treatment and save lives. For fast and accurate prediction of breast cancer in imaging and textual data, AI-based predictive algorithms can analyze breast cancer lesions microscopically [2, 21, 22]. Based on the exploited methodology, AI-based predictive algorithms are categorized either as ML-based approaches or DL-based approaches. Figure 8.1 represents the various categories of AI-based breast cancer prediction algorithms.

ML-based breast cancer prediction algorithms utilized various models namely, Naïve Bayes (NB), Decision tree (DT), Support vector machine (SVM), multi-layer perceptron (MLP), Logistic regression (LR), AdaBoost, XGBoost, and many others for analyzing textual data in the form of EHR [11, 23], numerical data in the form of.csv files [5, 24] and imaging data in the form of mammography [25, 26]. Based on the exploited deep neural network (DNN), DL-based breast cancer prediction algorithms are categorized either as UNet techniques [27–29] or non-UNet techniques. UNet-based breast cancer prediction algorithms utilized UNet and its variants such as TransUNet [2], Tubule UNet [13], Asymmetric UNet [8], UGGNet [19], and many others [20, 22, 28, 30] for making an accurate diagnosis. On the other hand,

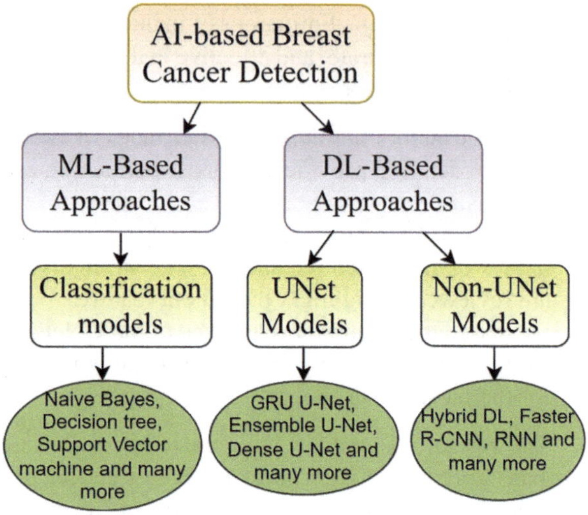

Fig. 8.1 Various categories of AI-based predictive algorithms for breast cancer detection

non-UNet based predictive algorithms exploited DL networks such as CNN [14], Faster R-CNN [31], adversarial networks [32], and LSTM [12] for faster and more effective diagnosis. The details about the representative work under each category are elaborated in the subsequent sections.

8.2.1 Risk Factors for Breast Cancer Prediction

Initially, breast screening can be done to assess the risk of cancer. Clinicals and researchers can identify the associated risk factors to predict the possibility of cancerous cells, especially in the breast of women. There are many associated genetic and non-genetic factors that if reviewed critically can reduce the chances of breast cancer to a great extent [33]. The potential risk factors are identified as depicted in Fig. 8.2.

Hormonal variations/Hormonal therapy: The changes in the estrogen levels during first live birth and at menopause, lead to the high chances for breast cancer. In addition, exposure to postmenopausal hormonal therapy also increases the chances of breast cancer in females.

Increased breast density: Breast density is also considered to be a vital parameter to determine the possibility of breast cancer in women. An increase in the breast density leads to a higher risk of breast cancer. Changes in breast structure due to milk production after first live birth in the glandular region of the breast also enhance the chances of cancer in women.

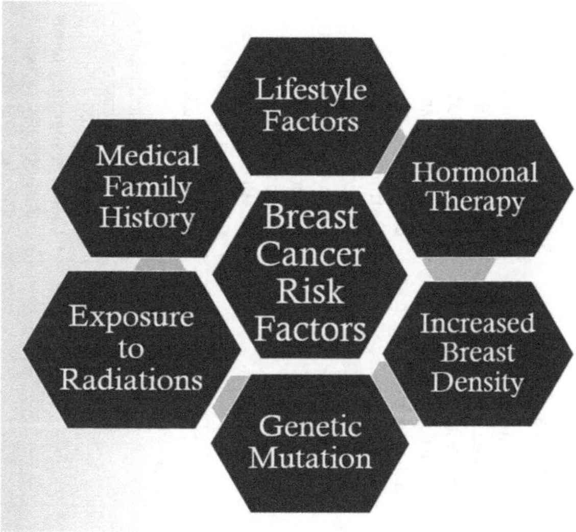

Fig. 8.2 Associated risk factors with breast cancer

Genetic mutation: It has been observed that the medical family history of the women is a well-known risk factor that increases the chances of occurrence of breast cancer in women. The risk of genetic mutations from infected mother, sister, or even from any male relative increases the chances of inheritance of diseased cells. Gene screening is recommended to predict the chances of cancer in such cases [34].

Exposure to radiations: Prolonged exposure to various radiations also increases the risk of breast cancer in women. Middle-aged females are considered to be at high risk.

Lifestyle factors: Obesity, reduction in physical activity, and smoking history in postmenopausal women increase the possibility of breast cancer in females. Increase in alcohol intake in pre and post-menopausal women is a potential parameter that enhances the chances of breast cancer.

8.3 Machine Learning-Based Algorithms for Breast Cancer Prediction

With the change in lifestyle and environmental conditions, there has been a sudden rise in the breast cancer cases for the last few years. ML-based algorithms such as LR, DT, SVM, XGBoost, and many more have been utilized to classify breast tumors either as benign or malignant [35, 36]. Table 8.1 tabulates the salient features of the representative work exploiting ML techniques for breast cancer prediction.

A lot of researchers have realized the potential of ML algorithms for the detection of breast cancer at an early stage. In this direction, Zhou et al. [37] have exploited

Table 8.1 Representative work in ML-based breast cancer prediction

Reference	Dataset description	Model utilized/proposed	Modality	Performance measures	Summary
Zhou et al. [37]	WDBC Benign/Malignant: 357/212	DT, SGD, RF, SVM, LR, KNN and AdaBoost	Numerical dataset	Spearman's correlation, p-value, Wilcoxon rank sum test, box plot, CM, PR, RE, F1-score, ACC and support	• Examined the relationship between feature subset data and labeled data using Spearman correlation • Distribution difference between different characteristics of breast cancer examined using Wilcoxon rank sum • AdaBoost+ LR achieved the highest ACC (98.96%)
Botlagunta et al. [11]	Total size: 25,652 entries in.csv file	LR, KNN, DT, RF, (linear, polynomial, radial) SVM, GB, and XGBoost	Textual EHR dataset	ACC, PR, RE, AUC-ROC, and F1-score	• EHR records with record numbers and histopathological reports utilized for breast cancer prediction • Cancer malignancy terms namely, biopsy, lymph node, and metastasis reviewed for breast cancer prediction • DT achieved superior ACC (83%) in comparison to other methods

(continued)

Table 8.1 (continued)

Reference	Dataset description	Model utilized/proposed	Modality	Performance measures	Summary
Bhattarai et al. [25]	114 Patient data from Nottingham City Hospital	NB, DT, DA, SVM, KNN, and ensemble	Serial mammography	HR, p-value, Kaplan–Meier survival curve	• KNN achieved ACC of 70% with three features namely Ki67, MI, and tumor size • Ensemble achieved ACC (70%) by including four additional features • Categorized the cohort either as slow-growing or fast-growing tumor subgroups
Dalal et al. [5]	Kaggle dataset	RT, LR, XGBoost, and MLP	Numerical dataset	ACC and AUC	• SimpleImputer function to handle the missing values in the dataset • Ensemble multiple models to achieve generalization and better accuracy • The ensemble model achieved the highest ACC (99.59%)
Chen et al. [24]	WDBC dataset	XGBoost, RF, LR, KNN	Numerical dataset	ACC, PR, RE, F1-score	• Utilized Pearson's correlation test to select 15 features out of 30 • Computed performance by dividing datasets as training and test (8:2 and 7:3) • Determined malignant and non-malignant cancerous cells

(continued)

Table 8.1 (continued)

Reference	Dataset description	Model utilized/proposed	Modality	Performance measures	Summary
Ferroni et al. [23]	• Real-world breast cancer dataset from Clinical database and Biobank project • Training/Testing: 318/136	MKL based on SVM, RO	Pathological characteristics	HR, ACC, AUC, SEN, SPE, CI, p-value	• Utilized multiple features such as socio-demographics, tumor information, and biochemical data • Threefold cross-validation for better classification accuracy • Relative weights in RO used for specifying feature importance
Kumar et al. [38]	• WDBC • Training/test: 70:30	KNN, RF	Numerical dataset	ACC	• Handled missing values by imputing mean, median, zero or constant • Determined distance using Manhattan and Euclidean methods • Predicted cancer at an early stage with ACC (83.33%)
Kurian and Jyothi [34]	• 4 datasets: 50,100,150, 200 sequences • Data size: 156, 303,454, and 667 • Training/Test: 80:20	LR, KNN, DT, NB, SVM, RF, AdaBoost, GB, LDA	Numerical dataset	RE, F1-score, ACC, CM,	• Considered three different sequences namely, Homo sapiens, BRCA1, and BRCA2 • Considered multiple feature extraction algorithms • Feature selection can be done by granular computing

(continued)

Table 8.1 (continued)

Reference	Dataset description	Model utilized/proposed	Modality	Performance measures	Summary
Mohammed et al. [21]	• Two datasets namely WDBC and breast cancer dataset • Total instance: 699 (WDBC), 286 (breast cancer dataset)	DT, NB, SMO	Numerical dataset	TP, FP, ROC, SD, ACC	• Resampling technique to address imbalanced datasets • Tenfold validation to improve accuracy • Classified tumors either as malignant or non-malignant
Naji et al. [39]	• WDBC • Total instances: 569 • Benign/malignant: 357/212	SVM, RF, LR, DT, KNN	Numerical dataset	CM, ACC, PRE, SEN, F-Measure, AUC	• SVM performed best in comparison to other models • SVM obtained ACC (97.2%), AUC (96.6%), F-Measure (0.96%), SEN (0.94%) and PRE (97.5%)
Nemade and Fegade [40]	• WDBC • Total instances: 569 • Benign/malignant: 357/212 • Training/Test: 80:20	NB, LR, SVM, KNN, DT, and ensemble (RF, Adaboost, XGBoost)	Numerical dataset	PRE, TP, FP, RE, F1-score, ACC, CM, AUC-ROC	• Exploited multiple ML and ensemble techniques for cancer prediction • DT obtained the highest ACC (97%), and LR highest AUC (0.99) • Ensemble technique XGBoost highest ACC (97%) and AUC (0.99)

(continued)

Table 8.1 (continued)

Reference	Dataset description	Model utilized/proposed	Modality	Performance measures	Summary
Rabiei et al. [26]	• Breast cancer DB from Motamed cancer institute, Tehran, Iran • Total instance: 5178 • Training/test: 75:25	RF, MLP, GBT, GA	Imaging	ACC, SEN, SPE, AUC, ROC, CM	• Considered multiple features such as 24 demographic, laboratory, and mammographic features • Weighted features based on importance • RF achieved the highest ACC (80%) and SEN (95%), GB highest AUC (0.59), and SPE (87%)
Rasool et al. [41]	• WDBC and BCCD • Total instance: 569 (WDBC) and 116 (BCCD)	Polynomial SVM, LR, KNN, Ensemble classifier	Numerical dataset	PRE, RE, F1-score, ACC, TP, TN	• Computed correlation between two features using PCC • Important features selected using recursive features elimination

(continued)

Table 8.1 (continued)

Reference	Dataset description	Model utilized/proposed	Modality	Performance measures	Summary
Sharma et al. [42]	• WDBC • Training/Test: 80:20	DT, Gaussian NB, MLP, AdaBoost	Numerical dataset	ROC, AUC, SPE, SEN, F1-score, ACC	• Tenfold cross-validation for better selection of the model • Ensemble modeling to enhance classifier performance • Proposed method with ACC (97.66%), PRE (92%), SEN (93.49%), SPE (91.07%), and F1-score (92.73)
Uddin et al. [43]	• WDBC • 569 instances	SVM, RF, KNN, DT, BV, LR, AdaBoost, GBT, MLP, Near cluster and Voting classifiers	Numerical dataset	Error rate, ACC, PRE, F1-score, RE, CM	• Feature scaling and feature optimization using PCA for robust feature extraction • GridSearchCV for tuning hyperparameters • Highest ACC (98.77%) for ensembled LR and SVM

EHR: Electronic health record, *WDBC:* Wisconsin diagnosis breast cancer, *BCCD:* Breast cancer Coimbra dataset, *NB:* Naïve Bayes, *DT:* decision tree, *DA:* Discriminant analysis, *SVM:* Support vector machine, *KNN:* K-nearest neighbor, *RF:* Random forest, *LR:* Logistic regression, *MKL:* Multiple kernel learning, *RO:* Random optimization, *LDA:* Linear discriminant analysis, *SMO:* Sequential minimal optimization, *MLP:* Multi-layer perceptron, *GBT:* Gradient boosting trees, *GA:* Genetic algorithm, *PCA:* Principal component analysis, *ACC:* Accuracy, *PR:* Precision, *RE:* Recall, *HR:* Hazard ratio, *AUC:* Area under the curve, *SEN:* Sensitivity, *SPE:* Specificity, *CI:* Confidence interval, *CM:* Confusion matrix, *TP:* True positive, *FP:* False positive, *ROC:* Receiver operating characteristics, *SD:* Standard deviation, *PCC:* Pearson's correlation coefficient seven ML algorithms namely, LR, AdaBoost, stochastic gradient descent, RF, SVM, KNN, and DT individually and jointly. Preprocessing step for feature scaling and imputation of missing values utilized for efficient results. The relationship between feature subset and labeled data was examined using Spearman's correlation and dissimilarity in feature data distribution was analyzed using the Wilcoxon rank sum test. Pearson correlation coefficient (PCC) for feature selection to classify the tumors into benign and malignant. Similarly, authors selected fifteen sensitive input features for breast cancer prediction using PCC for ML-modal [24]. Z-score based dataset standardization was adopted to improve the dataset quality. Authors utilized DT, and NB along with sequential minimal optimization for breast cancer detection [21]. During the preprocessing step, information with missing values was removed and data resampling was done to maintain the class distribution. Tenfold cross-validation was applied before the classification of breast cancer as benign or malignant. In [39], authors exploited five different ML techniques namely, SVM, RF, LR, DT, and KNN. Data cleaning, feature extraction, and selection steps were followed to prepare the datasets before processing them through ML algorithms. However, the handling of missing values and utilizing feature extraction techniques was not discussed.

Further, authors have exploited ML classification techniques such as NB, LR, SVM, KNN, and DT along with ensemble techniques such as RF, Adaboost, and XGBoost to obtain better accuracy [40]. Standard scaling for feature scaling and label encoding for converting categorical values to numerical were adopted during preprocessing steps to enhance dataset quality. Ensembled bagging and boosting techniques ensured better classification accuracy. Rabiei et al. [26] exploited ensembled techniques namely gradient boosting trees, RF, MLP along with genetic algorithms. Twenty-four features from demographics, clinical laboratory, and mammographic data were selected for effective prediction of breast cancer. Missing values were replaced either with maximum frequency or the same mod data. The class imbalance was addressed using synthetic minority oversampling techniques. Authors proposed an improved ML-based approach by including data exploration techniques namely, feature distribution, correlation, elimination, and hyperparameter optimization for efficient and effective breast cancer prediction [41]. After this, ML techniques namely, SVM LR, KNN, and ensembled classifier classified the tumor into benign or malignant. Similarly, ensembled approaches including, DT, AdaBoost, Gaussian NB, and MLP were examined for breast cancer prediction [42]. Tenfold validation was adopted for choosing the best model out of the considered ML techniques.

Uddin et al. [43] exploited eleven ML algorithms for cancer prediction. These algorithms were optimized using principal component analysis (PCA) and hyperparameter tuning was done using grid search. The most accurate and optimized algorithm was chosen to develop a webpage where real-time inputs can be taken for breast cancer prediction. However, the authors investigated two serial mammograms to predict the vivo rate of tumor growth [25]. The tumor traces were missing in the first mammograms and the time interval between the two mammograms was recorded for the proper diagnosis. Based on the outcomes, tumors were categorized as fast-growing and slow-growing in two subgroups. On the other hand, Botlagunta et al. [11] analyzed the blood profile data and socio-demographic data from the

EHR collection for predicting breast cancer. Multiple ML algorithms along with text mining were adopted to determine breast cancer patients so that intensive care can be extended to improve the survival rate. Next-generation sequencing from gene information was proposed for the prediction of breast cancer [34]. Nine ML-based algorithms were adopted to process the features extracted from gene sequencing data extracted from humans. The DNA sequences were analyzed using DL techniques to predict cancer early and save lives.

To summarize, ML-based techniques can act as a powerful tool for accurate and effective prediction of breast cancer. Multiple ML algorithms along with an ensembled approach are recommended and adopted by many researchers to detect breast cancer in breast lesions. However, most of the work utilized the same datasets and techniques to make predictions. The generalizability and real-time deployment of these models for clinical practices are limitedly addressed. In addition, imaging data is rarely adopted to detect cancerous cells.

8.4 Deep Learning Algorithms for Breast Cancer Prediction

There is plenty of medical imaging data namely, USd, MRI, and mammogram available for prediction of breast cancer. It is essential to analyze this data microscopically to fight breast cancer diseases. These images are of poor quality, with varying contrast, and resolution. The images are not clear and the segmentation of cancerous cells from the neighboring tissues is challenging. DL-based algorithms have been proposed to segment these images accurately and efficiently. For automatic segmentation of breast imaging for cancer prediction, DL-based algorithms are categorized either as UNet-based DL models or non-UNet-based DL models. The details about both of these categories are elaborated in the following sections.

8.4.1 UNet-Based Deep Learning Predictive Algorithms

UNet and its variant networks are widely explored for image segmentation in medical systems. UNet architectures are not only lightweight but also captures the contextual and spatial features efficiently. Table 8.2 tabulates the representative work exploiting UNet and its variants for breast cancer prediction. The utilized methodology, along with datasets and performance metrics details are also extracted to determine the potential of existing work.

For segmenting breast cancer image UNet [44] and its variants such as 3D UNet models with transfer learning [18], 3D inception UNet [29], attention dense-UNet [30], Asymmetric U-Shape network (Aym-UNet) [8], and many more [2, 13, 19, 20] are proposed by various researchers. In [2], authors improved the BGRD-TransUNet

Table 8.2 Representative work in UNet and its variants for breast cancer prediction

Reference	Dataset description	Model utilized/proposed	Modality	Performance measures	Summary
Chen et al. [6]	• Private (DB1), Public (DB2) • Total: 878 (DB1), 562 (DB2) • Training/Validation/Test:3/1/1	Attention gate and dilation UNet	USd imaging	DSC, AER, Hausdorff's error, MAE	• Attention gates in skip connections eliminate noise in spatial and channel features • Crucial contextual information captured by integrating attention dilation between encoder and decoder • Parameters reduction improved computational complexity and training speed
Ortega-Ruíz et al. [20]	• BCSS DB • Total images: 12,930 • Training/Validation/Test: 10,735/2194/15	Dilation, Residual, Dense block-UNet (DRD-UNet)	Histopathological images	DSC, Jaccard similarity index, ACC, SEN, and SPE	• Included dilated convolutions, residual connections, and dense layer to improve segmentation results • Multi-class classification of segmentation results for a better prognosis of breast cancer • Compared results with fifteen other UNet variations to prove the efficiency of the DRD UNet

(continued)

Table 8.2 (continued)

Reference	Dataset description	Model utilized/proposed	Modality	Performance measures	Summary
Ji et al. [2]	• BUSI (DB1), DB2 • Total: 780 (DB1), 163 (DB2) • Malignant/Benign/ Normal: 210/437/133 (DB1) • Training/Validation/Test: 8:1:1 (DB1) and 8:2 (DB2)	TransUNet	USd Imaging	DSC, mean IoU	• Residual multi-scale feature module for robust feature extraction from the DenseNet121 module • Boundary attentional feature fusion module for integrating edge information in extracted feature • Parallel channel and spatial attention modules for refinement of extracted features
Xia et al. [7]	• MRI DB • Total: 260 cases	Shape enhanced UNet with transformer encoder layer (SETE-UNet)	MRI imaging	Mean IoU, PR, ACC, DSC, SEN, SPE, AUC and Hausdorff distance	• Early-stage breast tumor MRI images for segmentation including men and women • Transformer encoder layer with the global self-attention mechanism for automatic segmentation • Shape boundary for accurate segmentation of tumor cells in the presence of background noises

(continued)

Table 8.2 (continued)

Reference	Dataset description	Model utilized/proposed	Modality	Performance measures	Summary
Liu et al. [8]	• BUSI (DB1), DB2 • Total: 645 (DB1), 163 (DB2) • Malignant/Benign/Normal: 210/437/133 (DB1)	Asymmetric UNet	USd Imaging	Jaccard similarity, DSC, ACC, RE, and PR	• Incorporated attention-based multi-branch residual encoder to improve network capabilities • Deep supervised boundary detection for accurate segmentation of tumor lesions • Multi-scale feature extraction to improve the diversity of feature maps
Minh et al. [19]	• Breast USd imaging DB • Total: 780 • Training/testing: 8/2	UGGNet	USd Imaging	ACC, RE, and F1-score	• Integrated the benefits of both UNet and VGG for accurate segmentation of breast cancer lesions • Exploited variants of UNet for feature extraction and variants of VGG for classifications • Achieved ACC(78.21%), RE(0.78) and F1-score (0.77) and quite fast training speed
Robin et al. [46]	• Kaggle dataset with 115 images • Training/testing: 8/2	UNet	Histopathological images	Learning curve and ACC	• Resizing done as preprocessing step to reduce processing time • Feature extraction and image segmentation using UNet model

(continued)

Table 8.2 (continued)

Reference	Dataset description	Model utilized/proposed	Modality	Performance measures	Summary
Li et al. [30]	• DDSM from the University of Florida • 400 mammography X-ray • Training/validation/test: 4/1/1	Dense-UNet	X-ray images	F1-score, mean IoU, SEN, SPE, ACC and AUC	• Automatic image segmentation using modified UNet with densely connected attentional gates • Network with encoder and decoder for robust feature extraction • Obtained ACC (78.38%), AUC (0.806), F1-score (82.24%), and SEN (77.89%), approximately
Zhuang et al. [45]	• 131 breast USd imaging from SUMC	GRA UNet	2D image slices	ACC, SEN, SPE, PR, mean IoU, DSC, and F1-Measure	• Utilized AWBUS images to segment the nipple and localize the tumor • Manually segmented ground truth of the imaging data • Modified UNet for better accuracy of nipple segmentation for tumor identification • 32 groups of convolution operations were applied to each input image

(continued)

Table 8.2 (continued)

Reference	Dataset description	Model utilized/proposed	Modality	Performance measures	Summary
Wang et al. [29]	• Automatic breast cancer USd from Sun Yat-Sen University Cancer Center • 559 volumes of data with 661 cancer regions	3D Inception UNet	USd Imaging	SEN, IoU, and Fp/ automatic breast cancer USd/volume	• 3D inception UNet consisted of two inception blocks and multiple deep supervised branches • Utilized asymmetric loss to balance FP and FN regions • Relatively faster methodology with 0.1 s for an automated breast USd volume
Vianna et al. [27]	• 2054 images from the National Cancer Institute • Benign/malignant: 1351/ 703 • Training/validation/test: 7/ 1/2	UNet and SegNet	USd Imaging	DCE, and p-value	• Implemented two CNN networks for segmentation of lesions for breast cancer detection • UNet accurate, and faster for lesion segmentation • Multiple kernel size and variable learning rate with different activation functions for model implementation

(continued)

Table 8.2 (continued)

Reference	Dataset description	Model utilized/proposed	Modality	Performance measures	Summary
Khaled et al. [28]	• 46 cases from The Cancer Imaging Archive	Ensemble UNet	4D DCE-MRI	DSC, Hausdroff distance, FPR	• Automated lesion segmentation for tumor detection • Combined inputs from three different methods for accurate segmentation results • Preprocessing and patch sampling for faster training results
Tong et al. [47]	• 316 USd images and 316 mask images from traditional Chinese Medicine hospital • Malignant/Benign: 145/171	UNet MALF	USd imaging	PR, RE, F1-score, AUC, ACC, Mean IoU, SEN, SPE, DCE, and Training time	• Combined residual block and mixed attention loss function for handling low brightness and contrast in USd imaging • Integrate four attention loss functions into cross-entropy loss functions for accurate tumor localization • Residual convolution and extended residual convolution module replaced the convolution module for robust feature extraction

(continued)

Table 8.2 (continued)

Reference	Dataset description	Model utilized/proposed	Modality	Performance measures	Summary
Meraj et al. [44]	• BUSI (DB1) and OASBUD (DB2) • Toal/Benign/Malignant: 647/437/210 (DB1) and 100/52/48 (DB2)	UNet with ICA	USd Imaging	ACC, PR, RE, F1-score, Kappa, DCE, mean IoU and CM	• Two step segmentation using UNet and quantization for accurate tumor lesion detection • Automatic fusion and extraction of features using ICA • Lasso method-based feature classification
Tekin et al. [13]	• WSI of breast cancer • 51 WSIs from 51 different patients • 30,820 polygonal annotated tubules in 8225 patches	Tubule-UNet	Patch-based imaging dataset	DCE, RE, and SPE	• Adopted tubal image segmentation for the detection of breast cancer • Mirror padding strategy to enhance boundary of patches to obtain accurate segmentation results • Three different tubule segmentation models to obtain patch-based semantic segmentation for reliable and fast results

(continued)

Table 8.2 (continued)

Reference	Dataset description	Model utilized/proposed	Modality	Performance measures	Summary
Byra et al. [48]	• UDIAT (DB1), OASBUD (DB2), and BUSI (DB3) • Total/Benign/Malignant: 163/110/53 (DB1), 100/48/52 (DB2), and 630/421/209	Selective kernel UNet	USd Imaging	DCE, Spearman's rank coefficient, ACC, AUC, and detection rate	• Selective kernel adjusted network receptive field automatically using an attention mechanism • Fused feature maps with dilated and conventional convolutional networks • Dilated convolution for segmentation of larger breast masses
Sivamurugan and Sureshkumar [9]	• Kaggle dataset with 324 mammography images	LSTM with UNet	Mammography Imaging	ACC, F1-score, PR, SEN, SPE, FPR, FDR, MCC, and NPV	• Two different methods to classify breast cancer • Hyperparameters such as epochs and activation function optimization using A-BWO • Optimized LSTM for better accuracy and fast efficiency

(continued)

Table 8.2 (continued)

Reference	Dataset description	Model utilized/proposed	Modality	Performance measures	Summary
Punn and Agarwal [22]	• BUSIS (DB1) and BUSI (DB2) • Images: 562 (DB1), 780 (DB2) • Training/testing: 7/3	RCA-IUNet	USd imaging	ACC, PR, RE, DCE, mean IoU, average Hausdorff distance, and MAE	• Employed long and short skip connections for binary segmentation mask of the tumor • Long skip connections in cross-attention filter to generate attention for decoding feature maps • Hybrid pooling operation for efficient pooling of feature maps in two modes
Khalil et al. [18]	162 images Total IDC patches: 277524 Positive/negative patches: 78786/198738	3D UNet	Histopathology images	ACC, and CM	• Bipolar interpolation for resizing the image for estimating pixel value • Intensity normalization for addressing intensity variations in different images • Equipped 3D UNet with max pooling, up-sampling, and transposed convolution for cancer detection

DB: Database, *BCSS:* Breast cancer semantic segmentation, *DDSM:* Digital database for screening mammography, *AWBUS:* Automated whole breast ultrasound, *USd:* Ultrasound, *WSI:* Whole slide images, *SUMC:* Shantou University Medical College, *BUSIS:* Breast ultrasound image segmentation, *BUSI:* Breast ultrasound images, *OASBUD:* Open access database of raw ultrasonic signals, *GRA:* Grouped-resaunet, *RCA-I:* Residual cross-spatial attention guided inception, *MALF:* Mixed attention loss function, *ICA:* Independent component analysis, *LSTM:* Long short term memory, *DCE-MRI:* Dynamic contrast enhanced magnetic resonance imaging, *IDC:* Invasive ductal carcinoma, *ACC:* Accuracy, *PR:* Precision, *RE:* Recall, *HR:* Hazard ratio, *AUC:* Area under the curve, *SEN:* Sensitivity, *SPE:* Specificity, *CI:* Confidence interval, *CM:* Confusion matrix, *IoU:* Intersection over union, *DSC:* Dice similarity coefficient, *AER:* Area error ratio, *MAE:* Mean average error, *MCC:* Matthew's correlation coefficient, *FPR:* False positive rate, *FP:* False positive, *FN:* False negative by replacing the backbone network ResNet50 with DenseNet121. Boundary guidance strategy was adopted to extract contour information from breast imaging efficiently. Robust features from different layers of DenseNet121 were extracted using residual multi-scale feature modules. The edge information from the boundary guidance strategy and extracted features were integrated using attentional feature fusion modules. Similarly, authors exploited boundary detection module to refine the lesion segmentation results [8]. This module refined the segmentation capabilities of the network. Multi-branch encoder and external attention module in the decoder were implemented for capturing discriminative features and filtering out noise in the final segmentation outcomes. Li et al. [30] proposed UNet based method for the automatic segmentation of breast mass from a mammography database. The encoder was feature extraction layer with dense CNN and the decoder was UNet with attentional gates. Attentional gates enhanced the efficiency of UNet during breast segmentation. On the other hand, nipple regions were segmented with high accuracy by using modifying UNet as Grouped-Resaunet (GRU) UNet [45]. The model architecture consisted of five encoders for extracting hierarchical features and a strong skip connection to fuse the extracted features with the corresponding decoder layer for precise segmentation of breast imaging for cancerous cells. The residual layer was deployed to reuse activations from the previous layer to the adjacent layer for learning weights.

Conventional breast imaging techniques represent the details in 2D which could not capture the whole breast details efficiently. In addition, these techniques required specialized and experienced operators to capture the infected regions accurately. To address the limitations of 2D imaging, 3D-view breast imaging is recommended. However, the analysis of 3D imaging requires thorough examination for detection of cancerous cells and is time-consuming too as the number of images is quite large. For fast and effective evaluation of 3D breast imaging for detection of cancerous cells various DL-based techniques are investigated [18, 29]. In [18], authors processed 3D histopathological data using 3D UNet models along with transfer learning for the classification of ductal carcinoma for breast cancer. During the preprocessing step, images were fine-tuned by normalizing the brightness and their size. However, authors have analyzed USd data using 3D Inception UNet with asymmetric loss [29]. Asymmetric loss balanced the false positive and negative regions in the DL network for improving accuracy for small cancerous lesions. Features were extracted at each layer and concatenated in deep supervision blocks to improve the prediction accuracy. However, the authors improved the UNet architecture by integrating mixed attention loss function for lesion segmentation in USd images [47]. Residual convolution and extended convolution modules were used for extracting the features and four loss functions were integrated into the texture consistency index of the feature map to improve the segmentation accuracy. On the other hand, authors exploited 4D dynamic contrast-enhanced MRI based on UNet for automatic segmentation of breast lesions [28]. Three different UNet with different combinations of input were adopted to improve the breast cancer diagnostic accuracy.

Breast lesion segmentation for the prediction of cancerous cells is challenging due to varying intensity levels. To address this, Meraj et al. [44] proposed quantization-assisted UNet for the exact segmentation of lesions in sonographic images. The methodology isolated the lesion for feature extraction using the independent component analysis and fused the extracted feature using DenseNet201. Different size dataset images were resized using the nearest neighbor method and data augmentation techniques such as flipping, shifting, sharpening and adjustments were adopted for transforming the images with better resolution and contrast. Authors exploited dual models in optimized long short-term memory (LSTM) with UNet for performing the breast lesion segmentation [9]. Mammography datasets were processed through three steps namely, median filtering, histogram equalization, and morphological operations to remove the noise in the images. Features were extracted by optimizing the UNet parameters using adapted-black widow optimization. Three deep models namely, VGG19, Resnet150, and Inception were used for feature extraction, and fused features were fed to dual model optimized LSTM for obtaining the prediction scores.

With advancements in DL technologies segmentation accuracy in biomedical imaging has increased. To provide enhanced segmentation accuracy on histopathological images for breast cancer prediction, authors modified UNet DL-based architecture as DRD-UNet (Dilation, residual and dense) [20]. This architecture was comprised of three blocks namely dilated convolution, residual connections, and dense layers. The performance was compared against sixteen others UNet architecture proved the effectiveness of the method. To improve the prediction speed with the minimum number of training parameters, the authors proposed residual cross-spatial attention-guided inception UNet (RCA-IUNet) [22]. The architecture utilized cross spatial attention filter to eliminate irrelevant features and residual inception for depthwise separable convolution with hybrid pooling layers with short skip connections. The model achieved an inference time of 18.75 ms for generating results. Authors utilized UNet and SegNet for automatic segmenting ultrasonography images for breast cancer prediction to reduce the number of biopsies [27]. Self-trained network was used to classify the pixels in the breast images for predicting the cancerous cells. On the other hand, Tekin et al. [13] exploited Tubule-UNet by using patch enhancement techniques to improve the input image quality and asymmetric encoder-decoder semantic segmentation model for segmenting breast cancer. The encoder used for feature extraction and comprised of three different asymmetric DL models namely, EfficientNetB3, ResNet34, and DenseNet161 while the decoder architecture was the same as UNet. The model was precise and accurately deployed for real-time validation on a webserver where users can upload images and segmented tubules are generated as outcomes. The next section will detail non-UNet based DL models for segmenting breast imaging.

8.4.2 Non-UNet-Based Deep Learning Predictive Algorithms

Table 8.3 tabulates the salient features of the representative work exploiting DL methods for detecting breast cancer. Accurate, automatic, and fast detection of breast cancer is very essential to prevent its spread in the body impacting the neighboring tissues. DL-based networks have great potential in detecting cancerous cells from various imaging data such as USd, MRI, and X-ray mammography and preventing painful biopsies. In this direction, authors classified breast cancers into eight different classes by analyzing breast MRI with five different fine-tuned DL models [4]. The models were pre-trained on the ImageNet database consisting of images with multiple magnifications. The model was trained, validated, and executed on multiple datasets to ensure its generalizability. Yala et al. [10] utilized a hybrid DL model comprising of patient risk factors from EHR and mammogram imaging data. The model analyzed the 5-year medical data of the patient, identified the patterns, and predicted the risk of breast cancer in future.

In another line of research for automatic detection of breast cancer, authors integrated patch-based learning in deep belief networks (DBN) [49]. The histopathological images were utilized for the classification of cancer using path-based DBN. Features were extracted automatically using unsupervised pre-training and supervised fine-tuning. After this, the model processed the images and classified them either as cancerous cells or background. The authors utilized fuzzy merging techniques with the Deep-CNN model for classifying the cells into benign and malignant [50]. Breast cancer tissues were segmented using multilevel saliency nuclei detection and the segmented regions were merged using fuzzy-based statistical regions. On the other hand, authors utilized whale optimization algorithms (WOA) in a DL-based network for accurate classification of cancerous cells into malignant and benign [51]. The images were preprocessed and adjusted for their processing through CNN. The parameters were optimized using WOA maintaining high classification accuracy and processing speed. However, the authors exploited Kera-Tuner optimization to optimize the deep RNN [52]. The optimization technique comprised Bayesian optimization, hyperband, and random search algorithms to tune the hyperparameters of RNN for optimized performance. Feature selection was done using three techniques namely, correlation methods, univariate feature selection, and recursive feature elimination. These techniques not only support in selection of robust features but also reduce the number of features to improve the model processing.

To address the limitations in the automatic segmentation of breast imaging for cancer prediction, authors extracted contextual information from conditional generative adversarial learning framework [32]. The texture features were integrated with contextual information to capture the spatial and semantic features efficiently. The essential features for tumor detection eliminating the effects of the artifact were selected using the channel attention with channel weighting mechanisms. For better classification accuracy, the background information was captured using the structural similarity index metric and L1-norm in the loss function. Siddiqui et al. [55] multimodal imaging data and decision-based fusion in a DL model. The model was trained on multiple datasets and decision fusion along with fuzzy logic to improve the classification accuracy in fused images. Also, the authors analyzed multimodal data for breast cancer prediction in two phases in the attention-based DL model [54]. During the first phase, stacked features were generated using sigmoid gated attention CNN while the second phase applied flattened, dense, and dropout layers

Table 8.3 Representative work in DL-based breast cancer prediction

Reference	Dataset description	Model utilized/proposed	Modality	Performance measures	Summary
Chai et al. [17]	• Seven datasets from GEO and TCGA • Total instances: BRCA(609), GSE2990 (125), GSE9195 (77), GSE11121 (200), GSE17705 (298), GSE19615 (115), GSE25066 (508) and BRCA_all (1932)	UISNet	Multi-modal	CI, AUC, SEN, Kaplan–Meier survival curve	• Uncertainty-based strategy along with integrated gradients of feature extraction to incorporate interpretability in the model • Binary adjacency matrix for representing sparse connections between genes and functional pathways • Computed gene importance to determine its contribution to cancer prognosis
Zeng et al. [15]	Dataset from three hospitals Total: 261 (DB1), 107 (DB2) and 72 (DB3)	Transfer learning-based CNN	Multi-modal	ACC, SPE, SEN, ROC, AUC and F1-score	• Feature extraction on pathological images with different magnifications • SVM classifier to predict pathological complete response for optimizing neoadjuvant chemotherapy • Clinicopathological features for designing the clinical model

(continued)

Table 8.3 (continued)

Reference	Dataset description	Model utilized/proposed	Modality	Performance measures	Summary
Wang [14]	Multiple datasets	CNN-based DL models	X-ray Mammography	ACC, SEN, SPE, AUC, PR, F1-score, MAE and MSE	• Examined various DL-based models for the detection of breast cancer • Model privacy, generalizability, and interpretability need to be addressed for potential DL models • Analyzed the potential of various DL models for the prediction of breast cancer
Mustafa et al. [12]	• METABRIC dataset • Total samples: 1980	CNN, DNN and LSTM	Multi-modal	SEN, SPE, PR, ACC, ROC, CM, and AUC	• CNN, DNN, and LSTM handled clinical data, copy number variations, and gene expression data, respectively • Utilized minimum redundancy and maximum relevance for dimensionality reduction • Incremental approach for robust feature selection
Zhang et al. [53]	GSE2990, GSE3494, GSE9195, GSE17705, and GSE17907	Ensembled DL model with LDA, and AE	Numerical	AUC, MCC, F-Measure, ACC	• Two-phase unregulated learning for feature extraction • Utilized ReLU activation and descent gradient as error function • Approach obtained overall ACC (98.27%)

(continued)

Table 8.3 (continued)

Reference	Dataset description	Model utilized/ proposed	Modality	Performance measures	Summary
Hirra et al. [49]	• Real-world cancer datasets: HUP, CWRU, CINJ and TCGA • Total images: 239 (HUP), 110 (CWRU), 40 (CINJ), 195 (TCGA) • Training/test: 75:25	Patch-based DL and DBN	Histopathology imaging	ACC, ER, SEN, SPE, FP	• Utilized patch matching to create probability estimation matrix • Mini batch for fine-tuning and Sigmoid for activation • Achieved ACC (86%), ER (14%), SEN (87.9%), SPE (84%) and FP (15.9%)
Kayikci and Khoshgoftaar [54]	• METABRIC DB • Total patients: 1980	DL-based CNN attentional bimodal	Multi-modal	AUC, ROC, RE, PRE, FP, SEN, ANOVA test	• Two phase model for stacking and flattening of features for robust prediction • Utilized ReLU activation function and GNN for assigning initial values to kernels • Achieved ACC (91.2%), AUC (0.95), PRE (0.841) and SEN (0.798)

(continued)

Table 8.3 (continued)

Reference	Dataset description	Model utilized/ proposed	Modality	Performance measures	Summary
Krithiga and Geetha [50]	• Real-World Cancer datasets: IUHPL, TMC, TCIA And Breakhis • Total image: 9109 From 82 patients • Benign/malignant: 2480/5429 images	Deep CNN with fuzzy merging	Histopathology imaging	ACC, SEN, SPE, PRE, IoU, MCC	• Image enhancement using anisotropic diffusion and multi-level saliency nuclei detection for reinforcing accuracy in segmentation • Utilized ReLU activation and expectation maximization for parameter optimization • Achieved ACC (98.6%), SEN (0.947), SPE (0.964), F1-score (0.935), MCC (0.913), and AUC (99.4)
Rana et al. [51]	• BreakHis dataset • Total images: 7909 from 82 patients • Benign/malignant: 24/58 patients	DL-based whale optimization	Histopathology imaging	PRE, RE, F1-score	• Applied data augmentation techniques such as shifting, rotation, zooming, and flipping • Softmax activation and whale optimization for optimizing parameters and automatic cancer prediction
Saleh et al. [52]	• WDBC dataset • Training/testing: 80:20	Optimized deep RNN	Numerical dataset	ACC, PRE, F-Measure, RE	• Correlation for univariate feature selection and recursive feature elimination • ReLu activation and Keras-Tuner optimization for parameter optimization • Achieved ACC (95.18%), PRE (95.44), RE (95.18), F-Measure (95.21) during the testing phase

(continued)

Table 8.3 (continued)

Reference	Dataset description	Model utilized/proposed	Modality	Performance measures	Summary
Siddiqui et al. [55]	• Image and feature dataset • Total samples: 7909/569 • Training/testing: 80:20	Deep extreme learning machine and CNN	Multi-modal	ACC, Miss rate	• Managed uncertainty and imprecision in dataset using fuzzy logic • Fusion-based multi-modal dataset for better prediction accuracy • Achieved ACC (97.97%) and Miss rate (2.03%) during the validation stage
Yala et al. [10]	Total: 88,994 from 39,571 women Training/validation/test: 71689/8554/8751	Hybrid DL	Mammography imaging data	AUC, CI, and p-value	• Included socio-demographics such as ethnicity, age, and other parameters in the analysis • Implemented risk-factor only model to assess patient's risk for cancer redevelopment • Single dataset and single mammogram impacted the generalizability of the model
Singh et al. [32]	Mendeley (DB1), UDIAT (DB2) Malignant/Benign: 150/100 (DB1) Total/tumor/healthy: 267/163/104 (DB2)	Deep adversarial learning network	Imaging data	DCE, mean IoU,	• Captured spatial and scale context using atrous convolution to handle varying sizes of tumors • Channel attention with channel weighting to promote specific tumor-relevant features • L1-norm loss function for capturing local context information from surroundings

(continued)

Table 8.3 (continued)

Reference	Dataset description	Model utilized/ proposed	Modality	Performance measures	Summary
Zhou et al. [31]	ABUS data from Peking University People's Hospital Total/tumor/healthy: 158/83/75	Faster R-CNN	3D Imaging data	RE, PR, FP/ Volume, and SEN	• Concatenated conv3 and conv5 layers features for generating feature maps • Normalized features before processing in the model • Fivefold cross validation for 2D and 3D detection for enhanced detection performance
Moon et al. [56]	SNUH (DB1), BUSI (DB1) Total/Benign/Malignant: 1687/953/ 734 (DB1) Total/Benign/Malignant/Normal: 697/ 437/210/133 (DB2)	VGGNet, ResNet, and DenseNet	USd Imaging from different machines	ACC, SPE, SEN, PR, RE, AUC, ROC and F1-score	• Ensembled architecture for obtaining the best accurate classification • CAD approach for tumor detection by image fusion • ROI extraction from the tumor region and generated 3-channels fused images from different images

(continued)

Table 8.3 (continued)

Reference	Dataset description	Model utilized/ proposed	Modality	Performance measures	Summary
Abunasser et al. [4]	BreakHis dataset from Kaggle Total/Benign/Malignant: 7909/2480/5429 Training/validation/testing: 6:2:2	Breast cancer CNN with Xception, InceptionV3, VGG16, MobileNet and ResNet50	MRI Imaging dataset	ACC, PR, RE, F1-Score, and training time	• Imaging dataset with varying magnifying factors • Ensembled approach with multi-class classification of breast tumor cells • Dataset boosting using GAN to increase dataset size to 40,000 images
Boulenger et al. [16]	Dataset from Peking Union Medical College Hospital Total images: 183 (from 145 patients) Training/validation/testing (images): 8(654)/1(92)/1(86)	CNN with VGGNet	USd imaging	ACC, SEN, SPE, AUC, CI, PPV, NPV and F1-Score	• Utilized adaptive histogram equalization for addressing varying pixel intensity in images • Undersampling of the dominant class to address class imbalance issues • Small dataset and collection from single center restricted the generalizability of the model

DB: Database, *DL:* Deep learning, *LDA:* Linear discriminant analysis, *CAD:* Computer-aided diagnosis, *AE:* Auto encoder, *DBN:* Deep belief network, *HUP:* Hospital of the University of Pennsylvania, *CWRU:* Case western reserve university, *CINJ:* Cancer institute of New Jersey, *SNUH:* Seoul National University Hospital, *TCGA:* The cancer genome Atlas, *IUHPL:* Indiana university health pathology laboratory, *TMC:* Tata medical center, *TCIA:* the cancer imaging archive, *BreakHis:* Breast cancer histopathological database, *GNN:* Glorot normal initializer, *RNN:* Recurrent neural network, *LSTM:* Long short term memory, *UISNet:* Uncertainty based interpretable deep neural network, *DNN:* Deep neural network, *ROI:* Region of interest, *AUC:* Area under the curve, *MCC:* Matthew's correlation coefficient, *GAN:* Generative adversarial network, *ReLU:* Rectified linear unit, *ACC:* Accuracy, *ER:* Error rate, *FP:* False positive, *SEN:* Sensitivity, *SPE:* Specificity, *IoU:* Intersection over union, *MSE:* Mean squared error, *MAE:* Mean absolute error for bi-model attention. The patient's socio-demographic data such as age and family history were also integrated to improve prediction accuracy.

Whole breast USd has a better prediction rate in comparison to traditional imaging. However, manual analysis of whole breast USd requires specialized expertise to detect subtle tumors. For this, authors proposed a faster R-CNN connected feature extraction network to detect tumor from 3D multi-view breast cancer USd volumes [31]. High-level features from conv3 and conv5 layers were integrated to generate fused features containing detailed descriptions for tumor prediction. However, Boulenger et al. [16] integrated CNN with the VGG network to predict triple-negative breast cancer in USd imaging with bad diagnosis results. The image contrast and intensity were enhanced by using multiple normalization and equalization algorithms to improve segmentation accuracy. For each patient images were classified independently to ensure model generalizability and t-distributed stochastic neighbor embedding analysis and saliency maps for visualizing the model interpretability. The authors utilized a DL network termed UISNet (uncertainty-based interpretable deep semi-supervised network) to interpret the important features [17]. Patient's heterogeneous information was considered to extract essential features and Monte Carlo dropout to improve the reliability of the extracted outcome. Sparse layer was introduced to process high-dimensional gene expression data for the prediction of breast cancer.

Ensembled learning integrates multiple models to improve the prediction accuracy and better classification rate [12, 53, 56]. In [12], authors ensembled three DL models namely, CNN, DNN, and LSTM. CNN extracted features from clinical modalities, DNN to handle copy number variations and LSTM to address high dimensional gene expression data. The outcomes from each individual network were integrated to predict the final accuracy. Also, Minimum redundancy maximum relevance were adopted for selecting effective features to improve model training performance. However, the authors exploited advanced ensembled classification approach to classify gene expression data for traces of breast cancer [53]. Linear discriminant analysis and autoencoder classifier to classify different features based on gene expressions for effective diagnosis of breast cancer. Moon et al. [56] ensembled different CNN networks namely, VGGNet, ResNet, and DenseNet to classify breast cancer into malignant and benign. Unweighted average, stacking weighted average, and voting were used as ensemble methods.

To summarize, nonUNet-based DL methods can classify the breast cancer from mammography, MRI and high dimensional data such as gene expression. These techniques also investigated ensembled approach by integrating multiple models to ensure the high diagnostic accuracy. The techniques were adopted for selecting essential features to ensure the fast-training capabilities of the model. The robust performance metrics ensured the effectiveness of these methods in classifying brain tumors into malignant and benign.

8.5 Comparison of Various Breast Cancer Prediction Techniques

We have reviewed breast cancer prediction techniques such as ML-algorithms, UNet algorithms, and non-UNet algorithms. The salient features of each of the techniques are tabulated in Table 8.4. The various parameters are discussed and reviewed, and performance is analyzed to provide the future perspective for each technique for breast cancer classification and segmentation.

It has been observed that methods under each category have their advantages and limitations. There does not exist any definite method for the prediction of breast cancer prediction but a combination of these methods exploiting patient's sociodemographic information and blood test reports in the form of EHR and imaging data collected using USd, MRI, and mammography can be used for early and accurate diagnosis of breast cancer. ML and DL-based models are helpful for radiologists to analyze imaging modalities at a faster rate [57–59]. The subtle cancer traces which may not be manually detected, can be easily detected with advanced ML and DL-based techniques. These techniques also addressed the limitations of lower contrast, poor resolution and intensity problems of medical imaging captured through various hardware devices. Breast abnormalities are easily detected which can be further classified into multiple classes to examine the severity of cancer microscopically.

Table 8.4 Similarities/differences of various breast cancer techniques

Attributes	Segmentation/classification techniques		
Methodology	ML-algorithms	UNet algorithms	Non-UNet algorithms
Feature extraction	✓	✓	✓
Feature selection	Less	Moderate	High
Dataset	Textual and numerical	Imaging, textual, and numerical	Imaging, textual, numerical, and gene expression
Model complexity	Less	Moderate	High
Automation	✓	✓	✓
Computational resources	Less	Moderate	High
Generalizability	Limited	✓	✓
Interpretability	Limited	✓	✓
Performance	Moderately accurate	Highly accurate	Highly accurate
Limitations	Not suitable for clinical deployment	Moderately suitable for clinical deployment	Highly suitable for clinical deployment

8.6 Summary

In this chapter, we have analyzed various ML- and DL-based breast cancer prediction algorithms for early prediction of breast cancer in textual and imaging modalities. To identify breast abnormalities, microscopic examination of breast lesions along with their neighboring tissues is crucial. The key to improving the survival rate in breast cancer diseases is early diagnosis and proper treatment is very crucial. The study of other risk factors such as family history, hormonal details, smoking habits, weight, and breast density are helpful in assessing the chances of occurrence of this aggressive disease.

AI-based algorithms can predict breast cancer survival rates in a very fast and effective way. It is a non-invasive method that can diagnose breast cancer by analyzing various modalities such as USd, MRI, and mammography. These algorithms can prevent painful biopsies in patients and detect delays that can cause severity to increase. It also prevents long-term exposure to radiation which can impact the other tissues in the body and increase complications. In addition, various image enhancement techniques and noise removal methodologies are also adopted to improve the imaging quality for accurate and clear diagnosis of cancerous lesions in the breast which may not be possible in case of manual analysis. It also addressed the limited availability of experienced and trained radiologists who can analyze imaging modalities and predict breast cancer.

AI-based techniques have shown remarkable performance in the prognosis of breast cancer to reduce the mortality rate. However, the generalizability and interpretability of these models to realistic deployment are yet to be proved. To ensure generalizability large and multiple datasets are utilized so that maximum scenarios can be covered. Various techniques are adopted to address the black box design of DL models and make the model interpretable for enhanced performance. However limited data availability, privacy and modification constraints are certain hurdles that restrict generic performance evaluation and clinical deployments.

In future, AI-based models require to be more generalizable and interpretable for its world-wide acceptability. Utilization of multi-model data is also recommended to ensure unbiased performance of these models. In addition, inclusion of more vital statistics, gene expression examination and histopathological images in the training data can also be very helpful in predicting the occurrence and reoccurrence of breast cancer in humans.

References

1. Nassif AB, Talib MA, Nasir Q, Afadar Y, Elgendy O (2022) Breast cancer detection using artificial intelligence techniques: a systematic literature review. Artif Intell Med 127:102276
2. Ji Z, Sun H, Yuan N, Zhang H, Sheng J, Zhang X, Ganchev I (2024) BGRD-TransUNet: a novel TransUNet-based model for ultrasound breast lesion segmentation. IEEE Access

3. Tariq M, Iqbal S, Ayesha H, Abbas I, Ahmad KT, Niazi MFK (2021) Medical image based breast cancer diagnosis: state of the art and future directions. Expert Syst Appl 167:114095
4. Abunasser BS, Al-Hiealy MRJ, Zaqout IS, Abu-Naser SS (2023) Convolution neural network for breast cancer detection and classification using deep learning. Asian Pacific J Cancer Prev APJCP 24(2):531
5. Dalal S, Onyema EM, Kumar P, Maryann DC, Roselyn AO, Obichili MI (2023) A hybrid machine learning model for timely prediction of breast cancer. Int J Model Simul Sci Comput 14(04):2341023
6. Chen J, Shen X, Zhao Y, Qian W, Ma H, Sang L (2024) Attention gate and dilation U-shaped network (GDUNet): an efficient breast ultrasound image segmentation network with multiscale information extraction. Quant Imag Med Surg 14(2):2034
7. Xia Y, Liu W, Yang D, Wang H, Wang H, Jiang M (2024) Early stage tumor segmentation in breast MRI using shape enhanced U-Net. Biomed Signal Process Control 93:106198
8. Liu J, Shao J, Xu S, Tang Z, Liu W, Li Z (2024) Asym-Unet: an asymmetric U-shape network for lesion segmentation of breast cancer. Available at SSRN 4676212
9. Sivamurugan J, Sureshkumar G (2023) Applying dual models on optimized LSTM with U-net segmentation for breast cancer diagnosis using mammogram images. Artif Intell Med 143:102626
10. Yala A, Lehman C, Schuster T, Portnoi T, Barzilay R (2019) A deep learning mammography-based model for improved breast cancer risk prediction. Radiology 292(1):60–66
11. Botlagunta M, Botlagunta MD, Myneni MB, Lakshmi D, Nayyar A, Gullapalli JS, Shah MA (2023) Classification and diagnostic prediction of breast cancer metastasis on clinical data using machine learning algorithms. Sci Rep 13(1):485
12. Mustafa E, Jadoon EK, Khaliq-uz-Zaman S, Humayun MA, Maray M (2023) An ensembled framework for human breast cancer survivability prediction using deep learning. Diagnostics 13(10):1688
13. Tekin E, Yazıcı Ç, Kusetogullari H, Tokat F, Yavariabdi A, Iheme LO et al (2023) Tubule-U-Net: a novel dataset and deep learning-based tubule segmentation framework in whole slide images of breast cancer. Sci Rep 13(1):128
14. Wang L (2024) Mammography with deep learning for breast cancer detection. Front Oncol 14
15. Zeng H, Qiu S, Zhuang S, Wei X, Wu J, Zhang R et al (2024) Deep learning-based predictive model for pathological complete response to neoadjuvant chemotherapy in breast cancer from biopsy pathological images: a multicenter study. Front Physiol 15:1279982
16. Boulenger A, Luo Y, Zhang C, Zhao C, Gao Y, Xiao M et al (2023) Deep learning-based system for automatic prediction of triple-negative breast cancer from ultrasound images. Med Biol Eng Comput 61(2):567–578
17. Chai H, Lin S, Lin J, He M, Yang Y, OuYang Y, Zhao H (2024) An uncertainty-based interpretable deep learning framework for predicting breast cancer outcome. BMC Bioinform 25(1):88
18. Khalil S, Nawaz U, Zubariah Z, Mushtaq S, Arif MZ, ur Rehman et al (2023) Enhancing ductal carcinoma classification using transfer learning with 3D U-Net models in breast cancer imaging. Appl Sci 13(7):4255
19. Minh TC, Quoc NK, Vinh PC, Phu DN, Chi VX, Tan HM (2024) UGGNet: bridging U-net and VGG for advanced breast cancer diagnosis. arXiv preprint arXiv:2401.03173
20. Ortega-Ruíz MA, Karabağ C, Roman-Rangel E, Reyes-Aldasoro CC (2024) DRD-UNet, a UNet-like architecture for multi-class breast cancer semantic segmentation. IEEE Access
21. Mohammed SA, Darrab S, Noaman SA, Saake G (2020) Analysis of breast cancer detection using different machine learning techniques. In: Data mining and big data: 5th international conference, DMBD 2020, Belgrade, Serbia, 14–20 July 2020, Proceedings 5. Springer
22. Punn NS, Agarwal S (2022) RCA-IUnet: a residual cross-spatial attention-guided inception U-Net model for tumor segmentation in breast ultrasound imaging. Mach Vis Appl 33(2):27
23. Ferroni P, Zanzotto FM, Riondino S, Scarpato N, Guadagni F, Roselli M (2019) Breast cancer prognosis using a machine learning approach. Cancers 11(3):328

24. Chen H, Wang N, Du X, Mei K, Zhou Y, Cai G (2023) Classification prediction of breast cancer based on machine learning. Computat Intell Neurosci 2023
25. Bhattarai S, Klimov S, Aleskandarany MA, Burrell H, Wormall A, Green AR et al (2019) Machine learning-based prediction of breast cancer growth rate in vivo. British J Cancer 121(6):497–504
26. Rabiei R, Ayyoubzadeh SM, Sohrabei S, Esmaeili M, Atashi A (2022) Prediction of breast cancer using machine learning approaches. J Biomed Phys Eng 12(3):297
27. Vianna P, Farias R, de Albuquerque Pereira WC (2021) U-Net and SegNet performances on lesion segmentation of breast ultrasonography images. Res Biomed Eng 37:171–179
28. Vidal J, Vilanova JC, Martí R (2022) A U-net ensemble for breast lesion segmentation in DCE MRI. Comput Biol Med 140:105093
29. Wang, Y, Qin C, Lin C, Lin D, Xu M, Luo X et al (2020) 3D Inception U-net with asymmetric loss for cancer detection in automated breast ultrasound. Med Phys 47(11):5582–5591
30. Li S, Dong M, Du G, Mu X (2019) Attention dense-u-net for automatic breast mass segmentation in digital mammogram. Ieee Access 7:59037–59047
31. Zhou Y, Chen H, Li Y, Wang S, Cheng L, Li J (2021) 3D multi-view tumor detection in automated whole breast ultrasound using deep convolutional neural network. Expert Syst Appl 168:114410
32. Singh VK, Abdel-Nasser M, Akram F, Rashwan HA, Sarker MMK, Pandey N et al (2020) Breast tumor segmentation in ultrasound images using contextual-information-aware deep adversarial learning framework. Expert Syst Appl 162:113870
33. Kim G, Bahl M (2021) Assessing risk of breast cancer: a review of risk prediction models. J Breast Imag 3(2):144–155
34. Kurian B, Jyothi V (2021) Breast cancer prediction using an optimal machine learning technique for next generation sequences. Concurr Eng 29(1):49–57
35. Kumar A, Ahluwalia R (2021) Breast Cancer detection using machine learning and its classification. In: Cancer prediction for industrial IoT 4.0. Chapman and Hall/CRC, pp 65–78
36. Kumar A, Jain R (2021) Behavioral prediction of Cancer using machine learning. In: Cancer prediction for industrial IoT 4.0. Chapman and Hall/CRC, pp 91–105
37. Zhou S, Hu C, Wei S, Yan X (2024) Breast cancer prediction based on multiple machine learning algorithms. Technol Cancer Res Treat 23:15330338241234792
38. Kumar BS, Daniya T, Ajayan J (2020) Breast cancer prediction using machine learning algorithms. Int J Adv Sci Technol 29(3):7819–7828
39. Naji MA, El Filali S, Aarika K, Benlahmar EH, Abdelouhahid RA, Debauche O (2021) Machine learning algorithms for breast cancer prediction and diagnosis. Procedia Comput Sci 191:487–492
40. Nemade V, Fegade V (2023) Machine learning techniques for breast cancer prediction. Procedia Comput Sci 218:1314–1320
41. Rasool A, Bunterngchit C, Tiejian L, Islam MR, Qu Q, Jiang Q (2022) Improved machine learning-based predictive models for breast cancer diagnosis. Int J Environ Res Public Health 19(6):3211
42. Sharma A, Goyal D, Mohana R (2024) An ensemble learning-based framework for breast cancer prediction. Decis Anal J 10:100372
43. Uddin KMM, Biswas N, Rikta ST, Dey SK (2023) Machine learning-based diagnosis of breast cancer utilizing feature optimization technique. Comput Methods Programs Biomed Update 3:100098
44. Meraj T, Alosaimi W, Alouffi B, Rauf HT, Kumar SA, Damaševičius R, Alyami H (2021) A quantization assisted U-Net study with ICA and deep features fusion for breast cancer identification using ultrasonic data. PeerJ Comput Sci 7:e805
45. Zhuang Z, Raj ANJ, Jain A, Ruban N, Chaurasia S, Li N et al (2019) Nipple segmentation and localization using modified u-net on breast ultrasound images. J Med Imag Health Inf 9(9):1827–1837
46. Robin M, John J, Ravikumar A (2021) Breast tumor segmentation using U-NET. In: 2021 5th international conference on computing methodologies and communication (ICCMC). IEEE

47. Tong Y, Liu Y, Zhao M, Meng L, Zhang J (2021) Improved U-net MALF model for lesion segmentation in breast ultrasound images. Biomed Signal Process Control 68:102721
48. Byra M, Jarosik P, Szubert A, Galperin M, Ojeda-Fournier H, Olson L et al (2020) Breast mass segmentation in ultrasound with selective kernel U-Net convolutional neural network. Biomed Signal Process Control 61:102027
49. Hirra I, Ahmad M, Hussain A, Ashraf MU, Saeed IA, Qadri SF et al (2021) Breast cancer classification from histopathological images using patch-based deep learning modeling. IEEE Access 9:24273–24287
50. Krithiga R, Geetha P (2020) Deep learning based breast cancer detection and classification using fuzzy merging techniques. Mach Vis Appl 31(7):63
51. Rana P, Gupta PK, Sharma V (2021) A novel deep learning-based whale optimization algorithm for prediction of breast cancer. Braz Arch Biol Technol 64
52. Saleh H, Alyami H, Alosaimi W (2022) Predicting breast cancer based on optimized deep learning approach. Computat Intell Neurosci 2022
53. Zhang X, He D, Zheng Y, Huo H, Li S, Chai R, Liu T (2020) Deep learning based analysis of breast cancer using advanced ensemble classifier and linear discriminant analysis. IEEE access 8:120208–120217
54. Kayikci S, Khoshgoftaar TM (2023) Breast cancer prediction using gated attentive multimodal deep learning. J Big Data 10(1):62
55. Siddiqui SY, Naseer I, Khan MA, Mushtaq MF, Naqvi RA, Hussain D, Haider A (2021) Intelligent breast cancer prediction empowered with fusion and deep learning. Comput Mater Continua 67(1):1033–1049
56. Moon WK, Lee Y-W, Ke H-H, Lee SH, Huang C-S, Chang R-F (2020) Computer-aided diagnosis of breast ultrasound images using ensemble learning from convolutional neural networks. Comput Methods Programs Biomed 190:105361
57. Kumar A, Singh D (Eds.) (2025) Revolutionizing Medical Systems Using Artificial Intelligence: A Breakthrough in Healthcare. Elsevier
58. Kumar A, Aelgani V, Vohra R, Gupta SK, Bhagawati M, Paul S et al (2024) Artificial intelligence bias in medical system designs: a systematic review. Multimedia Tools and Applications, 83(6):18005–18057
59. Maindarkar M, Kumar A (2025) Artificial intelligence based disease diagnosis using ultrasound imaging. In Revolutionizing Medical Systems using Artificial Intelligence, pp 147–161. Academic Press

Part III
Artificial Intelligence for Personalized Care

Chapter 9
Role of Artificial Intelligence in Immunology

Abstract Artificial intelligence (AI) has become an inevitable part of the healthcare industry. The disease ontology is expanding at an alarming rate, and it is close to impossible for medical practitioners or healthcare professionals to deal with vast data in clinical records, submit lab reports, medical imaging data, drug targets, phenotypes, and genomics data, or find a concrete justification or treatment as per individual variability in disease management. AI can learn feature patterns from these huge datasets and can aid these professionals with robust and reliable predictions. With the integration of AI in immunology, the enhancements in diagnosis, drug discovery, and personalized therapy have significantly improved healthcare outcomes. The application of AI in immunology can be exploited for antigen-specific vaccine design, as well as the prioritization of potential immune epitopes from bacterial pathogens that activate human T cells.

Keywords Artificial intelligence (AI) · Immunology · Computational immunology · AI in biomedical research · Intelligent immunoanalytics · Immunoinformatics

9.1 Introduction

Immunology is a complex and versatile field. The intricate immune system protects the human body from diseases, but the field's perplexing concepts and mechanisms mean that researchers and clinicians alike need to think creatively and champion innovation to strive for an up-to-par level of biomedical advancement. Breakthroughs in immunology will also result in transitioning from a reactive model of care to a more proactive one, thereby ushering in the era of precision and personalized medicine [1]. Artificial intelligence is predicted to be an infrastructure technology that will impact several sectors in the years to come. The importance of AI and how it treads paths in fields like computational biology, computational chemistry, and machine learning is also on the rise, playing major roles in supporting logistical operations and in fields like drug discovery and disease diagnostics [2]. The marriage of AI and immunology

A. Kumar and D. Singh, *Artificial Intelligence in Modern Healthcare System*,
Transactions on Computer Systems and Networks,
https://doi.org/10.1007/978-981-96-6703-1_9

amplifies this leap of innovation in terms of speed, accuracy, and efficacy. The ability to sift data for patterns that were once considered 'random' and inapparent is one of the contrasting abilities of AI [3].

Immunology and its related terminology stem from the Greek terms; immunity and the concepts of antigens and antibodies find their roots in the word antigen. The study of the body's inbuilt defense mechanisms is known as immunology. The collection of cells, tissues, and organs that together attempt to resist foreign invaders and restore normalcy in the event of injury is the immune system [4]. The players involved include various cells, soluble factors, tissues, and organs, making this study an interdisciplinary one. The average lifespan of infectious diseases is rising, and researchers now believe the inborn immune components are much less versatile than we once thought [5]. The high-dimensional, unstructured nature of data with more noise is a computational burden that paves the way for AI to be included in immunological proteomics. AI holds all the necessary tools to increase the performance of analyzing, storing, interpreting, and actuating the mined complex data into pure, useful insights, giving us a new perspective and a better edge over the age-old standard computational times and strategies [6]. The key Contribution of this chapter is as.

- This chapter identifies the potential of AI application in the context of genomics, serology, and immune monitoring in the case of vaccine development.
- Further, integrated approaches of AI adoption with immunological concepts can contribute to the efficacy of vaccines and several other immunotherapies. We acknowledge that there are some limitations to the adoption of AI in immunological research.
- Additionally, the correct application of AI for disease vaccination purposes requires resolution of challenges such as the generalizability of AI vaccines, explainable AI, ethical and data sharing concerns, and AI model integration with immunological concepts, all of which are discussed in further detail.
- We close the chapter with an overview of opportunities and future developments that are on the edge of translational immunology.

The rest of the chapter is organized as follows. Section 9.2 elaborates on the AI-based applications of Immunology. In addition, AI-based techniques for Immunology to analyze Immune system models are categorized into ML-based and DL-based techniques in the Sect. 9.3. Section 9.4 details compared to analyze the merits, demerits, and limitations in each category of existing AI model for immunology. Future directions and opportunities of AI-based techniques are in Sect. 9.5. Lastly, the concluding remarks and future directions are sketched in Sect. 9.6.

9.2 Applications of AI in Immunology

Artificial intelligence (AI) has the potential to significantly improve medical diagnostics and treatment [7, 8]. This concept can certainly be applied to the field of immunology for numerous exclusive benefits. There are a significant number of applications currently emerging that involve using AI in the field of immunology. This section aims to explore these different areas and discuss how, in particular, AI might have the capability to assist in the immunology of the future [2].

Application Areas of AI in Immunology One of the most proposed advantages of utilizing AI in a medical context is the potential for an increase in diagnostic rates and disease prognostication. Enhancing and providing a more robust explanation for the data analytics that occur in the background can achieve this. The ability for machines to have smooth and consistent data provision is likely to bring data analytics higher in terms of accuracy and coverage [9]. Drug discovery and development has been another major area into which AI has been invested. The combination of AI and drug discovery aims for an increase in the gravitational pull for a particular candidate drug of interest. An enhancement in yield is produced if a client-liaised approach between the companies and the involved patients can be established. An immense need for personalized medicine has planted roots in today's world, and this phenomenon is no exception. In this highly customized age, industries are exploring a novel and niche approach to personalized medicine. Cells in the body have a very unique immune status, and there is a trend to investigate most of the abnormalities from these very immune-compatible cells [3]. In the context of personalized medicine, therefore, cells could be classified in accordance with the immune program within them. AI, if harnessed, could potentially be used to revolutionize personalized medicine and develop devices or software that diagnose the immune profile of a patient and classify these cells based on the unique immune program that belongs to them. This machine learning technology can be applied to human cells, animal cells, etc., in order to diagnose the disease. In conclusion, there are many exciting projects in the field of AI and immunology. The predictions are very promising in this area, and the applications are limitless. From the bench to the bedside, AI has the potential to forever change the face of healthcare [2].

9.2.1 Disease Diagnosis and Prognosis

Artificial intelligence (AI) has been extensively applied in disease diagnostics and prognosis, a domain known as medical diagnosis [10, 11]. In immunology, such techniques allow for the analysis of complex and multidimensional data extrapolated from immunological tests, patient history, and physical exams so the prognosis can be taken one step further for immune-related diseases. Inside AI, a considerable number of techniques have been described, including generative models, feature selection techniques, and a wide variety of machine learning (ML) and deep learning

(DL) approaches, which aim to enhance the capability obtained from simple pattern or marker analysis [12]. Diagnosis improves diagnostic accuracy or enhances the performance of a diagnostic protocol. As a prognosis starts when a disease marker is identified in a diagnostic scenario, we describe all diagnostic techniques, irrespective of the AI approach, here in this subsection. There are already multiple studies available where different ML and DL algorithms were used to diagnose a disease based on clinical or omics data. AI has even been applied in predicting infection or an autoimmune-like disorder in the early or latent stages of a disease [13]. AI systems have the potential to be developed and implemented directly into a professional's workflow in computer-assisted diagnosis systems [14, 15]. This integration could help clinicians make more definitive decisions, based not only on the information they gather but also on the results of tests or procedures conducted by AI. Ideally, such a system could allow a combined diagnosis to see the patient's test results and the generated diagnosis using artificial intelligence [13].

Researchers have used AI to identify and validate a panel of biomarkers in the blood of children or adults that can be used to diagnose a systemic condition, and studies of periodic blood-based protein levels in patients using bioinformatics methods have shown AI can predict either a patient's level of symptoms or disease progression. IDD diagnosis can still be improved, and working AI systems with the ability to directly impact the field of immunology will need to be released. One AI approach can hold the potential to revolutionize traditional diagnostic practices [16].

9.2.2 Drug Discovery and Development

Drug discovery and development is a costly and lengthy process. Artificial intelligence algorithms provide new approaches to analyze vastly available techniques that can identify small molecules and candidates. AI programs, including deep learning, can analyze larger sets of data to make the data analysis more accurate and efficient due to the lack of important datasets to be trained, to study drug mechanisms quickly and more economically, and hence to provide a better output [17]. The integration of these concerns with the new data also reduces the implemented algorithms of data into the strong requirements for computational power. Companies are developing large datasets to extend their usage in their workflows. A few examples showcasing the successful application of these approaches in discovering immunotherapeutics include AI technology developed in collaboration with a member of the Roche Group [18]. Nevertheless, AI algorithms have been recognized as one of the potential multifactorial design considerations for drugs. Major challenges of the application of AI approaches in this domain pertain to the quality of the data. The current data available for treatment and drug development, as well as operational and clinical data, could be integrated into a single patient file [19]. Moreover, most clinical data are textual records that express the doctor's decisions, which are characterized by social, organizational, and individual contexts. However, identified research opportunities in this domain include the development of deep learning AI approaches in patient immunity profiling and personalized drug therapy, where immune responses are measured on a single subject visit time series with good dependent variables. Ethical challenges in AI applications concern automated drug design and treatment, including the responsibility for developing drugs, distributive justice, and test versus control subjects' decision-making [20]. In contrast, AI applications in design could automate recommendations that will accelerate method development, and the decision is supported by the investigator's intuition for a more effective study. In conclusion, AI and machine learning have very promising roles in expediting multiple advances in drug design and treatment. The transformative potential, benefits, promise, and success of these tools in drug treatment and trials are increasingly concerning the integration of AI/ML tools into designs to improve their efficiency for testing or to reduce their costs [21].

9.2.3 Personalized Medicine

Personalized medicine is an attempt to use data from individual patients to guide disease treatment. In the case of immunology, immune responses vary widely between individuals. AI tools can be used to detect detailed immune responses in vaccinated individuals or detect sophisticated immune responses to infection [22]. Personalized immune-based drug treatments may involve the integration of various types of data such as genomic data, plasma or serum biomarkers, and data on tumor biology. Physicians treat patients individually based on the results of these tests. For example, by genetically detecting people who have impaired interferon responses and are thus likely to suffer from persistent, chronic, and debilitating infections and those who may respond with autoimmune symptoms to vaccines [19]. A similar approach has been proposed to optimize response to pharmaceutical countermeasures and exposure to viruses (The Applications of artificial intelligence in immunology is shown in Fig. 9.1).

Personalized medicine strategies are likely to be more complex, less intuitive, and more high-risk than traditional clinical practice [23]. Nevertheless, the treatment of immunotherapy in the field of immune lung cancer is closer to a personalized medicine strategy based on extensive genetic and molecular diagnostics. Case studies of personalized immunotherapies that use genetic immune response data to identify patients to exclude or functionally activate in patients exposed to anticancer checkpoint inhibitor therapies support the feasibility of the concept. AI and ML tools can also integrate and help predict and guide the effects and outcomes of complex personalized immunotherapy strategies. AI can therefore be considered to support decision science in personalized medicine to overcome the problem of the low number of test patients. Importantly, having access to and using individual health and gene data to develop personalized therapies also raises privacy and ethical concerns, as well as direct commercial use [19, 23].

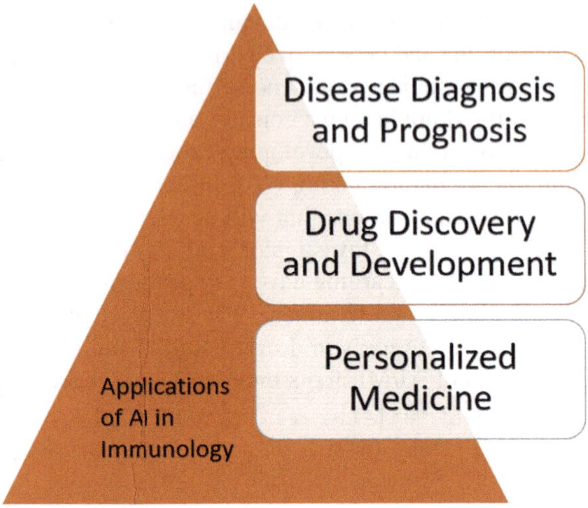

Fig. 9.1 Applications of artificial intelligence in immunology

9.3 AI Techniques and Models in Immunology

Artificial intelligence techniques, which are primarily used in immunological studies, mainly depend on the objectives of the studies. To process the existing complex and multifactorial immunological data, most studies are based on supervised, semi-supervised, and unsupervised learning algorithms. Additionally, machine learning techniques are widely used to elucidate interactions, changes in subpopulations, and patterns [13]. For the analysis of high-dimensional immunological data, deep learning architecture is utilized, including different types of CNN, RNN, autoencoders, and GANs. These AI models capture different types of features according to the study's priorities. As a result, the selection of best-fitted algorithms depends upon the tasks to be performed. The successful improvements in immunological research achieved via AI-based approaches are significant [24].

Therefore, it is crucial to select a suitable model according to the objective of the study. The list of more complex models, such as CNN, RNN, autoencoders, and GANs, requires additional consideration regarding the model design, parameter tuning, evaluation, computation, and reporting parts, particularly in the studies conducted for bench-to-bedside treatment discovery [25]. Immunological data is mostly high-dimensional and noise-oriented, requiring efficient analyses through sparse feature learning models, factorization-based models, nonlinear-based models, neural network-based models, ensemble-based models, and unsupervised models, depending on the subdomain type and size of data, and challenges. There is a strong desire to gain translatable AI-based knowledge required to understand host defense and infectious diseases, autoimmune and inflammatory responses, and immunoresistance development. As a result, the future direction suggests the selection of AI model types, with support from surrogate system models, for the promotion of immunological research [25, 26].

9.3.1 Machine Learning Algorithms

Enormous work has been done in developing data analysis and prediction tools using machine learning in immunology. Machine learning algorithms essentially work by learning patterns underlying the given input data and making predictions or discovering new insights from them. The algorithms can be broadly categorized into classification and regression methods, which are widely used in immunological studies as well. Classification techniques are used to categorize or position the data into specified classes or clusters and have also been applied to the prediction of new disease diagnoses. Examples of classification algorithms are decision trees, support vector machines, k-nearest neighbors, ensemble methods such as random forests, and neural networks. Regression, a supervised learning algorithm that finds the relationship between independent features and dependent outputs, is also used in quantitative modeling in immunological studies to predict outcomes such as drug response and patient survival, personalized treatment. Feature selection is often performed to eliminate irrelevant or redundant features in machine learning models, which are important in immunological applications [13].

Many reviewed papers discuss the application of various classifiers in different areas of immunology. Machine learning models have not only been used in different areas of immunology to make new findings and discoveries, but have also been applied in clinical studies where data from real-world clinical settings have been utilized to improve diagnostic precision, patient stratification, and facilitation of personalized medicine [27]. The application of artificial intelligence algorithms in solving diseases has been thoroughly reviewed, which underlines the application of artificial intelligence, including machine learning models, in the diagnosis of diseases associated with immunology such as asthma, hepatitis, AIDS, tuberculosis, and cancer, as well as the search for new drug targets and compounds [28]. However, the challenges associated with translational value in the feature space and the ongoing research should be addressed in future to make more robust and accurate immune system disease prediction models. In addition, data normalization and data bias are also key challenges in the application of machine learning in immunological classification problems that should be paid attention to [29].

9.3.2 Deep Learning Architectures

Deep learning, a subfield of machine learning, is an artificial intelligence architecture that learns layered representations of input data. Through neural networks, deep learning can identify essential characteristics of complex and specialized tasks, thereby allowing researchers to understand previously obscured layers of immunological data. Initially, deep learning is fed input data. Inside the neural network, the input data is transformed into different and highly specific forms using several mathematical operations [30]. As the operation is performed consecutively, mathematical operations rarely imitate the handmade form as the previous model architectures for artificial intelligence, known as artificial neural networks, and other models. Many of the complex characteristics of the immunological data must be evaluated with the original data after the completion of the output layer [31]. However, the intermediate results are considered to have the most important characteristics that reflect the functional information of the immune system. Therefore, deep learning can reveal functional immunological characteristics in body fluids and tissues or spatial immunological structures and immune molecular patterns by capturing corrected data required for further immune analysis [32].

There are many models in deep learning, and some are used in immunological fields. Notable in immunology, deep learning methods of image recognition tasks use convolutional neural networks. CNNs have been incorporated in numerous commercially available software packages that are capable of providing powerful and accurate quantitation for stained slides [33]. While numerous CNN-based models exist, very few have been used for applications in immunology. CNNs have demonstrated significant improvement in diagnosis and prognosis in various fields, including meningioma classification, skin cancer recognition, breast cancer subtyping, and prostate cancer detection. However, the most influential and widely recognized deep learning model in the medical field is natural language processing-based transformers, which are used for the analysis of medical records, inferring relationships between diseases, or for predicting the clinical outcomes of patients [34]. Although such examples are very rare in the field of immunological research, this is likely to become a focus for future studies. Despite its usefulness, deep learning has certain limitations. Because deep learning is deeply connected, the model may only capture patterns in the training dataset, and deep learning may require a large amount of reference data [35]. In certain cases of deep learning, a larger number of related studies are required to yield improvements for immunological data such as genomics and biomedical images regarding the limited number of specific immunological and clinical data.

In terms of practical applications, however, there is still little existing research that uses ANNs or deep learning for immune analysis [36]. Generally, there is little ethical controversy, but it often accompanies excessive benefits and restrictions, particularly in medical, diagnostic, and disease treatment fields; the interpretability of poor results is associated with ethical risks. Interpretation in immunology and cancer investigation is highly significant; deep learning in immunological analyses holds potential and is expected to have a significant impact. As the deep learning model becomes more sophisticated and larger, the application of deep learning and immunology is expected to further expand [37].

9.4 Challenges and Limitations of AI in Immunology

To conclude, although AI presents ample opportunities to explore the immunological landscape, there are numerous challenges that must be overcome in order to realize these ambitions. The production of high-quality human immunological data is currently limited and hence restricts the creation of reliable AI models. The modeling of biological systems is also complex and unique to each patient due to epistasis. In addition, integrating AI into clinical practices is problematic due to the resistance among many practitioners to this new technology [38]. Various ethical issues also need to be resolved, such as patient data privacy in addition to bias in the models. Regulatory procedures for AI, particularly in a clinical context, have not yet been solidified and hence constitute an additional barrier to widespread use. Some even suggest that the spotlight on AI may be exaggerated and premature given the string of recent failures of AI models across various industries [39]. There are various subdomain of Immunology as shown in Fig. 9.2.

Given the obstacles that have been highlighted, the ambition to run immunological studies through AI is a tough sell. Despite the advances that have been made in the application of AI to immunology, we see no foreseeable future where AI can replace traditional immunological studies, at least for the next 10–15 years [40]. However, this does not mean there is no place for AI in the field. More realistic goals include patient stratification using a combination of AI, genomics, and proteomics, as well as predictions of treatment responses with disease progression based on medication intake and lifestyle or environmental factors. These tools, when combined, could

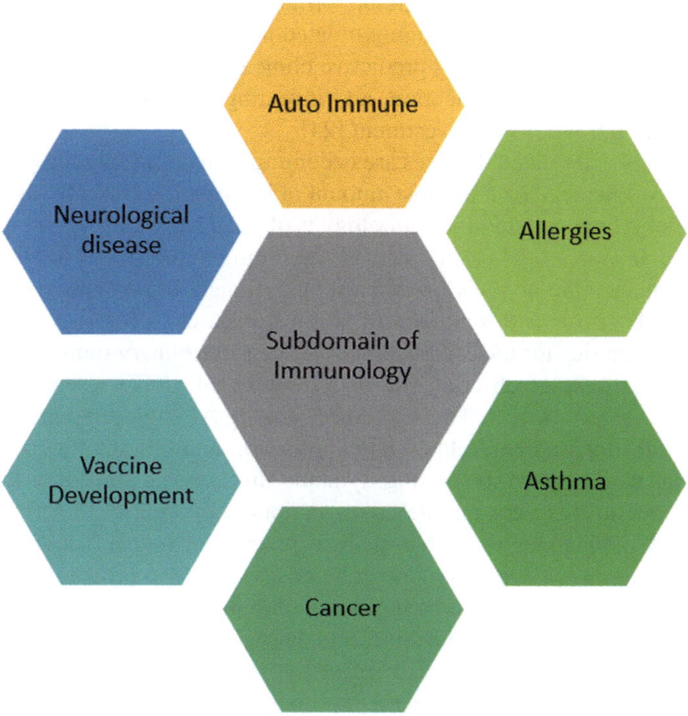

Fig. 9.2 Subdomain of immunology

inform practitioners as to the prognosis of their patients and hence inform therapeutic decisions, representing a more realistic and feasible future for AI in immunology [18].

9.5 Future Directions and Opportunities

We see several future possibilities created by AI to revolutionize the fields of immunology and immune-mediated diseases. Large volumes of complex data are increasingly generated in immunology from genomics, epigenetics, single-cell analyses, and other advanced technologies. AI will be increasingly utilized to facilitate efficient data analysis and deep learning, assist in experimental design, enable applications for more personalized medicine, and serve as an integral part of drug and diagnostic discovery [41]. Real-time predictive analytics to enhance patient care management and provide precise and optimal drug or therapy mining for autoimmune diseases, cancer immunotherapy, immune monitoring, or vaccine design are some conceptual predictions of where future AI applications can be widely developed [25]. We believe that the major advance in AI innovation is expected to come from interdisciplinary crossover between AI experts, machine learning, robotics, and experts in immunology, together with the formulation of immune theory and immunologists. Investment in AI for immune-related medicine and healthcare, including companies working on designing predictive clinical AI, using AI to mine immune-mediated diseases, as drug detection, and inventing AI-based detection kits, may again provide high returns on investment [24].

Immunology and related diseases are becoming an increasingly important part of healthcare worldwide. Given the vast amount of information currently available, the emergence of AI in the field of immunology is of great significance, and the outlook is optimistic. In close collaboration with clinical immunologists, AI has the potential to greatly enhance the application of basic and clinical immunology research and improve the diagnosis and treatment of various diseases in future. Globally, there are also opportunities for the establishment of interdisciplinary immunology-driven AI and healthcare research consortia or a network of global research communities with this shared interest. These alliances among AI engineers, artificial intelligence institutes for immune-related human disease, and immunological experts in the field can work toward generating dynamic discussions, spreading knowledge, and experience in this very unique cross-sectional opportunity to further develop the proposed applications to enable medical practice in future. The conception of future crossover interdisciplinary research agendas can also provide a platform for joint worldwide grant applications in AI data mining for immune-related diseases. Development of professionals and educational programs in AI and machine learning for immunopathology is essential to meet the growing medical industrial needs for the future. It can also provide new international career opportunities and long-term training in interconnected areas of imaging and clinical immune pathogenesis [25, 40, 41].

9.6 Summary

In closing, the potential of AI in immunology is transformative. If used in the right way, it can further some of the great clinical progress achieved in recent years. Building useful machine learning models, applying deep learning to image classification, and searching for patterns in omics data at scale are just a few of the evidenced successful applications of AI in immunology. The promising research emanating from these projects ought to be matched by responsible innovation. Since the integration of AI into complex immunology models is rife with technical, ethical, and adoption challenges, a more in-depth diagnostic survey and an assessment of the ongoing state of AI integration is needed to identify the pervasive challenges and provide a unified view of the landscape across disease states.

Early investments in the computational foundations required for such AI-first models, best practices for integrating AI into scientific processes, and exploration of the efficacy of the AI-standard immune health promotion vehicles on patient outcomes are the essential next steps. Finally, this work should only be done in an environment of interdisciplinary collaboration across immunology, computational science, ethics, and regulatory disciplines. Immunology stands at a pivotal cross-section where AI could be deployed across a range of available data, methods, and application areas. There are, without doubt, limitations and a need for improvements in both the data and methodology. Moreover, proving clinical utility is hard, and ethical, and micro- and macroeconomic models of implementation are not to be neglected. Still, progress is being made in the reconstruction of more complex models capable of accommodating much of our molecular understanding. Immunology is potentially one of the great fields of application for AI, given the proper imperatives to define questions and develop methodology. However, it represents a spectacular challenge to machine learning developers.

References

1. Zhang Z, Wei X (2023) Artificial intelligence-assisted selection and efficacy prediction of antineoplastic strategies for precision cancer therapy. In: Seminars in cancer biology. Elsevier
2. Xu Z et al (2021) Applying artificial intelligence for cancer immunotherapy. Acta Pharmaceutica Sinica B 11(11):3393–3405
3. Sahu M et al (2022) Chapter three—artificial intelligence and machine learning in precision medicine: a paradigm shift in big data analysis. In: Teplow DB (ed) Progress in molecular biology and translational science Academic Press, pp 57–100
4. Pimenov DY et al (2023) Artificial intelligence systems for tool condition monitoring in machining: analysis and critical review. J Intell Manuf 34(5):2079–2121
5. Li T et al (2023) Artificial intelligence in cancer immunotherapy: applications in neoantigen recognition, antibody design and immunotherapy response prediction. In: Seminars in cancer biology. Elsevier
6. Himeur Y et al (2023) AI-big data analytics for building automation and management systems: a survey, actual challenges and future perspectives. Artif Intell Rev 56(6):4929–5021

7. Kumar A et al (2023) 6G-enabled IoT and AI for smart healthcare: challenges, impact, and analysis. CRC Press
8. Kumar A, Sareen P, Arora A (2023) Healthcare engineering using AI and distributed technologies. Smart distributed embedded systems for healthcare applications. CRC Press, pp 1–14
9. Kaplan A et al (2021) Artificial intelligence/machine learning in respiratory medicine and potential role in asthma and COPD diagnosis. J Allergy Clin Immunol Practice 9(6):2255–2261
10. Kumar A et al (2023) Security and privacy issues in smart healthcare using machine-learning perspectives. 6G-enabled IoT and AI for smart healthcare. CRC Press, pp 41–56
11. Kumar A, Mamgai R, Jain R (2023) Application of IoT-enabled CNN for natural language processing. In: IoT-enabled convolutional neural networks: techniques and applications. River Publishers, pp 149–177
12. Kumar Y et al (2022) A systematic review of artificial intelligence techniques in cancer prediction and diagnosis. Archives Computat Methods Eng 29(4):2043–2070
13. Peng J et al (2021) Machine learning techniques for personalised medicine approaches in immune-mediated chronic inflammatory diseases: applications and challenges. Front Pharmacol 12:720694
14. Kumar A, Bhalla S (2025) Exploring brain tumor detection through artificial intelligence. In Revolutionizing Medical Systems using Artificial Intelligence, pp 93–120. Academic Press
15. Kumar A, Walia GS, Sharma K (2020) Recent trends in multicue based visual tracking: a review. Expert Syst Appl 162:113711
16. Kaur S et al (2020) Medical diagnostic systems using artificial intelligence (ai) algorithms: principles and perspectives. IEEE Access 8:228049–228069
17. Yang Y et al (2022) Artificial intelligence for prediction of response to cancer immunotherapy. In: Seminars in cancer biology. Elsevier
18. Xie J et al (2023) Advances in artificial intelligence to predict cancer immunotherapy efficacy. Front Immunol 13:1076883
19. Xu Y et al (2021) Technological advances in cancer immunity: from immunogenomics to single-cell analysis and artificial intelligence. Signal Transduct Target Ther 6(1):312
20. Ghaffari Laleh N et al (2023) Facts and hopes on the use of artificial intelligence for predictive immunotherapy biomarkers in cancer. Clin Cancer Res 29(2):316–323
21. Gao Q et al (2023) The artificial intelligence and machine learning in lung cancer immunotherapy. J Hematol Oncol 16(1):55
22. Ai-ris YC et al (2021) Immunogenicity of COVID-19 mRNA vaccines in pregnant and lactating women. JAMA 325(23):2370–2380
23. Sharma A et al (2022) Artificial intelligence-based data-driven strategy to accelerate research, development, and clinical trials of COVID vaccine. Biomed Res Int 2022(1):7205241
24. Elemento O et al (2021) Artificial intelligence in cancer research, diagnosis and therapy. Nat Rev Cancer 21(12):747–752
25. Baxi V et al (2022) Digital pathology and artificial intelligence in translational medicine and clinical practice. Mod Pathol 35(1):23–32
26. Feng R et al (2021) Knowledge gaps in immune response and immunotherapy involving nanomaterials: databases and artificial intelligence for material design. Biomaterials 266:120469
27. Quazi S (2022) Artificial intelligence and machine learning in precision and genomic medicine. Med Oncol 39(8):120
28. MacEachern SJ, Forkert ND (2021) Machine learning for precision medicine. Genome 64(4):416–425
29. Iqbal MJ et al (2021) Clinical applications of artificial intelligence and machine learning in cancer diagnosis: looking into the future. Cancer Cell Int 21(1):270
30. Pertseva M et al (2021) Applications of machine and deep learning in adaptive immunity. Annu Rev Chem Biomol Eng 12(1):39–62
31. Germain RN et al (2022) Understanding immunity in a tissue-centric context: combining novel imaging methods and mathematics to extract new insights into function and dysfunction. Immunol Rev 306(1):8–24

32. Li L-S et al (2023) From immunology to artificial intelligence: revolutionizing latent tuberculosis infection diagnosis with machine learning. Mil Med Res 10(1):58
33. Yang Y et al (2021) Identification and validation of efficacy of immunological therapy for lung cancer from histopathological images based on deep learning. Front Genet 12:642981
34. Ripoll DR, Chaudhury S, Wallqvist A (2021) Using the antibody-antigen binding interface to train image-based deep neural networks for antibody-epitope classification. PLoS Comput Biol 17(3):e1008864
35. Pandya S et al (2021) A study of the recent trends of immunology: Key challenges, domains, applications, datasets, and future directions. Sensors 21(23):7786
36. Sharma S et al (2022) Performance evaluation of the deep learning based convolutional neural network approach for the recognition of chest X-ray images. Front Oncol 12:932496
37. Mochalova EN et al (2022) Imaging flow cytometry data analysis using convolutional neural network for quantitative investigation of phagocytosis. Biotechnol Bioeng 119(2):626–635
38. Girdhar A, Kapur H, Kumar V (2022) Classification of white blood cell using convolution neural network. Biomed Signal Process Control 71:103156
39. Franklin MM et al (2022) A deep learning convolutional neural network can differentiate between *Helicobacter pylori* gastritis and autoimmune gastritis with results comparable to gastrointestinal pathologists. Arch Pathol Lab Med 146(1):117–122
40. Abbasi N, Nizamullah F, Zeb S (2023) AI in healthcare: using cutting-edge technologies to revolutionize vaccine development and distribution. JURIHUM: Jurnal Inovasi dan Humaniora 1(1):17–29
41. Kumar A, Singh D (Eds.) (2025) Revolutionizing Medical Systems Using Artificial Intelligence: A Breakthrough in Healthcare. Elsevier

Chapter 10
Managing High-Risk Surgery Using Artificial Intelligence

Abstract High-risk surgical procedures concern complex surgical procedures requiring substantial postoperative care, such as those in cardiothoracic, urosurgical, and neurosurgical disciplines—in particular, those involving medical significance or the perception of elevated associated risk. Due to rapidly rising healthcare costs, research into recent medical advancements has been expedited by pressure from national and economic factors to provide an explanation of entitlement to tax-funded service suppliers. Quality assurance management systems have been implemented to serve as an adjunct to the development of high-performance healthcare systems. Artificial intelligence is currently being developed and applied to all aspects of medicine with improvements in computational ability. AI might be an important contributory element due to its expertise in deciphering text and images, especially concerning surgeons who enter into the tougher excision results for diagnostic classification purposes. However, any AI application helping in important surgical procedures necessitates a strong accreditation strategy that supports qualified professionals' concerns regarding the effect of peripheral hyper-regulation and de-scaled protection, also ensuring that a technology adherer can comply with new professional requirements and encourage the proliferation of already scarce surgical specialists. High-risk surgical processes are tailored here to these issues, which are capital-intensive and profoundly affect patient outcomes. We maintain that AI technologies are built as a 'third operating hand' to support mental and physical exercise, and not as independent research systems. Every move to build private surgical processes must also follow AI regulation, including current legal health regulations.

Keywords High-risk surgery · Surgical assistance · AI in surgery · Smart operating rooms · Robotic-assisted surgery · Surgical automation

A. Kumar and D. Singh, *Artificial Intelligence in Modern Healthcare System*, Transactions on Computer Systems and Networks,
https://doi.org/10.1007/978-981-96-6703-1_10

10.1 Introduction

Surgical robots have become an irreplaceable equipment in today's operating rooms, thanks to their skillfulness. They are widely applied in operations that are at high risk of high postoperative complications or surgery difficulties, such as in thoracic surgery, delicate hepatopancreatobiliary and minimally invasive surgeries, assisting in the operations [1]. With the assistance of these intelligent technologies, patient safety can be effectively improved. The reoccurrence of disease could be reduced, concurrent co-magnetic of complex operations could be avoided, and the psychiatric pressure of young surgeons can be relieved that stems from the heavy load of practice learning. These advantages make artificial intelligent robot-assisted surgery high-risk operation patients elective [2]. Currently, robots operate in a manner of "imitation" rather than "thinking"; however, the story takes a turn in the world of Artificial Intelligence (AI), as AI-based robots are the "main characters", equipped with their distinct advantage, with the capacity for "thinking" [3]. It is more interesting to discuss how AI is incorporated to assist these robots performing their operations [4]. As a newly-emerged field, surgical AI addresses capturing the capabilities of the latest advancements in bioinformatics and biomedical computation that could provide suggestions to verbalize insights, diagnostic, or procedural performance of surgical robotics systems, while bearing the specially-developed clinical practice guidelines and engineering standards in mind. Although augmented by the most advanced clinical and engineering technologies, it is challenging for surgical AI to achieve a fully autonomous intraoperative surgical procedure, which also ignites a rebound discussion on the unique role of individual surgical robots [5, 6].

10.1.1 Background and Significance

The beginning of the twenty-first century has seen the rise of a dynamic, globally connected technology and services-based economy propelled by digitization and telecommunication. Despite the now daily experience of virtual meetings, much of the genuine disruptive changes toward such a connected economy are still on the finance, create, transfer, operate, and monetize information and knowledge [2]. In healthcare, even the simplest disruptive technologies based on the combination of fast web-based access to molecular diagnostic kits, fast electronic processing of sensor's biomedical readings, and web protocols and rights management to have the diagnosed data being processed and the results being sent back within a relatively short time space are still not deployed [7]. Although it is possible to combine high-performance diagnostics with the currently very established non-invasive surgery, this leads to ambulatory or even real-time surgeries in an operating theater environment with immense benefits [8]. Only the speed of service and the lack of economical models leveraging the low-cost access to the process to support the pioneer establishment of robotic systems at very high service costs. To support critical argumentation on the effectiveness of surgery, some concepts must be agreed upon to comprehend the role of surgery in healthcare [9].

Surgery continues to be the central and decisive form of therapy for most cancers. Although substantial progress has been made in the development of anticancer drugs and radiotherapy, successful outcomes from surgery still remain essential in most cases. To avoid recurrence/metastasis and to improve the overall prognosis for patients with cancer, radical en bloc resections are required when possible [10]. In addition, successful and assiduous minimally invasive surgeries are necessary to ensure the firm establishment of more than 20 years' progress in endoscopic diagnosis and biopsies, as well as various other endoscopic diagnoses. Unfortunately, since 1965, curricula thin fiberscopic technologies have not had significant continual progress in enabling endoscopic diagnosis, biopsy, and surgery to be combined and conducted in a single board-certified non-invasive machine like endoscopic diagnosis and biopsy [11].

10.1.2 Purpose and Scope of the Study

This research aims at developing a model to assist high-risk surgery using artificial intelligence (AI) techniques such as deep learning and big data. In the developed model, data from intelligent medical devices will be collected and analyzed, as well as existing patients' data from medical diagnostic images already known, which can reflect therapeutic outcomes in order to reduce the risk of re-exploration and improve prognosis [12]. The intelligent medical device is composed of existing biosensors, biochemical sensors, and an AI software program [13]. The AI software program will be able to early detect abnormal signs and symptoms of the patient by monitoring changes in patients' information and inform the healthcare provider of the results in a timely manner. This device can bring both instant improved quality of treatment and convenience for patients and healthcare providers, being an eco-friendly medical device. The first solution was that the biosensors were 3D printed for easy application to the body [14]. For the second solution, commonly used biosensors (Pulse oximetry) and the method of applying the peripheral venous lines were included in order to create a more comprehensive solution [15]. Needed high-efficiency materials were determined for each solution. The results of the pre-experiment were validated to confirm the excellence and compatibility of the results. The 3D printed sensors and applications were presented. Finally, the value and limitations were discussed for commercialization [12].

This research aims at developing a model to assist high-risk surgery using artificial intelligence (AI) techniques such as deep learning and big data. In the developed model, data from intelligent medical devices will be collected and analyzed, as well as existing patients' data from medical diagnostic images already known, which can reflect therapeutic outcomes in order to reduce the risk of re-exploration and improve prognosis [16, 17]. The significant theme mentioned in the Purpose and Significance of the Study chapter was expanded in this study. The latest research and progress at

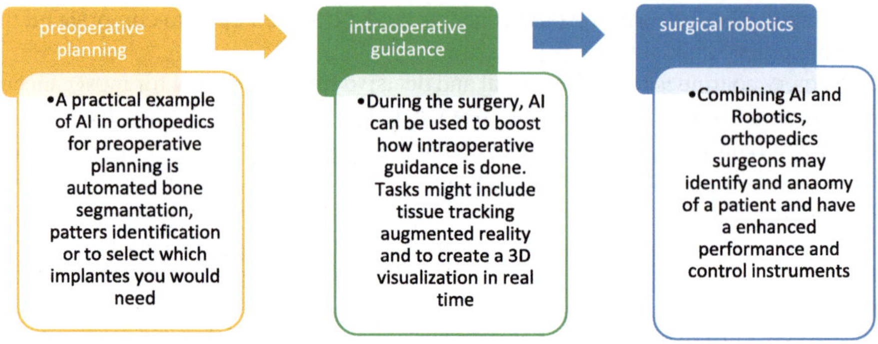

Fig. 10.1 Challenges in high risk surgery

home and abroad related to this theme were mentioned in an introduction chapter, and the foundation of the theory was introduced in a matter chapter. Moreover, this study was expected to make a positive contribution to the sustainability of healthcare policies by using big data and AI, as per an expectation chapter. The key contributions in this chapter are as follows:

- The risk factors that lead to the chances High Risk Surgery occurrence are elaborated in detail to provide awareness pointers in order to prevent its spread among humans.
- We have discussed applications of AI-assisted Surgery, Limitations of traditional methods and advantages of AI assisted High Risk Surgery.
- Emphasizes the importance of AI based surgery, Ethical and legal considerations and innovations for AI-assisted surgery.

The rest of the chapter is organized as follows. Section 10.2 elaborates the background and current challenges for AI assisted High risk surgery. In addition, Applications of AI-based surgery, image analysis detection and robots assisted surgery are discussed in Sect. 10.3. AI based Decisions, results, Improved Precision and Accuracy and advantages of AI in high risk surgery. Ethical and Legal considerations, Patient Consent and Autonomy, Liability and Accountability Issues are discussed in Sect. 10.5. Future trends and innovations in AI assisted surgery are mentioned in Sect. 10.6. Lastly, the concluding remarks and future directions are sketched in Sect. 10.7 (The Challenges of high risk surgery is shown in Fig. 10.1).

10.2 Current Challenges in High-Risk Surgery

Even though technical assistive devices and technical surgical procedures are harmonized during surgery, patient clinical status during surgery varies with the volume lost for replacement, and therefore these patient parameters should be adjusted.

Analyzing the surgical video, each patient has a unique video preoperative and intraoperative profile [18]. However, each patient has both enjoyable periods recognized by decreased parameter deviation and unforeseen circumstances, leading to overwhelming workload. The anesthesiologist team must increase patient safety control quickly. Artificial intelligence for medical purposes is considered in the era of anesthesiologist support [19]. It is expected to provide information for better decision-making in variable patient clinical situations, based on high level performance both in intensive care unit monitoring and surgical condition assessment.

Facing high operative risk and struggling against time to save life during surgery, technical surgical skills are crucial. But many factors can unexpectedly influence surgical outcomes. Not all patients with the same acute disease need surgical intervention [20]. On the other hand, the difference in the quality of technical skills among surgical team members becomes evident during surgery, in terms of ease and time to accomplish key parts of the surgery with less bleeding. Currently, selection of patients for high-risk surgery and the choice of surgical team for high-risk surgery are not standardized [21]. Focusing on medical practice in developed countries, a new approach using artificial intelligence in this surgical decision-making problem to better inform both patients and doctors is needed. Here, we explore how frequent surgical complications that emerge during high-risk surgery are intertwined with the technicalities of the surgical procedure and the patient's clinical situation from a general surgical specialist perspective and propose directions for future refinement of practical artificial intelligence support [20].

10.2.1 Risk Factors and Complications

While human beings are prone to test error, fatigue, or stress, among others, machines are less deficient in many of these areas and their performance level is consistent over time depending on their programming. Although machine-learning models do make mistakes, they are less likely than human errors when the technology is well-maintained and parameters of operation are obeyed. While the concept of placing people's lives in the hands of machines may be unattractive to some, it's still important to remember that humans designed methods, set their requirements, and instructed them in their tasks [22]. AI will allow healthcare professionals to develop different technologies to help improve patient outcomes with the implementation of supportive computer programs through transforming the structure of their jobs. These technologies will create opportunities for health professionals to work in a supported role in a manner that allows for consistent best care [19].

10.2.2 Limitations of Traditional Surgical Techniques

Despite the significant improvement and advancements in both anesthesia and surgical techniques during the last decade, they cannot ensure a zero-risk intervention for patients. As I stated above, a considerable number of patients face a high-risk surgical intervention. Those are cases where the high-risk is mostly due to the specific clinical condition of the patient and concerns mostly two points. The one is the general condition and functionality of vital organs (heart, lungs, liver, kidney, etc.), while the other refers mostly to the circulatory system itself due to stenoses, ruptures, malfunctions, etc. of arteries [23]. This is a case where Artificial Intelligence can make a significant additional contribution, allowing the vascular surgeon to ensure a higher level of intervention with a minimum risk, especially for the functionality of major organs during the operation. In this chapter, after the first assessment of the AI in surgery itself, an analysis of the high-risk surgical interventions is made, while research conducted during the last years for scientific research and prototypes concerning the role of AI in aortic surgeries in an AI-oriented society is presented [24, 25].

10.3 Applications of Artificial Intelligence in Surgery

Preoperative test prediction tools are developed to predict preoperative test results using individual patient's information. These predictions can help surgeons make a more personalized surgical planning, lower the risks, or guide cautioning the post-operative care. There are also various decision-making assisting tools. Instead of employing the general instant information during the surgery, these tools can provide more precise and individualized recommendations [26]. The other important group is the tutorial tools for simulation or guided surgery. Coordinative virtual environment software promotes group work within surgery to integrate information from recorded surgeries and provide advanced communication between the surgery team members. While many artificial intelligence technologies in surgery are introduced to optimize the surgical workflows by enhancing the four stages of data collection, data management, data analysis, and data display, distinct artificial intelligence technologies may have distinct dominant stages in which they can demonstrate their unique benefits [27]. However, most of these technologies are still in the conceptual stages. With the rapidly expanding capabilities of artificial intelligence, they are expected to mature and be helpful for numerous surgeons presently.

There are numerous potential applications for artificial intelligence in surgery, serving different purposes and administering to distinct parts of the workflow associated with this specialty. The following paragraphs detail individual uses of the technology in surgery [28]. The existing correlation between these outlined uses provides a glimpse at one of the unique trends in surgery practice where artificial intelligence is more prominent. These assistive technologies are concurrently user interactive and automated, complementing and guiding but not substituting the surgeon. Some of these assistive tools even offer explanations that the surgeon can interpret, further empowering the latter to make a well-informed decision [29].

10.3.1 Image Analysis and Interpretation

One of the best-known applications of AI technology in the medical field is the ability to analyze images to assist with diagnostic tasks [30, 31]. A good portion of current AI applications is used in various diagnostic methods, making predictions based on the available electronic data has resulted in significant advances in personalized medicine [3]. In high-risk surgeries, these methods have a smaller but still widely used role in the form of medical imaging and informative pre-operative counseling. Using MRI scans and other imaging technologies, AI can provide assistance with pre-operative planning. Information provided by the AI can be used to weigh different techniques and individualize operative strategies before the patient arrives in the operating room [32]. A review of the recent impact on different AI imaging tools in high-risk abdominal incidentalomas surgeries has been studied in 2021. Collecting high volumes of data directly from patients as they move through the hospital can expand the data utilized by AI models [33]. Limitations of medical images include low reproducibility and weak correlation to objective functional parameters.

10.3.2 Robot-Assisted Surgery

Advantages of robot-assisted systems include magnification and stereoscopic vision, high-precision instruments, tremor reduction, wide range of movement, stable camera, minimal invasion, and reduced operator fatigue. The features of robot-assisted devices are visual imaging systems; servo actuators; three-dimensional sensor; and either the data storage system, American Standard Code for Information Interchange (ASCII) interface, Universal Serial Bus (USB) interface, display; input systems or haptic interfaces [34]. The robotic surgical process consists of moving the robot to the location of the surgical site, placing the robot precisely where the incision is to be made to help the surgeon visualize the surgical site, and then make the incision while assisting the surgeon.

Robot-assisted surgery is an application of robots. Robots are machines that can be programmed to carry out very complex personal tasks or can be manipulated by general-purpose machines. Robots that assist surgeons are in the form of an articulated arm [35]. The robots have different roles in surgery; they include laparoscopic surgery, cataract surgery, orthopedic surgery, cardiovascular surgery, and many different high-risk and minimally invasive surgeries. The disadvantages that exist in robot-assisted surgery are the lack of dedicated training, the high cost of human surgery for initial testing, and easy access. As a result, surgeons need to be educated about robotic systems, which tools are best suited to their procedures, and how to support personnel in order to ensure effective robotic system use [36].

10.4 Benefits and Advantages of AI in High-Risk Surgery

One of the surgeons who has studied the use of AI-assisted liver surgery most is Positano from the San Giovanni di Dio and Ruggi d'Aragona Hospital in Salerno, Italy. His experience is that the use of AI in high-risk surgery improves the likelihood that the patient will stay overnight in the ICU rather than having to stay there for a longer period [37]. AI-assisted high-risk surgery could mean a great advantage in terms of cost–benefit relationships. Researchers in computer science disagree to some extent with this rather optimistic finding from surgeons. They believe that improvements in the use of AI-generated imaging for the guidance of complex operations increase the likelihood of the patient having to stay overnight in the ICU, rather than the opposite. They argue that improving clinical outcomes in the form of a higher number of patients being discharged from the ICU in the afternoon can be a side effect of AI that investigators, ethics committees, surgeons, hospital administrations, and payers perhaps should focus on more, i.e., the avoided complications and extra days in the ICU [38].

Artificial intelligence (AI) can be used to help surgeons in the planning phase as well as during the procedure itself. The AI is fed a large number of images to help it learn what the patient's own anatomy looks like, which is then used to assist the surgeon in the operating room by providing guidance when the difficult, high-risk sections of the operation are performed. It can be a way for the surgeon to feel more reassured and confident by making the 'blind' parts of the procedure 'visible' [39, 40].

10.4.1 Improved Precision and Accuracy

High-risk surgery often necessitates reconstruction or curative treatment. This often involves procedures with a narrow margin for error, both in terms of the initial accuracy of the procedure and the subsequent timeframe. For example, if a new graft is not accurately joined during microvascular surgery, then blood flow must be quickly restored to the tissue to avoid ischemic damage. Or if a 'marginal' donor organ is not given away, the liver transplant must be completed within a limited window before the organ becomes unusable [41, 42]. AI's inherent capabilities in data processing, pattern analysis, and real-time decision-making make it ideally suited to support precision and accuracy in challenging surgical scenarios. This can be at a coarse level through the automation of important pre- and intra-operative tasks, such as organ localization, resection, and suturing. At a more refined level, AI can also support the successful execution of tasks that might be possible for a user but which are highly sensitive to fatigue or stress-induced hand tremor, such as the accurate separation of co-joined vasculatures. Finally, AI can also help prevent errors. The introduction of advanced imaging modalities, such as embedded near-infrared fluorescent angiography to track blood flow, can provide valuable in situ guidance to ensure the intended outcomes of high-risk procedures are being achieved in a timely fashion [43].

10.4.2 Enhanced Decision-Making Support

How can AI help individual complex, high-stress decision-making in high-risk surgical teams? Can we better provide each operating room's team with the necessary information, confidence, and trust they need to effectively work together, in order to (if not to cure the disease) reduce the chances of possible critical care and long-term death and minimization of the patient's family's pain and suffering? In Appendix I, I describe HIPAA-safe procedures to appropriately develop and use specific hospital electronic health data records. A great opportunity provided by today's electronic health record technology is the "check what's needed" stage [44]. Today, it is possible to effectively learn from previously recorded results of what happens when experienced human professionals create, say, the type of physical interventions on certain type each undergoing certain type initial after a certain type long-term. The first step is for prestigious and diverse medical domain expert confederacies to develop such custom-designed, essential datasets. Communication channels should include using private and public knowledge on favorite surgeon interventions and brands. Recall the famous line in Butch Cassidy and The Sundance Kid film about the danger of knowing real-time knowledge of the new Union Pacific safe. Such applications should be made to not only surgical team decision-making but also emergency department triage problems [45].

Can decision support systems enhance decision-making in high-risk surgery? For over a generation, it has been clear that experienced human surgeons, soliciting data from the patient's case and historical outcomes with similar cases, can't cover all the potential pathways and possibilities of further trouble. In the heat of the moment, psychology studies show we suffer from cognitive biases when facing risky decisions [45, 46]. I might even call this the case that Mother Nature feels regret. The majority of talk and work in medical decision support systems is falsely focused on replacing the experienced human, who solicits advice when new and unexpected special circumstances or problems arise. Such a focus misses the opportunity to effectively support. For complex care such as high-risk surgery, recognizing, understanding, and accommodating these notable differences between different human surgeons are prime opportunities for AI and system-based tools [47].

10.5 Ethical and Legal Considerations

Moreover, in addition to these basic principles, other issues will be important for policy and regulatory purposes [48, 49]. There is not necessarily an alignment between personal legally held information and legally derivable data. It is important that the majority of analyses from educated guesses are not instrumental in the discussion of change but are important to guide how the future unfolds. There is a need to develop different forms of supportive technologies to address what the best techniques are for human augmentation [50]. The well-being and dignity of the human person are highlighted as one of the first objectives of the international organization. It is not sufficient to argue that increased states are good instruments because they are getting better. The working group needs to work out a rationale that ensures human operations are not reproductive [51].

This research underlined the variety of ethical and legal considerations that are important in the development and integration of autonomous surgical procedures in the surgical theater. These considerations are important for ongoing progress in technology, the refinement of policy or regulatory responses, and the development of guidelines or best practices for surgical automation. It is important to build trust with patients and medical professionals [52]. These considerations are important for ongoing progress in technology, the refinement of policy or regulatory responses, and the development of guidelines or best practices for surgical automation. Transparency and the development of appropriate public dialogue are important to address the issues that arise. There remain substantial areas of difference across various health-care systems over a number of these considerations without international consensus [53].

10.5.1 Patient Consent and Autonomy

To the present time, the growth of health accession capabilities and health democratic awareness with AI has often been a disappointing failure; AI-based diagnosis is actually a frequently unreliable robust sensitivity, upgrades in supply techniques have, over the decades, created none of the actual technological knowledge specified, and state regulation impairments have privileged the technical destructions and prohibited acquisition resistance by the relevant human organizations. Contrast the unrealistic claims about the independent empowerment of AI-based operation due to infallibility with the Japanese operation disaster of the Kongo, which resulted from arrogant faith in the machine Infal [54].

Just as AI can impact patient accuracy, so can it be used to track whether a patient has really given valid, informed consent for a high-risk operation. AI can be used both to compare and contrast data recorded because someone has seen a doctor and believes a surgical situation requires immediate action, but the patient also said the operation data has the legal valid consent written. Even after demonstrating transmitted discussed issues in multiple-patient populations, it should be agreed that the best use of AI is in promoting the ethical and cognitive enhancement of human professionals [55].

10.5.2 Liability and Accountability Issues

But again, AI liability is a very new territory that the current law is struggling to handle. There are also questions of how the law needs to be further developed to deal with these future AI issues. The liability should be legally designated to strengthen the performance of AI when put into use, as the company that has an increased risk of being sued is responsible for making sure that the AI passes safety tests. Therefore, the company involved may consider dedicating resources, software engineers, as well as psychiatrists, radiologists, CRAs, or MRIs, and compliance staff to reduce exposure to lawsuits [56].

In terms of liability issues, this will depend on how the physician views the AI. If they view the AI as a tool or virtual assistant, the physician shall bear the liability in the event that errors, malfunctions, bugs, or the like have occurred during the medical operation. However, if they view the AI as a trusted clinical diagnosis, whereby the responsibility of the operation lies on the AI, the AI may probably decide to opt for the best course of action [57]. As such, the AI and the company that created the AI could be held liable and accountable for the actions that the AI decided to take and implement.

10.6 Future Trends and Innovations in AI-Assisted Surgery

AI and machine learning can be applied to improve perioperative patient care and enable remote telemedicine. This success is achieved through mass-market connected sensors, smart home monitoring, and deep neural network virtual complication metrics developed from thousands of electronic health records for real-time assessment and pre-disease prediction. Virtual reality, connected operating rooms, and high network bandwidth support the spatially aware mixed reality of AI and augmented human capabilities [58]. This relies on an ultra-low latency 5G private network combined with edge and cloud processing concepts to overcome the limitations of centralized cloud processing, such as the speed of light, packet overhead, and radiowave spectrum congestion [59]. This allows for remote high-power AI computing for the human cyber-physical systems who must make high-stakes decisions with blood on their hands, either inside or on the patient, at a specific time, place, and manner.

Thus far, AI and machine learning applications have focused on pattern recognition tasks such as imaging, labeling, and predicting solutions using known methods (e.g., titrating medication or identifying complicated anatomical structures to optimize device placement) [60]. They have also been used after surgery to plan robotic surgeries, including coordinating ports and placing arms. The potential for AI and machine learning extends to connected cyber capabilities, ranging from remote cloud computing to real-time sensing and commanding of AI within an augmented reality projected from remote access. This allows for the capture, embedding, translation, and display of vital signs from the body, the surgical environment, and virtual machine learning models directly into the surgical field of view [56].

10.6.1 Integration of Machine-Learning Algorithms

With respect to surgery, a large portion of tasks ranging from image analysis in radiologic tests to preoperative evaluation of patients and even artificial intelligence (AI) to improve the outcomes of the surgery can be made better in the direction of personalized surgery support systems [61]. Since surgery is one of the last resolutions to maybe the worst problem that a patient encounters, personalized assistance and supportive systems should also enable such operations to result in the best possible outcome [62]. This section aims to discuss the role of artificial intelligence with an inclusive approach in the captioned surgery process in a comprehensive way [59]. Everyone or every organization knows that breaking the barriers created by the standard procedures will cost a lot of money and time; however, in the end, the gigantic gains will be taken over.

In today's modern world, it is not difficult to find the interaction of robots with humans with daily examples. If one looks forward, personalized systems that are tailor-made just for ourselves come into play. In this matter, the integrated system provides feedback and forth between two or more systems. In this way, personalized systems that are integrated naturally with humans are made possible. With the advantage of fast digitizing world and the capability of humanity having an immense knowledge of data storage capacity, personalized complex systems that understand us, learn along with our interactions, providing us with their help will be implemented.

10.6.2 Advancements in Surgical Robotics

There has been a range of projects that aim to provide improvements to surgical robots, such as minimally invasive surgery, teleoperated fingertip control and manipulation, and minimally invasive and natural orifice surgery [63]. Surgical robotics have been developed to enable minimally invasive surgery (MIS), which features small incisions, less trauma to patients, faster recovery, and lower postoperative complications in comparison to conventional open surgery. In the development of surgical robotics, several types of robot systems have been used in surgery such as RAS (robotic-assisted surgery), RAV (robot with augmented visualization), RAVS (robot with augmented visualization and semi-automation), and RPlanningV (robotic planning and virtual navigations with real-time visualization) [64]. Experimental results of the surgical robot have demonstrated intuitive RAVS capabilities and trustworthiness of the system for effective clinical use. The conducive use of the robot platform for MIS robotic surgeons and clinical translation enables the potentials of supplying worldwide high-quality, high-standard surgeons to expand the reach to emerging and low resource settings. By aiding the earliest stages of laparoscopic surgery, such a "robot-augmented" surgeon may democratize MIS for procedures such as cholecystectomy and appendectomy [65]. The benefits of robotic dexterity offer minimally invasive natural orifices trans luminal endoscopic surgery (NOTES) on the jaw, pharynx, lung, rectum, and vagina. Teleoperated miniaturized robotic surgeries (T-shaped robot, independent anchor hooks) and dexterous endoscopic options offering advantages like greater degrees of freedom of the joint and the distal tip of the instrument could be useful tools in the field of robotic surgery [66].

10.7 Artificial Intelligence in Medical Imaging

Artificial intelligence (AI) techniques have taken an expanded position in various fields, including medical imaging. AI can perform a wide range of activities with the assistance of digital medical imaging, which comprises the automation of diagnosis and image analysis. By considering the various difficulties, AI has become more popular. For implementing AI with medical imaging, the computation needs to be performed at different stages [67]. AI performing with medical images requires preprocessing, feature extraction, segmentation, image post-processing, feature selection, and feature classification. Extracting features and performing multiple sequences will increase classifiers' efficiency. This ensures that pattern recognition plays a predominant role in all medical imaging methodologies [68].

AI includes numerous methods for computer-aided identification and imaging interpretation. These various methodologies are used in medical imaging systems, such as computational methods and AI methods. Mainly, the studies utilize AI to expand the use of medical imaging methods, incorporate image databases, and 3D image reconstruction. Feature extraction methods are used to remove non-related low-frequency values and enhance the image potential. The various extraction techniques used in medical imaging are also dependent on the nature of the medical images and the type of object to be segmented. Preprocessing techniques can greatly improve the intrinsic performance of image processing tasks such as image restoration, feature extraction, feature enhancement, and noise reduction [69].

10.7.1 Overview of AI in Healthcare

In healthcare, AI is referred to as using complex algorithms and software to emulate human cognition. AI is articulately deployed in the healthcare field. Machine Learning (ML) and Deep Learning (DL) are two substantial categories used to create AI models. ML deals with searching and analyzing data using statistical methods. ML models are capable of recognizing hidden patterns within the database [70]. ML can be divided into three groups: supervised, unsupervised, and hybrid learning. The supervised model implements input-output pairs using a ground truth dataset. In unsupervised learning, novel properties and relationship structures are accounted for. In the hybrid method, it involves both the techniques of supervised and unsupervised learning [71].

Deep Learning is a universal method used in computer vision and image recognition. One of the favorable aspects of DL over ML is that it can effectively handle large and complex data. The complex dataset has a deep learning model. Convolutional Neural Networks (CNN) are a popular form of DL techniques. CNN captures spatial and temporal patterns across images; these patterns are also known as high-level features of the images. Training an AI model requires large datasets and expert service support to build high-quality models [72]. The development of such software must reach validated metadata, follow explicit rule standards, and handle sensitive and interpretable high-quality data. Nowadays, AI has transformed healthcare and revolutionized it with a drastic increase in data. The AI model identifies potential cells for specific areas such as cancer detection, cancer treatment, and diabetic complications [73].

10.7.2 Applications in Medical Imaging

Deep learning-based models are reinventing medical imaging procedures. The conventional imaging-based diagnostic methods suffer from several shortcomings, prompting a rethinking of their formulation for quicker and more accurate diagnosis. However, the burden of sifting through a volume of medical imaging files for grasping subtle and sometimes imperceptible visual clues remains dominant. It seems that the golden phase of radiomics and texture analyses is rapidly exhausting patience as well as our big data acquisition capability [74]. The AI community is finding in generative adversarial networks, as well as deep learning-based neural networks, particularly CNNs, an unopposed power substrate to complement radiological advancements. CNN is indeed unleashing the full potential of a wave of revolution in making computer-aided diagnostic capabilities a reality for robust interpretation of medical imaging, ophthalmological examination, clinical pathology, and finally radiological reporting [75].

Artificial intelligence (AI) has made its pioneering footprints in the domain of medical image capturing, storage, and transmission by helping medical experts view health disorders in a natural and easy way. The human visual system is dumbfounded with the tasks of aggregating, integrating, synthesizing, and interpreting the surface visual structures of an anatomic plane as well as their color composites to achieve correct visualization answers. AI-based computational models are currently duplicating the visual interpretational capacity of the human visual system, sometimes even surpassing it [76].

10.8 Conclusion and Recommendations

At this point, the unnaturalness brought about by the simulation evaluation is part of the technical development process. However, due to cost, accessibility issues, ethical and safety oversight issues, and other factors, the current simulation system is not widely used and cannot effectively assist the surgeon's re-education and training. Therefore, providing tools that can improve surgery with feedback systems, restoring touch, and simulating tactile information integration will be very meaningful. Touch feedback will allow artificial intelligence to fully assist the surgeon in completing any procedure or any step within the operation.

Surgical handheld robotics is the first clinical application of artificial intelligence in digital surgery. It has opened a new era for digital laparoscopic surgery, allowing surgeons to perform high-precision and reliable operations at a low cost. It is of great significance for surgeons to use handheld surgical robots more efficiently to carry out training, and it solves a long-standing construction problem of digital surgery - loss of touch. The fact that no touch has been considered necessary when developing digital surgery tools has created problems for many surgical techniques that are highly dependent on touch, ultimately leading to a lack of tactile input during digital surgery. This will affect the quality of surgery, increase the difficulty of surgery, and ultimately affect the effect of surgery. This will be transformed when surgeons learn to perform delicate examinations or palpation skills, visualize the endoscope based on the digital platform, and use the surgical robot to perform high-precision dissections. We believe that it is now possible to restore touch and further digitize one or all of the surgical procedures within the framework, and artificial intelligence trained based on touch function is becoming closer, performing real-time feedback to help us obtain surgical feedback data and data fusion in the surgical area weapon-wheel.

10.9 Summary of Key Findings

Fortunately, the majority of AI uses in surgery are the assistance of expert and experienced surgeons and support their decisions. High-risk surgeries specific AI algorithms can play an indispensable role in assisting surgery. With the help of AI algorithms, the preoperative preparation of surgeons and doctors can be dramatically improved. Using AI assistances, high-risk surgeries can be considerably shortened and the complications can be dramatically reduced. These are enacted thanks to the fact that AI can analyze, perceive, and comprehend complex and large data sets more effectively compared to humans. Patient care has to be individualized and molecular biology will play a more important role in this sense. AI technology will support solving the problems such as personalized patient care, preventative approach, individualized planning also in before and after surgery periods, and contribute to personalized surgery in all aspects.

Artificial intelligence (AI), a technology that mimics certain functions of the human mind, also has the potential to play an important role in surgery and the consolidation of the surgery. AI systems are producing useful tools and applications and in future, there may be future AI-assisted and AI-directed autonomous surgeries. To this end, surgeons should become aware of the effect of AI and learn the necessary knowledge and skills for surgery.

10.9.1 Recommendations for Future Research

Given that high-risk surgery may have a huge impact on individual outcomes and costs, it is rare that research has been conducted into machine learning interventions in this field. We present an exploratory report about the potential utility of predictive models designed to ease the transition toward high-risk surgery. When present, the studies mainly concentrated on the use of traditional statistical methods and analyses used to develop said machine learning models. However, the evidence is insufficient to outline the impact and incentives for using such predictive models in clinical use. Such a standard would pave the way for valuable decision-making resources for patients, families, and doctors, thereby offering society important clinical benefits. Despite the great strides that have been made in the last century in surgery, it is still among the most dangerous procedures. Surgery has helped numerous human beings worldwide but high-risk operations sadly increase the risks and place strain on healthcare providers and their money. Recent advances in AI have taken us to the cusp of unprecedentedly predictive power, and researchers are quick to investigate how this may affect surgery. In this report, AI is herein presented in the context of high-risk surgery to assist surgeons. Until now, AI has contributed little to high-risk surgery. Future study should concentrate on how modern technology can lower the threat of high-risk surgery when being closely integrated with clinical workflows. This includes on how to make predictive models that are readily programmable in EPRs.

References

1. Fosch-Villaronga E et al (2023) The role of humans in surgery automation: Exploring the influence of automation on human–robot interaction and responsibility in surgery innovation. Int J Soc Robot 15(3):563–580
2. Luțenco V et al (2024) New horizons of artificial intelligence in medicine and surgery. J Clin Med 13(9):2532
3. Messaoudi N, Gumbs AA (2024) Artificial intelligence surgery: how autonomous actions and surgiomics can decrease risk in high-risk surgery. Recent strategies in high risk surgery. Springer, pp 605–620
4. Kunkes T et al (2021) Influence of hierarchy on risk communication during robot-assisted surgery: a preliminary study. Surg Endosc 1–7
5. Koch A et al (2023) Operating room team strategies to reduce flow disruptions in high-risk task episodes: resilience in robot-assisted surgery. Ergonomics 66(8):1118–1131
6. Thanh Thai M et al (2020) Advanced intelligent systems for surgical robotics. arXiv e-prints, arXiv: 2001.00285
7. Robertson S (2020) A novel web application for rapidly searching the diagnostic case archive. J Pathol Inf 11(1):39
8. Thomasian NM, Kamel IR, Bai HX (2022) Machine intelligence in non-invasive endocrine cancer diagnostics. Nat Rev Endocrinol 18(2):81–95
9. Jurado M et al (2022) The role of oncovascular surgery in gynecologic oncology surgery. Int J Gynecol Cancer 32(4)
10. Hlozek J et al (2022) Genetic changes in thyroid cancers and the importance of their preoperative detection in relation to the general treatment and determination of the extent of surgical intervention—a review. Biomedicines 10(7):1515
11. Bausys A et al (2022) Prehabilitation in modern colorectal cancer surgery: a comprehensive review. Cancers 14(20):5017
12. Deng J, Zhong Q (2022) Analysis of prognostic factors for in-hospital mortality in patients with unplanned re-exploration after cardiovascular surgery. J Cardiothorac Surg 17(1):82
13. Jin X et al (2020) Artificial intelligence biosensors: challenges and prospects. Biosens Bioelectron 165:112412
14. Qureshi R et al (2023) Artificial intelligence and biosensors in healthcare and its clinical relevance: a review. IEEE Access
15. Babu M, Jesudas T (2022) An artificial intelligence-based smart health system for biological cognitive detection based on wireless telecommunication. Comput Intell 38(4):1365–1378
16. Spadaccio C et al (2023) Early Re-exploration versus conservative management for postoperative bleeding in stable patients after coronary artery bypass grafting: a propensity matched study. J Clin Med 12(9):3327
17. Tran Z et al (2021) Failure to rescue after surgical re-exploration in lung resection. Surgery 170(1):257–262
18. Kitaguchi D et al (2022) Artificial intelligence-based computer vision in surgery: recent advances and future perspectives. Ann Gastroenterol Surg 6(1):29–36
19. Mofatteh M (2021) Neurosurgery and artificial intelligence. AIMS Neurosci 8(4):477–495
20. Varghese C et al (2024) Artificial intelligence in surgery. Nat Med 30(5):1257–1268
21. Bjerring JC, Busch J (2021) Artificial intelligence and patient-centered decision-making. Philos Technol 34(2):349–371
22. Argyle EM et al (2021) Physiological indicators of task demand, fatigue, and cognition in future digital manufacturing environments. Int J Hum Comput Stud 145:102522
23. Dyas AR et al (2022) Using the surgical risk preoperative assessment system to define the "high risk" surgical patient. J Surg Res 270:394–404
24. Wan MA et al (2022) Can the risk analysis index for frailty predict morbidity and mortality in patients undergoing high-risk surgery? Ann Surg 276(6):e721–e727

25. Meersch M et al (2017) Prevention of cardiac surgery-associated AKI by implementing the KDIGO guidelines in high risk patients identified by biomarkers: the PrevAKI randomized controlled trial. Intensive Care Med 43(11):1551–1561
26. Gumbs AA et al (2023) Surgomics and the artificial intelligence, radiomics, genomics, oncopathomics and surgomics (AiRGOS) project
27. Benzakour A et al (2023) Artificial intelligence in spine surgery. Int Orthop 47(2):457–465
28. Pakkasjärvi N, Luthra T, Anand S (2023) Artificial intelligence in surgical learning. Surgeries 4(1):86–97
29. Loftus TJ et al (2023) Artificial intelligence–enabled decision support in surgery: state-of-the-art and future directions. Ann Surg 278(1):51–58
30. Kumar A et al (2023) 6G-enabled IoT and AI for smart healthcare: challenges, impact, and analysis. CRC Press
31. Kumar A, Sareen P, Arora A (2023) Healthcare engineering using AI and distributed technologies. Smart distributed embedded systems for healthcare applications. CRC Press, pp 1–14
32. Guni A et al (2024) Artificial intelligence in surgery: the future is now. Eur Surg Res
33. Balch JA et al (2024) Integration of AI in surgical decision support: improving clinical judgment. Glob Surg Educ J Assoc Surg Educ 3:1–6
34. Walker RJ et al (2024) Hospital learning curves for robot-assisted surgeries: a population-based analysis. Surg Endosc 38(3):1367–1378
35. Püschel A, Schafmayer C, Groß J (2022) Robot-assisted techniques in vascular and endovascular surgery. Langenbecks Arch Surg 407(5):1789–1795
36. Xu Z et al (2022) Robot-assisted surgery in total knee arthroplasty: trauma maker or trauma savior? A prospective, randomized cohort study. Burns Trauma 10:tkac034
37. Lopez CD et al (2023) Machine learning can accurately predict overnight stay, readmission, and 30-day complications following anterior cruciate ligament reconstruction. Arthroscopy J Arthroscopic Relat Surg 39(3):777–786. e5
38. Bishara A et al (2022) Postoperative delirium prediction using machine learning models and preoperative electronic health record data. BMC Anesthesiol 22(1):8
39. Abraham J et al (2023) Integrating machine learning predictions for perioperative risk management: towards an empirical design of a flexible-standardized risk assessment tool. J Biomed Inform 137:104270
40. Hasanin AM, Mostafa MM, Abdulatif M (2021) Comment on: "Perioperative goal-directed therapy in high-risk abdominal surgery. A multicenter randomized controlled superiority trial". J Clin Anesth 75:110557
41. Kumari M (2023) Artificial intelligence and machine learning-driven decision-making in spinal disease treatment. J Theor Appl Inf Technol 101(24)
42. Saravi B et al (2022) Artificial intelligence-driven prediction modeling and decision making in spine surgery using hybrid machine learning models. J Pers Med 12(4)
43. Bajgain B et al (2023) Determinants of implementing artificial intelligence-based clinical decision support tools in healthcare: a scoping review protocol. BMJ Open 13(2):e068373
44. Tertulino R, Antunes N, Morais H (2024) Privacy in electronic health records: a systematic mapping study. J Public Health 32(3):435–454
45. Woldemariam MT, Jimma W (2023) Adoption of electronic health record systems to enhance the quality of healthcare in low-income countries: a systematic review. BMJ Health Care Inf 30(1)
46. Pal A et al (2024) Human factors in pelvic exenteration: themes in high-performing teams. Colorectal Dis 26(1):95–101
47. Gogalniceanu P et al (2024) Surgical leadership in a culture of safety: an inter-professional study of metrics and tools for improving clinical practice. Am J Surg 228:32–42
48. Kumar A et al (2024) Artificial intelligence bias in medical system designs: a systematic review. Multimedia Tools Appl 83(6):18005–18057
49. Kumar A et al (2023) Security and privacy issues in smart healthcare using machine-learning perspectives. 6G-enabled IoT and AI for smart healthcare. CRC Press, pp 41–56

50. Gros B, García-Peñalvo FJ (2023) Future trends in the design strategies and technological affordances of e-learning. Learning, design, and technology: an international compendium of theory, research, practice, and policy. Springer, pp 345–367
51. Bednar P, Welch C (2020) Socio-technical perspectives on smart working: creating meaningful and sustainable systems. Inf Syst Front 22
52. Ursin F et al (2024) Intraoperative application of mixed and augmented reality for digital surgery: a systematic review of ethical issues. Front Surg 11:1287218
53. Bellini V et al (2024) Artificial intelligence in operating room management. J Med Syst 48(1):19
54. King MR (2023) The future of AI in medicine: a perspective from a Chatbot. Ann Biomed Eng 51(2):291–295
55. Moy S et al (2024) Patient perspectives on the use of artificial intelligence in health care: a scoping review. J Patient Cent Res Rev 11(1):51–62
56. Shaik T et al (2023) Remote patient monitoring using artificial intelligence: Current state, applications, and challenges. Wiley Interdisciplinary Reviews: Data Mining and Knowledge Discovery 13(2):e1485
57. Golpayegani D, Pandit HJ, Lewis D (2023) To be high-risk, or not to be—semantic specifications and implications of the AI act's high-risk AI applications and harmonised standards. In: Proceedings of the 2023 ACM conference on fairness, accountability, and transparency. Association for Computing Machinery, Chicago, IL, USA, pp 905–915
58. Kumar A (2023) Augmented reality as a boon to disability. Smart distributed embedded systems for healthcare applications. CRC Press, pp 77–92
59. Bates DW et al (2021) The potential of artificial intelligence to improve patient safety: a scoping review. npj Dig Med 4(1):54
60. Zuhair V et al (2024) Exploring the impact of artificial intelligence on global health and enhancing healthcare in developing nations. J Prim Care Commun Health 15:21501319241245850
61. Khalighi S et al (2024) Artificial intelligence in neuro-oncology: advances and challenges in brain tumor diagnosis, prognosis, and precision treatment. NPJ Precis Oncol 8(1):80
62. Vollmer A et al (2023) Performance of artificial intelligence-based algorithms to predict prolonged length of stay after head and neck cancer surgery. Heliyon 9(11)
63. Barua R (2024) Innovations in minimally invasive surgery: the rise of smart flexible surgical robots. In: Emerging technologies for health literacy and medical practice. IGI Global, pp 110–131
64. Barua R et al (2023) Advances of the robotics technology in modern minimally invasive surgery. In: Design and control advances in robotics. IGI Global, pp 91–104
65. Chen K et al (2022) Robot-assisted minimally invasive breast surgery: recent evidence with comparative clinical outcomes. J Clin Med 11(7)
66. Masetti M et al (2022) Minimally invasive treatment of colorectal liver metastases: does robotic surgery provide any technical advantages over laparoscopy? A multicenter analysis from the IGoMILS (Italian Group of Minimally Invasive Liver Surgery) registry. Updates Surg 74(2):535–545
67. Kumar A, Singh D (Eds.) (2025) Revolutionizing Medical Systems Using Artificial Intelligence: A Breakthrough in Healthcare. Elsevier
68. Hardy M, Harvey H (2020) Artificial intelligence in diagnostic imaging: impact on the radiography profession. Br J Radiol 93(1108):20190840
69. Zhou SK et al (2021) A review of deep learning in medical imaging: Imaging traits, technology trends, case studies with progress highlights, and future promises. Proc IEEE 109(5):820–838
70. Bekbolatova M et al (2024) Transformative potential of AI in Healthcare: definitions, applications, and navigating the ethical landscape and public perspectives. In: Healthcare, MDPI
71. Tariq M et al, Principles and perspectives in medical diagnostic systems employing artificial intelligence (AI) algorithms. Int Res J Econ Manag Stud IRJEMS 3(1)
72. Kelso MR, Jaros M (2024) Improving wound healing and infection control in long-term care with bacterial fluorescence imaging. Adv Skin Wound Care 37(9):471–479

73. Zakir M et al (2023) Cardiovascular complications of diabetes: from microvascular to macrovascular pathways. Cureus 15(9)
74. Allam A et al (2021) Analyzing patient trajectories with artificial intelligence. J Med Internet Res 23(12):e29812
75. Kumar A, Bhalla S (2025) Exploring brain tumor detection through artificial intelligence. In Revolutionizing Medical Systems using Artificial Intelligence, pp 93–120. Academic Press
76. Mansour RF et al (2021) Artificial intelligence and internet of things enabled disease diagnosis model for smart healthcare systems. IEEE Access 9:45137–45146

Chapter 11
Benchmark Datasets for Analysis in Medical Systems

Abstract One of the vital areas within medical systems analysis is the benchmark dataset. Currently, many related research fields are using AI methods for more accurate and quicker diagnosis, prognosis, and generalization in many areas such as neurology, cardiology, retinal image processing, and so on. To make a decision, we need accurate data to support our analysis. Consequently, we need to consider performance and decide more efficiently. There are many neurological disorders, ranging from neurodegeneration to trauma, or other related pathologies, in which all of these illnesses have their own sub-type disorders. However, we will treat a patient, and most treatment will be either cognitive or medicinal therapy after we diagnose the disorders. The fundamental operations for disease analysis are image labeling, preprocessing, and feature extraction from a given dataset. In addition, after we have the dataset for each type of related disease, we will apply image regeneration-based techniques to learn the main principal components, reduce overfitting, and possibly improve the classification methods. We provide detailed benchmarks for subsequent analysis, namely diagnosis and prognosis for fatty liver classification, diabetic retinopathy, different neurological disorders, various mood disorders, cardiac-related treatment courses, and initiating early strokes and survival analysis. The paper is structured as follows: we briefly provide an overview of related data and methodologies for each dataset. The paper presents results of a series of benchmarks of different datasets in medical systems. The dataset is obtained from different sources and medical image data based on different diseases like strokes, diabetic retinopathy, and also based on spatial domain image data for neurological behavior diseases. It offers a performance analysis including precision, recall, F1-score, and the AUC of a machine learning paradigm. The result indicates a general trend in most of the datasets.

Keywords Benchmark datasets · Medical datasets · Healthcare data · Medical data repositories · Public health datasets · Standardized datasets · Electronic health records (EHR) · Disease classification · Patient health monitoring · Predictive modeling · Medical image analysis · AI model evaluation

A. Kumar and D. Singh, *Artificial Intelligence in Modern Healthcare System*,
Transactions on Computer Systems and Networks,
https://doi.org/10.1007/978-981-96-6703-1_11

11.1 Introduction to Medical Data Analysis

Today, the requirement for a proper medical data analysis technique has been recognized as critical to the success of diverse healthcare systems. It can handle (1) the enormous amount of medical information gathered from assorted sources such as clinical practices, bioinformatics, and imaging functions, delivering data styles for facilitating prognosis and treatment following specialized education and narrow research, (2) big data analytics for shedding light on genes and environments invoking divergent reactions in human biology, and (3) free medical records obtained from natural clinical practices that applicants cannot systematically manage for scientific analysis purposes. Accordingly, there exists an urgent call for data analytics where the amalgamation of required characteristics of the practice and measurements of patient parameters from many hospitals is not feasible with just one. Data extracted from diverse domains play a crucial role in developing enhanced healthcare with improved quality services [1].

Currently, medical data sets are numerous from diverse bioinformatics applications and other physiological and biomedical systems. New data resulting from advancements in technology and empirical valuation with detailed bibliographies are important to guide development in the desired direction. In addition, a huge amount of data collected for research purposes also includes public, federal, clinical, and ill patients, which constitutes a subset of the healthy population and hence shows less valid identification of patterns. As such, investigators find it hard to contribute to one whole, mainly because of the volume and quality of the data sets. Hence, the need for benchmark sets for new comparative research is a must. Many of the tools for prognosis, diagnostics, risk prediction, and clustering have already appeared in the recent past for the alliances of diabetes. Moreover, a relatively large number of tools and methodologies for diabetes prediction have appeared in the last few years, and the majority do not justify or appreciably cater to any significantly different application [2].

11.2 Importance of Benchmark Datasets in Medical System Analysis

Benchmark datasets are becoming more crucial for the analysis of medical systems. As a result, research has attained a degree of standard in developing new methods and algorithms. Benchmark datasets are required to evaluate a variety of applications making use of medical systems. Both their importance and potential applications are discussed. The study is intended to persuade the medical community to produce more quality benchmark datasets for researchers in any area to use. A medical data-based methodology for creating benchmark datasets is also described.

At the heart of extensive research, the importance of reproducibility and comparability cannot be overstated. Benchmark datasets aid in the quantification of various medical systems for the purposes of evaluating algorithms and methodologies. As a result of the availability of these data, numerous applications and techniques have been able to compete against each other. Additionally, benchmark datasets help researchers develop more advanced algorithms and techniques. Benchmark datasets play a vital role in the training of deep learning models and ensure the development of more accurate and improved models [3]. To substantiate these viewpoints, we must continue working and looking for new advancements in medical systems. A recommendation is provided as the final point: a plan to upgrade and change the present benchmark datasets in order to match the latest medical devices or systems. In order to detect medical systems training and validation, it is important to update the benchmark databases frequently, as the efficacy of some equipment changes according to their updates [4]. Researchers will no longer require a large amount of data to collaborate and develop deep and intelligent systems as a result of these databases. Data acquisition for medical monitoring devices must always be done ethically in order to protect the privacy of patients and researchers. It is critical to maintain patient values and norms when designing these new database frameworks. Ethical considerations can alter and reflect a person's situation or where they are living. In the near future, it is hoped that researchers will validate and repair problems in benchmark research through these datasets. Real-life applications, such as prediction or disease diagnosis, could be made more cost-effective or efficient. New techniques and applications are required to accomplish this. We need to continue to make advances in order to keep up with these disease prediction strategies [5].

11.3 Overview of AI Applications in Medical Diagnosis and Prognosis

The modern healthcare sector is marked by an advance toward the use of artificial intelligence (AI) as a tool for the improvement of clinical decision-making. Among the numerous areas of data-driven applications, machine learning and its special case, deep learning, are widely used primarily in diagnostic and prognostic modeling tasks. The application of AI techniques to medical tasks is aimed at enhancing diagnostic accuracy, reducing the time of obtaining a final diagnosis, and increasing disease detection sensitivity and negative test conversion. The number of AI applications in medical decision support is continuously increasing [6]. AI technology has demonstrated state-of-the-art performance on numerous tasks, improving validity and shortening the time of analysis of medical tests of different natures, from microscopic image cancer screening to ECG evaluations. Several case studies have demonstrated the capability of AI technology to outperform the greatest clinicians on specific medical tasks in terms of diagnostic accuracy [7]. Among various approaches to AI, deep learning, a subfield of machine learning, is making AI accessible to a broad range of applications with substantive potential for societal impact. Its prime strength lies in its ability to automatically detect features from raw data. Despite its immense potential, deep learning also brings its own set of challenges, particularly when moving to real-world settings, especially in medicine where the ethical use of AI should be crucial [8].

11.3.1 Neurological Disorders and Mental Illness Diagnosis and Prognosis

Diagnosis and prognosis of mental and neurological diseases is very complicated [9]. Currently, for clinical diagnosis and treatments, the evaluation of symptoms, clinical organization, cognition, metabolic status, brain physiology, brain anatomy, and genetic tools are used. The diagnosis of multiple neurological and psychiatric diseases is important for choosing the right treatment, estimating the disease progression, and distinguishing the causes of the symptoms accurately. The diseases are actually multifactorial complexes where the symptoms overlap and the individual's brain anatomy and physiology of a normal individual show various alterations during the execution of the same task. Despite the complex nature of mental and neurological diseases, and to develop predictive algorithms possibly considering the brain age, many studies have tried to classify subjects from functional, structural, and fMRI images with good generalization performance across different datasets, laboratories, and countries according to cognitive measures related to classifications in order to develop predictive and personalized diagnostic tools for clinicians [10]. In the following, we give some examples of diseases that should be investigated in terms of structure, function, and connectivity for developing benchmark datasets to identify the disease in the early stages before advancing to the point where they become irreversible in terms of treatment. There are different mental and neurological diseases, and each of them needs to be investigated in a separate study for developing benchmark datasets in terms of structure and function to be used for training (or testing) more accurate diagnostic methods. Diseases where many people face the diagnosis and prognosis include Alzheimer's disease, Parkinson's disease, Huntington's disease, autism, attention deficit hyperactivity disorder, schizophrenia, major depression, bipolar disorder, PTSD, traumatic brain injury, epilepsy, obsessive–compulsive disorder, Tourette syndrome, and severe anxiety. The following is the average age of onset and prevalence per year from major disorders. Correct diagnosis may help to treat people and manage the symptoms in order to effectively control them [11].

11.3.1.1 Key Challenges in Neurological Disorders and Mental Illness Diagnosis

Automated Diagnosis in Neurological Disorders and Mental Illnesses: A Critical Analysis of Benchmark Datasets; Evaluating the diverse range of manifest symptoms and formulating a diagnosis is one of the greatest challenges in neurological and mental illness conditions. Often, observers need to classify symptoms into various disorder categories, for instance, motor disorders, mental illness symptoms, comorbid symptoms, and symptoms of functional loss. Subjective interpretation of symptoms is inevitable, except for a few that are diagnosed objectively with specific biomarkers [12]. However, for most of these types of diagnoses, no definitive biomarkers or parameters are available for objective diagnosis. It is also seen that manual interpretations often lead to interobserver variability and misinterpretation due to bias errors. Additionally, some pertinent factors complicate disease diagnosis, including the stigma of mental illness, which requires precise diagnostic decision formulation to avoid subjective biases. The availability of poor drug targets also necessitates specific diagnosis.

One feasible way to provide a precise basis would be to create a collective effort of experts in the fields of neurology and psychology, along with the addition of patient history as an important interdisciplinary approach to make accurate diagnoses of mind-brain-sensorium conditions. The need for objective, continuous, and real-time examination of evolving mental disorders is currently being enabled using radiological imaging techniques, supported by other advanced tests including biochemical, electrophysiological, and cardiovascular assessments [13]. Various tools have been used in the field, including clinical scales and screening questionnaires that are used in primary and secondary care to distinguish among neurological and mental illness conditions. On the other hand, cognitive assessment tools for diagnosing mental disorders have been developed systematically to adhere to psychiatric comorbid assessments at the bedside. It is evident from the current literature survey that most existing diagnostic tools are suicidal psychometric evaluation techniques. These are often seen as self-reporting and interview questionnaires that loosely differentiate among affected persons and groups, or specifically, any two similar categories. This is sufficient to predict the disorder type or category. In future, comprehensive data mining techniques and artificial intelligence applications will play important roles as advanced decision-making tools for disease diagnosis [14].

11.3.1.2 Benchmark Datasets for Neurological Disorders and Mental Illness Analysis

In systems dealing with neurological and mental health diagnosis, it is of utmost importance to have an external validation of the outcome given by the systems. Access to benchmark datasets is the main hub in diagnostic algorithm development, where several pathologists diagnose diverse diseases in relation to clinical or imaging supervised assessments aiming at a medical curriculum. Benchmark datasets will ensure the reproducibility between different experiments and algorithms that claim to diagnose the illnesses or disorders of interest, facilitate early detection of patients who are possibly developing the disease, and might lead to personalized therapy, predict the treatment outcome, and promote collaborative algorithm development, among others. There is no free large dataset of brain tumors in pediatric patients; in that context, public repositories can be used, focusing on data of children with brain tumors [11].

A huge number of public websites offer numerous datasets; they are partially listed for reference purposes. Some datasets are available with patients' clinical assessments and imaging files, and others have an extensive range of available data with readily downloadable genetic information. Intellectual disability and attention-deficit hyperactivity disorder assessments are not within the gender or touchstone analysis public datasets and usually have a relatively low number of assessments in the ones found [15]. This is probably due to the complexity of the diagnostic process and patient confidentiality constraints. However, sufficient image uploads for the comparison of algorithms can be found online. In contrast, cancer data usually have tens of thousands of highly curated patients for benchmarking deep learning algorithms, but validity constraints [13].

11.3.2 Brain Tumor Prediction Using Brain Imaging Segmentation

Prediction of tumors in the brain is one of the most significant studies in medical applications [16]. On investigating the entire literature, we find that the details present on tumor prediction through brain imaging are limited. This could be an inoperative source or matter, and we suggest that only a few publications have worked extensively on tumor prediction. Tumor detection and prediction are done with the help of segmentation of the contributed abscess in preoperative and postoperative brain imaging. Segmentation of predicted biopsy collected data from an MR image report is the need of the pathology for detecting the tumor separately. Several studies have attempted brain imaging for tumor prediction by using classification or prediction models. The principal contribution behind choosing brain imaging is its primary function [17]. It appears to confirm persistent symptoms and is a rather essential test result that provides detection of infarction, tumor, and other abnormal activities of bleeding to guide evidence-based treatment. The human brain is imaged with the help of diagnostic imaging techniques like MRI and CT scans. Segmentation is also preferred in the medical imaging reports through the data of computational brain imaging segmentation since normal patients can differ in different report scans [18].

Classifications or models that predict the tumor are mainly dependent on single or multiple imaging modalities. There has been interest for more than a decade in using magnetic resonance imaging to improve tumor segmentation. There is significant clinical interest in recent years in MR and CT imaging for developing statistical atlases of normal human brain anatomy and quantitative assessment of brain tumors. The prominent five main modality units responsible for the examination of MR are T1, T1-weighted, T2, T2-weighted, T1c, and T1-weighted post-contrast. The segmented masks consist of whole brain, white matter, gray matter, tumor cyst, edema, necrosis, and tumor-enhancing lesions for relevant classifications. Several studies support independent model predictions, random forest classifiers, and multiple features with reduced modality, achieving 91–97% accuracy of a large dependent classification model. Medical data implies that integrating clinical data and imaging data sets may yield better predictions than multi-dimensional systems; clinic computed tomography data was fused with imaging data sets, and it showed enhanced model data, resulting in better results and accuracies. Research in neuroimaging has now shown tremendous interest in further brain tumor studies that combine clinical and molecular data [19, 20].

A large number of recent neuroinformatics spectroscopy writings and pharmacogenomics therapies enable active machine learning and analog methods research. Convolutional neural networking or deep learning for topology such as UNet can detect and search and design the intensity of the lesions. For example, U-adds can have pre-and post-fused modalities to improve complete search results, and high-end models increased testing response in a day model. If this model is pre- and post-fused MRI imaging using MRI CT scan that is known for the last complete tumor and brain management, it can help the model learn the name of search areas and further improve search accuracies. This involves artificial intelligence and aids in transforming the patient's treatment plans [19].

11.3.2.1 Significance of Brain Tumor Prediction

The prediction of brain tumors, being one of the most common tumor sites in the body, is of paramount importance. Although brain tumors are relatively rare, their high morbidity and mortality pose major public health challenges. The prognosis of brain tumors is very poor because the tumors grow very rapidly and the surrounding structures of the brain are affected by the tumors. In such cases, early intervention is essential to save the patients. The choice of treatment modality depends on the complete grading of the brain tumor tissue. The availability of efficient brain tumor prediction methodologies in the clinic is the need of the hour. Early detection of brain tumors is the key to saving patients' lives because the 5-year survival rate can be improved with surgical procedures. When growing in the major portion, there is a high possibility of impacting the patient's personal abilities. These psychological imbalances can be well managed and can avoid such strain by providing predictive capabilities if we plan to save the brain tumor patient [18].

The prediction of brain tumors is the clinical challenge facing most brain tumor clinicians. The CNN model with the VGG-16 architecture achieves the highest classification. It is possible to get clearer details, diagnostics, and accurate predictions of brain tumors using AI and MRI features. Prediction of brain tumors can avoid unnecessary operations and save patients' lives. We need to standardize the prediction protocols used in medical systems, whether they are experimental or analyzed in an actuarial technique. The balance of increased patient satisfaction and care is affected by the knowledge of relevant outcomes and predictions. For fair and ethical reasons, it is important to recognize any vulnerabilities related to the trivial size associated with brain tumor patients when designing such predictive models. New strategies to improve care quality need to be discovered so that informed decisions can be made [17] (Table 11.1).

11.3.2.2 Benchmark Datasets for Brain Tumor Prediction

The datasets containing brain tumor imaging information are collected from different hospitals and various imaging instruments. Their wide range of variations has implications for pathological diagnoses and treatment plans. To make a reliable evaluation, benchmark datasets were developed. These benchmark datasets include data from one or more hospitals and also include different magnetic resonance (MR) imaging video sizes and different diagnoses. They contain both preprocessed data such as T1c, T2, T2-Flair, pre-Gd, and GD-transformed tumor images such as T1. Normative datasets also include some clinical parameters, data sources, and data timestamps, which can better test and analyze the data. Researchers have formed several robust benchmark datasets as an overview of the key hurdles such datasets are required to be distinguished. This includes the absorption of multimodal images consisting of various imaging sequences along with clinical parameters associated with brain tumors [28].

Each dataset is prepared to make the process of data analysis robust rather than scattering different datasets from distinct sources in order to illustrate the advanced domain in conjunction. Each dataset provides a different set of instructions according to the scope of a dataset. Some datasets include data varying with respect to age and diagnosis for a single group of patients at a single hospital. Other datasets include different patients from other sources. Importantly, these datasets differ in size and consist of 2D and 3D data. Some of these datasets were established in the past decade to include the most comprehensive imaging data in the medical field. The advanced models that predict the body part of the dataset can be evaluated [29].

The need to improve predictive techniques and their models in the brain tumor field has led to the creation of several datasets. Nowadays, despite the presence of different collections, they assist in one or several tasks such as disease classification, survival analysis, habitability prediction, tumor segmentation distribution, and surgical therapy. Many new complementary collections that appear in two or more tasks are not yet public in nature and are being continuously investigated and duplicated by other professionals. In our current dataset, in contrast to the view of previous datasets, the path that distinguishes the centers and tumor types in the case of multicenter data was also used to find a basis for analyzing data from multiple centers. Also, comparable global sums of abnormal tumor systems are given at small tumor points. New significant progress has been made in building unique ground facts that

Table 11.1 Description of brain imaging datasets

Datasets	Dataset type	Number of samples	Key features
BRATS (Brain Tumor Segmentation Challenge) [21]	MRI scans	3000 +	Multi-modal images, tumor annotations
	Patient types	500 +	Gliomas and meningiomas
TCIA (The Cancer Imaging Archive) [22]	Brain tumor images	10,000 +	Various imaging modalities
	Clinical data	Varies	Includes associated patient metadata
CAMP (Cancer and Aging in Male Patients)	MRI scans	500 +	Focus on aging-related tumors
	Patient demographics	Available	Includes age, gender, and treatment data
Brain tumor dataset (Kaggle) [23]	MRI images	500 +	Labeled images of tumors
	Class labels	4 Type	Includes benign and malignant tumors
Multi-modal brain tumor segmentation [22]	Multi-modal images	1000 +	T1, T2, Flair modalities
	Annotations	Available	Segmentation masks for training
NCI Genomic Data Commons (GDC) [24]	Imaging studies	20,000 +	Includes clinical and genomic data
	Tumor types	Various	Covers multiple cancer types
OASIS (Open Access Series of Imaging Studies) [25]	MRI scans	1000 +	Focus on aging and dementia
	Patient information	Available	Include demographic data
IXI dataset [26]	MRI scans	600 +	Multi-modal imaging
	Annotations	Limited	Mainly for research use
DICOM from RSNA [27]	DICOM images	1000 +	Rich clinical imaging data
	Metadata	Available	Includes patients and study information
DeepLesion [22]	Medical images	32,000 +	Includes brain and other lesions
	Annotations	Available	For segmentation and detection tasks

not only exclude those facts but also have a look that is still open. The new datasets have the potential to identify the patient as multi-integrated trust signals in hospitals and to have an exceptionally large palm configuration measurement data source. More than 10 doctors have been asked to assess the integrated multispectral images of such good hand data for the complete statistical stroke treatment for thousands of patients [19].

11.3.3 Predicting Diabetic Retinopathy Using Retinal Imaging Segmentation

Diabetes is the leading cause of blindness, with the number of people facing vision loss being 1.1 billion. Diabetic retinopathy, a general complication of diabetes, is asymptomatic in the initial stages. As a result, detecting diabetic retinopathy in advance of clinical symptoms is essential to controlling it [30]. We categorize the methods of retinal imaging to identify diabetic retinopathy based on fundus photography, identifying the mild and severe stages of the disease over angiographic modalities. Using retinal imaging techniques, we consider the usefulness of captured images. This imaging is extracted from multiple signs into a technique called retinal imaging segmentation and reconstruction [31].

Retinal imaging can diagnose diabetic retinopathy, but until now, the predictive capability of the diagnosed retinal changes to those involving diabetes has not been established. Many trials of the segmentation method of retinal imaging background and retinal imaging segmentation technique behind diabetic retinopathy are presented. Researchers perform the task of classifying diabetic retinopathy by supervised learning following a retinal imaging digital enlargement signal. Segmentation may produce a disease-related area that is more accurate in detecting subtle disease and improving the system's interpretation regarding the characteristic features after separation. Segmentation has been a trend in the process of detection for the identification of changes in pathology, so it is concluded that the performance level reaches a prediction value of 99%. The right timing or concept is needed for the implementation of the signal available in the previous segmentation method. However, the proposed method developed in this chapter does not detail the segmented area as the main focus of further capabilities to improve the system [32].

11.3.3.1 Understanding Diabetic Retinopathy

Diabetic retinopathy (DR) is a common microvascular complication of both type 1 diabetes mellitus (T1DM) and type 2 diabetes mellitus (T2DM). The pathophysiology of DR is a complex system of retinal inflammation, vasculature, and neuronal damage. Chronic hyperglycemia is the prime stimulus for vascular changes in the retina, classified as either microaneurysms or compromised circulation. Prolonged hyperglycemia first dilates the blood vessels; however, the walls of the vessels become cauterized, leading to capillary occlusion. This leads to hypoxia-inducible factor-1 activation within the retina, aiding in angiogenesis via VEGF, which creates both neovascularization and vascular leukostasis. Nonproliferative stages were described and later expanded based on initial severe levels of grading. Four stages were recognized, from minimal to severe, and proliferative DR was further classified by severity and risk of vitreous hemorrhage for prognosis [33] (The description of CVD diabetic retinopathy dataset is in Table 11.2).

The actual prevalence can be higher because a significant proportion of diabetics remain undiagnosed. Diabetic patients have about a 2% annual risk of becoming blind. Because of the reversible nature of early microaneurysms, it is important to detect capillary dilation in the early stages of the disease; therefore, healthcare providers recommend annual exams to detect early signs of DR. Photocoagulation treatments such as argon or xenon laser and, recently, pharmacologic therapy with VEGF inhibitors have been used to reduce macular edema and loss of vision by promoting the regression of pathological fibrovascular membranes. The current standards of care are the most effective in treating the vision-threatening consequences of DR and rebleeding. Moreover, systemic anti-VEGF drugs may even alter the natural history of DR by not only being a therapeutic treatment but also a diagnostic tool in early prediction [43].

11.3.3.2 Benchmark Datasets for Diabetic Retinopathy Analysis

Analysis of disease focuses on the computer-aided prediction and diagnosis of several eye-related diseases, including diabetic retinopathy. To validate the robustness and accuracy of the developed predictive model, many researchers use public data as benchmark datasets. Thus, several benchmark datasets for the analysis of diabetic retinopathy have been made available. These datasets not only offer a standardized set of ground truth but also relatable data favoring the detection model when the researcher aims to develop a prediction model [44].

The development of new diabetic retinopathy datasets is essential because they offer diverse, high-quality image data and relevant clinical information. This dataset's clinical data provide a patient's medical history, including diabetes status, with additional ocular diagnoses giving a comprehensive analysis of risk prediction factors. Owing to the fact that focus is shifting toward the computer-aided prediction of disease, it is critical that retinal images and eye exam results are not only collected consistently from different medical institutes but also that firmware upgrades and new conventions in imaging are conducted. The annotations for diabetic retinopathy datasets are produced by employing a multicriteria system approach [45]. This is the approach used by various datasets designed for diabetic retinopathy. The databases designed to assess retinal fundus images with regard to the presence of diabetic retinopathy are available for public use. The datasets are organized and updated to promote public health by encouraging research and publications in the development of automated algorithms for the detection of diabetic retinopathy. Sharing

Table 11.2 Description of CVD diabetic retinopathy datasets

Datasets	Dataset type	Number of samples	Key features
Kaggle diabetic retinopathy detection [34]	Fundus images	35,000 +	Labeled for stages 0–4
	Image resolution	3000 × 3000	High-quality images for analysis
DRIVE (Digital Retinal Images for Vessel Extraction) [35]	Fundus images	40	Includes annotations for vessel segmentation
	Annotations	1000 + pixels	High-resolution masks
STARE (STructured Analysis of the Retina) [36]	Fundus images	400	Contains a variety of retinal conditions
	Annotations	Vascular and lesion masks	Detailed structure for analysis
EyePACS dataset [37]	Fundus images	88,702	Diverse patient demographics
	Image resolution	Varies	Images from various sources
APTOS 2019 blindness detection [38]	Fundus images	4000	Labeled for stages 0–4
	Image resolution	2048 × 2048	High-resolution for detailed analysis
RetiSpec dataset	Fundus and OCT images	Varies	Multi-modal imaging data
	Annotations	Varies	Comprehensive labels for conditions
OCT (Optical Coherence Tomography) dataset [39]	OCT images	3000 +	Useful for cross-modal analysis
	Annotations	Varies	Segmentation labels for various conditions
Indian Diabetic Retinopathy Image Dataset (IDRiD) [40]	Fundus images	5000 +	Labeled for segmentation and classification
	Annotations	Available	Detailed masks for lesions
MESSIDOR (MEdical Screening of Diabetic Retinopathy) [41]	Fundus images	1200	Includes images for training and validation
	Annotations	Available	Annotations for diabetic retinopathy features
RETINA image database [42]	Fundus images	Varies	Includes multiple conditions
	Annotations	Available	Labels for various retinal diseases

retinal images is a global effort and goes beyond national boundaries. Government bodies and private organizations have been funding public health research by sharing the dataset within the scientific community [46]. The availability of the dataset is an essential real-world condition, and the success of public health depends on the automated systems in place. In this section and the following subsections, we review various datasets and clinical analytics applications on fundus images. We also review the publicly available databases for diabetic retinopathy across the world in various repositories.

11.3.4 CVD Risk Stratification Using IVUS Imaging

Cardiovascular disease (CVD) remains one of the major causes of death worldwide and demands systematic risk stratification. The presence and degree of atherosclerotic burden are crucial for risk estimation. A study has been designed to investigate the role of high-resolution imaging and advanced computation in predicting future cardiovascular events in low to intermediate risk individuals with subclinical coronary atherosclerosis. This imaging approach allows the acquisition of detailed real-time, precise, high-resolution cross-sectional images of the artery wall and direct visualization of patterns of atherosclerotic modifications, wall-lesion interactions, and arterial function, particularly in the coronary arteries where arterial wall distortions and motion artifacts may occur less frequently [47].

Effective risk stratification tools are needed in order to identify patients at high risk of cardiovascular events who truly deserve medical therapy to prevent silent myocardial infarction (MI). Indeed, event prediction with a single risk factor in light of weak benefits from preventive therapy is less important. A variety of methods are commonly used to leverage imaging data, including physician examination of artery images, manual or computer-aided measurements of arterial dimensions, echo signal interpretation, or computation based on various metrics derived therefrom for multi-group or individual measurements. Combining clinical and imaging findings has been of increasing importance in clinical patient management and could predict future ischemic events when patients undergo coronary angiography [48]. Generating a full patient profile integrating clinical, laboratory, carotid fine B-mode echo, and imaging allows for refined ischemic risk prediction in medium and high-risk patients. Developments regarding plaque composition in non-culprit coronary arteries might predict culprit lesions when moderate predictors are already elevated in a heterozygous familial hypercholesterolemia population. Moreover, B-mode echo combined with artificial intelligence confirmed the predictive role for future ischemic events of carotid remodeling evolution in a follow-up clinical trial in which normal to prevent coronary artery disease was examined. In diabetic patients, who are at increased risk of cardiovascular events, plaque strain rate assessing mechanical properties in culprit and non-culprit arteries was able to predict 36 months for plaque rupture extensions, clearly linking positively to the vascular system. More frequent updates are gradually proposed in which machine learning was employed to better refine the clinical risk of cardiovascular events during the broader follow-up [49].

11.3.4.1 Cardiovascular Diseases and Risk Stratification

More than 80% of citizens of economically developed countries worldwide die of cardiovascular diseases. CVD occurs in different forms, and coronary heart disease and stroke are the most common. Today's advanced diagnostic capabilities enable timely diagnosis and the development of appropriate treatment strategies. It is essential for medical specialists to identify patients at both low and high risk of CVD. The smaller the probability of developing the disease, the more circumspect and cautious specialists should be about starting preventive and treatment procedures, balancing potential benefits with possible ill effects. In contrast, a more severe patient prognosis requires a more accurate analysis of the patient, as well as changes in their daily habits and lifestyle [50]. One of the major patient risks in developing CVD is considered to be the presence of various cardiovascular risk factors contributing to the deterioration of metabolic and functional parameters [51].

From a clinical perspective, the most important factor for the development of the stratification strategy and the selection of the optimal treatment is the risk of adverse CVD outcomes, such as cardiovascular mortality and morbidity. Achieving high sensitivity and specificity in these pathologies is the primary role of long-established traditional cardiovascular risk factors; for example, hypertension, diabetes, hyperlipidemia, and cigarette smoke have played a key role in the assessment of atherosclerotic cardiovascular risks over the past three decades. The final stage of the atherosclerotic process, as well as the most recent and decisive step for cardiovascular consequences, is an acute critical event: the plaque rupture, activating the coagulation chain, clot formation, and the development of an acute crisis with occlusion/hemodynamic collapse and the precipitating symptoms. In this context, an increased risk of CVD outcomes also means assessment of an increased risk of acute cardiovascular mortality/morbidity in the short term. Early intervention in these patients could significantly reduce potential mortality and morbidity. The evolution of emerging technologies plays a major role in improving clinical atherosclerotic patient risk phenotyping and, thus, in improving the accuracy of the risk assessment [52] (The description of cardio vascular disease dataset is in Table 11.3).

One of the major challenges of CVD stratification is handling the extreme variability of the population regarding different included factors. The combination of most of these known and new risk factors in a Power-Scoring Risk Equation is somewhat useful, but robustly evaluating the interacting risk factors for the evolution of cardiovascular diseases in a highly variable population, by influencing sometimes several linked cardiovascular pathways categories, could be very difficult in defining the evolution of the cardiovascular process in every single categorized patient heterogeneity. Indeed, there are different external and internal factors that interfere in this process, contributing to diversity within those included in the same category [54].

Table 11.3 Description of cardio vascular disease datasets

Datasets	Dataset type	Number of samples	Key features
IVUS imaging dataset from the American College of Cardiology (ACC)	IVUS images	1000 +	Includes diverse patient demographics
	Clinical data	Available	Comprehensive patient health information
The Cardiovascular Imaging Study (CAVIS) [53]	IVUS images	Varies	Focus on coronary lesions
	Patient outcomes	Available	Longitudinal data on treatment efficacy
The IVUS-QCA database	IVUS and QCA images	500 +	Detailed lesion characterization
	Clinical data	Available	Integration of imaging and clinical outcomes
Cardiovascular Disease Imaging Repository (CDIR)	Multi-modal images	2000 +	Includes IVUS, angiography, and more
	Patient demographics	Available	Comprehensive clinical information
The Rotterdam study	IVUS images	1200 +	Focus on aging and cardiovascular health
	Longitudinal data	Available	Detailed patient history and outcomes
The BioImage study	IVUS and MRI images	800 +	Comprehensive cardiovascular imaging
	Patient profiles	Available	In-depth clinical evaluations
The AtheroPoint IVUS dataset	IVUS images	600 +	Focused on atherosclerotic lesions
	Clinical data	Available	Patient risk factor information
IVUS and OCT imaging dataset	IVUS and OCT images	400 +	Detailed comparison of imaging modalities
	Annotations	Available	Includes segmentation and feature labels
The MACE (Major Adverse Cardiac Events) IVUS dataset	IVUS images	300 +	Focus on complications and outcomes
	Clinical outcomes	Available	Detailed tracking of adverse events
The Vascular Quality Initiative (VQI)	IVUS and clinical data	10,000 +	Comprehensive data on vascular interventions
	Quality metrics	Available	Includes patient outcomes and follow-up

11.3.4.2 Benchmark Datasets for CVD Risk Stratification

To create, update, or compare CVD risk stratification analysis, benchmark datasets have been adopted. These are representative in terms of both the participants and related clinical outcomes, yet also include large volumes of data on comorbidities and relevant standard imaging where available, as well as longitudinally acquired ECG data. In such datasets, candidate predictors can be used to create or validate the algorithms present with clinical outcome labels and be comparatively evaluated against one another. Benchmark datasets explicitly designed for the development and validation of risk algorithms for the prediction of CVD are presented here. People can evidence new risk stratification approaches that are both successful and groundbreaking through careful studies of each dataset's individual qualities and the pooling of larger datasets in combination with secondary proof of organizational measures also provided at the dataset level [55].

The research is broad and focused, considering most risks related to clinical outcomes, related biomarkers, and imaging details. Four datasets with longitudinal ECG data and another three with specific clinical staff-level familiarity with ECG quality assurance provided detailed imaging-related information, consolidating a total of 96,695 participants, comprising 39.9% women and representing a large number of relevant developed countries. For the standardization and derivation of universally well-represented risk algorithms for the prediction of imminent CVD, collaboration vehicles dedicated to public forums are required when benchmarking on these datasets [56].

11.3.5 Breast Cancer Prediction

Background: In women, breast cancer is the most prevalent cancer, with early detection and diagnosis resulting in successful therapy. Despite the availability of effective approaches to treatment and diagnosis, breast cancer has become a life-threatening disease. Breast cancer is the most prevalent cancer in women, accounting for almost 25% of all cancers, or one million new instances. Tumors that form in the breast region of a woman's body are referred to as breast cancer. Tumor growth in the cells of the female breast, which then grows to the surrounding tissue, is how breast cancer is defined. It grows and invades different layers of surrounding tissue that cause substantial harm afterward. The age of onset is decreased to 30 to 45 years in breast cancer occurrence in women.

Clinical significance: Breast cancer, a malignant type of cancer, is accountable for millions of fatalities in women. When the surface grows milk-producing gland cells, various kinds of risk factors might result in breast cancer's etiology. There are diverse kinds of breast cancer with varying degrees of invasion, impingements, and complexities directly associated with personal lifestyle and genetic background. Advancements in molecular and genetic research have resulted in precise breast cancer analysis approaches. To date, clinicians have conducted diagnostic research into just a diagnosis, a guiding clinic, and effective breast cancer enhancement which can take it smoothly with further guidance. Fortunately, the recent developments in soft computing and computer-aided diagnostic technology have allowed increased accuracy in predictive investigation for better diagnosis and subsequent treatment. Many current predictive methodological research has integrated clinical information, digital imaging, and proteomics, promising to elucidate more biological and clinical queries. Similarly, several latest synthetic methodologies optimize execution by hand from the low-command adjustment and separate learning. This chapter starts from the section that discusses the recent technical advancements in breast cancer using the latest clinical remarks. Following that, it goes through the origin part, which discusses the breast cancer consequences. These findings provide a methodological basis that demonstrates the necessity for a more accurate means of discovering the fields and anomalies. There is also a lot of modifications to the individual technique.

11.3.5.1 Breast Cancer Statistics and Impact

In 2020, breast cancer was the most commonly diagnosed cancer in nearly all regions and the leading cause of deaths from cancer worldwide. In females, breast cancer is the most frequently diagnosed cancer, comprising nearly a quarter (24.5%), and about 1 in every 6 women diagnosed with breast cancer would not survive. The incidence rate has been decreasing in the past few years, with real-time statistics available. Between 2009 and 2010, the global estimate of the age-standardized incidence rate of breast cancer was found to be 43.0 per 100,000 women. The incidence rate of breast cancer is influenced by age, and worldwide, there is a trend of increasing age-specific incidence rates of breast cancer among women aged 40–74 years, except for the age group 45–49 years. For instance, the rates in Brazil start at 87 per 100,000 women in the age range of 40–44 and go as high as 1420 per 100,000 women in the age group of 70–74. Also, for a higher age of up to 79 years, the rates decrease to 1420 per 100,000 women, but the decline is not consistent. As with many cancers, the incidence of breast cancer is higher in older women compared with younger women in the Brazilian population [57] (The Description of breast cancer prediction dataset is in Table 11.4).

Impact for women worldwide, it is reported that one in every eight women could be diagnosed with breast cancer in their lifetime. Experiencing breast cancer is extremely stressful, and the treatments, including surgery, chemotherapy, and radiotherapy, have side effects leading to a change in lifestyle, which can alter body shape and image, causing a loss of self-esteem, emotional strain, divorce, depression, and frequently, even suicide. In addition to this, the breast cancer diagnosis can cause financial strain, as adults diagnosed with cancer would likely reduce their hours of work, resulting in average earnings loss each year. It is reported that early diagnosis increases the chances of survival, and hence, the development of techniques in the early detection of breast cancer would reduce mortality and improve outcomes [58]. Furthermore, it is reported that one in three cases of approximately 6.38 million cases related to economic burden and nearly 71,000 new cases of breast cancer could

Table 11.4 Description of breast cancer prediction dataset

Datasets	Source	Description	Key features
Breast cancer Wisconsin (diagnostic) dataset	UCI machine learning repository	Contains 569 samples with 30 features derived from FNA images, focusing on benign vs. malignant classification	Cell nucleus characteristics: radius, texture, perimeter, area, smoothness, etc.
Breast cancer Wisconsin (original) dataset	UCI machine learning repository	An earlier version with 699 samples and 10 attributes, also for benign vs. malignant classification	Mean values of radius, texture, area, perimeter, etc.
METABRIC (Molecular Taxonomy of Breast Cancer International Consortium)	cBioPortal	Extensive genomic and clinical data from over 5,000 patients, facilitating understanding of molecular characteristics and treatment outcomes	Genomic data, clinical attributes, subtype classifications
NCI Genomic Data Commons (GDC)	National cancer institute	Provides access to diverse genomic, transcriptomic, and clinical data related to various cancers, including breast cancer	Genomic datasets: mutations, expression data, clinical information
SEER database (Surveillance, Epidemiology, and End Results)	National cancer institute	Population-based cancer statistics, providing insights on incidence, treatment outcomes, and demographics	Patient demographics, tumor characteristics, treatment data, survival statistics
Kaggle breast cancer dataset	Kaggle	Multiple datasets available for analysis, often related to competitions on breast cancer classification and prediction	Varies widely; includes clinical, genomic, and imaging data
Breast Cancer Surveillance Consortium (BCSC)	BCSC	Data on breast cancer screening, diagnosis, and outcomes, focusing on risk factors and the effectiveness of screening methods	Screening history, biopsy results, tumor characteristics, follow-up outcomes

(continued)

Table 11.4 (continued)

Datasets	Source	Description	Key features
The Cancer Genome Atlas (TCGA)	National cancer institute	Comprehensive datasets for various cancer types, including genomic, epigenomic, and clinical data for breast cancer	Genetic mutations, expression profiles, clinical outcomes, patient demographics
Iris dataset for breast cancer (using cytological features)	Educational resources	Focuses on cytological features for breast cancer samples, mainly used for educational classification exercises	Cytological characteristics of cells
Molecular profiles of breast cancer (GEO)	Gene expression omnibus	Contains gene expression profiles associated with breast cancer subtypes, aiding in subtype classification and biomarker discovery	Gene expression data, clinical attributes, subtype classifications

be avoided. In other words, it has the potential to prevent around 30% of cases. The most common modifiable risk factor for breast cancer is alcohol, followed by obesity, which may be associated with other modifiable lifestyle risk factors. Many disparities in health status regarding the actual incidence of breast cancer and mortality rates are evident in various countries and population groups, including restricted physical functioning, quality of life, and psychological distress, as the severe impact of the disease on functional status is more frequent in breast cancer patients [59].

11.3.5.2 Benchmark Datasets for Breast Cancer Prediction

Datasets play a crucial role in the advancement of knowledge in medicine, medical systems, and breast cancer prediction analytics. A dataset is important due to the need for comprehensive data for predicting breast cancer. Researchers in the area of research develop, validate, and test the predictive models using datasets. Benchmark datasets are available freely on various platforms. Collaborative efforts of researchers and scholars are initiated for curating meaningful benchmark datasets and making them publicly available. Healthcare practitioners and breast cancer patients can benefit from benchmark datasets if predictive models developed using these datasets have high performance metrics. To aid breast cancer research in discovering novel aspects, benchmark datasets are very important [60].

DSO are evolving nodes in analytics where such efforts are pivotal. Quality datasets are very important to predict the phenotype of breast cancer with high accuracy. Quality data require a rich and diverse dataset. Diverse data include imaging, histopathology, radiogenomics, cancer markers, clinical data, physiological states, and genetic information. Publicly available benchmark datasets contain image data, genetic or clinical data, cancer information, gene expression data, and histological data for training, validating, and testing. Several benchmark datasets are available. Users can use publicly available benchmark datasets on various platforms [61]. Healthcare practitioners or scientists have ensured that all the available benchmark datasets are validated by clinicians and that all the indicators explained above are realized to perform any experiments in healthcare with confidence. This section provides detailed information on seven benchmark datasets. Collaborative investments are initiated to update the benchmark datasets periodically due to the continuous emergence and outcomes of research activities. Breast cancer prediction research in the world will have benefits collectively if such scholarly energy is amalgamated and fostered [62, 63].

11.3.6 Diabetic Foot Ulcer Prediction

The prevalence and incidence of diabetic foot ulcers (DFUs) have increased recently, given the global rise in diabetes. DFUs frequently lead to osteomyelitis, cellulitis, and amputation. Infection and subsequent amputation can cause additional mortality and morbidity. Amputations pose more severe consequences, with an estimated five-year survival rate of 50%. Up to 85% of diabetic amputations are preceded by foot ulceration, suggesting that early preventive strategies are central to minimizing the threat of lower-extremity amputation [64]. The human and economic costs of DFUs continue to rise, necessitating a better approach to predicting ulcer development to prevent them. Hazards directly related to DFUs include neuropathy, deformity of the foot and ankle, poor circulation, and a foot ulcer's history. Given these realities, identifying those at highest risk for developing a foot ulcer is important, as intervention may help avert a chronic wound from forming. The concurrent use of validated clinical assessment tools with predictive algorithms could be advantageous when developing outcome measurements. New research and technology continue to create enthusiasm for advances in our capability to identify those at greatest danger effectively. There has recently been an explosion of interest in abnormal foot pressure associated with plantar ulceration. Interestingly, a recent survey showed that only 15% of clinic visits involve the measurement of foot pressure. There is also a suggestion that the integration of numerous instruments may enhance predictive capabilities [65]. In the era of evidence-based medicine, it is pivotal to move current practice forward from the use of population parameters to identify hospital admissions and threats for amputation to the use of the latest research on developing foot ulcer predictors. The application of emerging artificial intelligence-driven models may reveal increased predictive capabilities.

11.3.6.1 Challenges in Diabetic Foot Ulcer Prediction

The incidence of diabetes is growing at an alarming rate and is found to be the cause of deaths in every 6 s. The chance of developing diabetic foot ulcers (DFU) in an individual living with diabetes is around 25%. Prevention of foot ulcers is highly required because once developed, treatment is complicated and contributes to healthcare expenses. The prediction of diabetic foot ulcers is a challenging task because of the multifactorial nature of their development. Even with the known significant risk factors like neuropathy, peripheral arterial disease, foot deformities, and self-reported loss of protective sensation, these can only explain some variations, and they do not predict foot ulcer development on an individual basis. This has resulted in several limitations in epidemiological studies and prediction models, with no agreement on the evaluation of each risk factor in the clinical setting [66].

Studies have indicated some inherent limitations in current prediction models. Firstly, the commonly used risk factors are associated with the inception of foot ulcers on a quantitative scale. Secondly, neuropathy is one of the earliest risk factors and can confound the predictive model at the initial population when starting to follow up with patients. Furthermore, even with a high risk of foot ulcers shown in individuals with neuropathy, these can suffer from poor reporting by clinicians, leading to delayed referrals to expert foot services [67]. This is because, in the early course of neuropathy, years can go by before pain initiation. Thirdly, missing foot data can affect the overall satisfaction of the study. It is not always feasible to have complete foot data; some patients will avoid removing their shoes for clinical assessment, and in addition, patient data can frequently be moved across different settings. Fourth, relying on neuropathy as the point of entry for a multidisciplinary foot care team is not appropriate because the progression of peripheral arterial disease and foot deformity can vary between individuals, and this is the point of early intervention, especially in acute settings, to save a leg or a life. The current studies indicate a 7.5-year cumulative incidence of foot ulceration. Given current research interaction with family physicians, providing this figure is not enough to raise an eyebrow. The current landmark studies on foot ulcer prediction do not consider the majority of patients experiencing this from a specific perspective. The impact of the 7.5-year cumulative incidence in a typical general practice foot clinic can be higher, given recent evidence that referral of high-risk patients can reduce risk by up to 74% [68].

11.3.6.2 Benchmark Datasets for Diabetic Foot Ulcer Prediction

The relevance of complex datasets with comprehensive clinical and imaging data develops benchmark datasets. These datasets include extensive recent clinical history data and multiple imaging data. In total, 94 variables are included. The data comes from nine different departments and represents different international data for external validity. The main challenge of including these additional and complementary imaging data is the increasing time to harmonize and validate them. By expanding the number of patients, we improve the generalizability (including multiple centers) and complex input predictors via varying input data types (including clinical data) plus ease of replication due to the use of clinical variables [69].

The benchmark dataset comprises data from 3000 patients with identified foot complications: 1597 with a previous diabetic foot ulcer, 794 with a current diabetic foot ulcer, and 609 with a history of a major extremity amputation. Moreover, 80% have active osteomyelitis; 15% have received lower extremity amputation. The 3,000 patients determined eligible for the benchmark dataset generation were drawn from three country populations: the USA, Spain, and the Netherlands. The exclusion recall percentage for each country included the following exclusions: death, refusal to participate further in the study after index recruitment, absence of confirmed eligibility, physician-assessed miscarriage of eligibility, and consent contact status not yet determined. The resulting percentages excluded per country are as follows: Netherlands 18%; Spain 19%; USA 42%; Total 31% [70] (The Description of diabetic foot ulcer dataset is in Table 11.5).

11.3.7 Benchmark Datasets for Immunology

The twenty-first century has witnessed significant advances in artificial intelligence (AI) technologies, which are greatly expediting the pace of healthcare. AI and machine learning (ML) tools have already augmented accurate diagnostic capability in healthcare. Since its inception, AI has become the cost-effective engine of several scientific disciplines. Translated AI has already benefited diagnosis, treatment, and monitoring of most diseases as it enables healthcare providers to merge large sets of patient data with complex disease pathways. This is especially effective in fields of genomics, radiogenomics, precision medicine, and immunotherapy. The immune system is the central mechanism of modern medicine where all the pathological and intrinsic modifiers produce their action. Advances are being made in AI-radiomics-based immune profiling of cancer and also in analyzing cancer immunotherapy and predicting the survival of patients with various cancer types [71].

Unsupervised and supervised ML models are trained on immunome data and used for understanding the cell biology behind leukocyte differentiations. Data-driven AI models perform immune phenotyping, disease mapping, and patient monitoring. AI has the capability to classify, compensate, and enumerate leukocytes at high speed and in real time from PAP stains, blood smears, bone marrow aspirates, and lymphoid excisions. It also enables one to monitor and understand changes in the cellular composition and phenotype states [72]. With technological advances, it is not difficult to estimate the risk of transplant rejection based on a single-cell transcriptomics data by defining immune cell states. Serial antibody-based single-cell cytometry and high-throughput single-cell sorting experiments fulfill the high-resolution lineage tracing requirements of AI. All these strong points enable one to study the complexities of the immune system and decide the best strategy for logistics when a pandemic or mass infection pressure crops up, in the shortest span of time. The aim of this review

Table 11.5 Description of diabetic foot ulcer dataset

Datasets	Dataset type	Number of samples	Key features
Kaggle diabetic foot ulcer detection	Fundus images	2500 +	Labeled for ulcer classification
	Image resolution	Varies	High-resolution images
DFU images dataset	Ulcer images	1000 +	Annotated for different ulcer types
	Image resolution	Varies	Standardized for analysis
Gumuchian dataset	Ulcer images	800 +	Images from clinical settings
	Annotation types	Multi-class	Detailed labeling for analysis
Podiatry network image database	Foot condition images	500 +	Diverse range of podiatric conditions
	Clinical metadata	Available	Associated clinical data for analysis
American Heart Association (AHA) DFU dataset	Ulcer images	1200	Clinical context for each image
	Associated data	Available	Detailed patient demographics
DFU segmentation dataset	Segmented ulcer images	500 +	Detailed segmentation masks
	Annotation format	Binary masks	Useful for training segmentation models
Mayo clinic DFU dataset	Ulcer images	1500	High-quality clinical images
	Clinical data	Available	Comprehensive patient health information
Diabetic foot ulcer dataset from UCI machine learning repository	Clinical records	300 +	Detailed patient records
	Data format	CSV	Suitable for various ML applications
Diabetic Foot Image Dataset (DFID)	Ulcer images	1000	Focused on various ulcer presentations
	Image quality	High	Useful for image analysis
DICOM database from RSNA	Medical images	Varies	Includes various imaging modalities
	Clinical context	Available	Associated data for research

is to enlighten the reader about the capabilities, advances, and power of AI-based immunology [73] (The Description of immunology datasets is in Table 11.6).

Table 11.6 Description of immunology datasets

Datasets	Dataset type	Number of studies	Key features
ImmPort	Clinical trials	100 +	Longitudinal data, diverse disease
	Gene expression	1000 +	Includes immune cell profiling
TCIA (The Cancer Immunome Atlas)	Tumor samples	10,000 +	Immune cell infiltration data
	Expression data	Varies	RNA_Seq data available
Human protein atlas	Tissue profiles	24,000 +	Includes immune tissues
	Cell type profiles	12,000 +	Protein localization data
Cibersort	Expression profile	Varies	Estimates immune cell fractions
	Validation datasets	3000 +	Includes various cancer types
Gene expression omnibus	Gene expression	200,000 +	Diverse immunological studies
	Array data	10,000 +	Includes microarray and RNA seq
ImmuneSpace	Transcriptomic data	2000 +	Integration of various studies
	Protein data	Varies	Cross-study comparisons
ImmGen (The Immunology Genome Project)	Mouse immune cells	1000 +	Profile of various immune cell type
	Expression data	Varies	Includes differentiation states
The Human Immune System Database (HISDB)	Immune cell types	500 +	Functional response data
	Response profiles	Varies	In-depth immunological analysis
ArrayExpress	Gene expression	20,000 +	Comprehensive immunological datasets
	Multi-omics data	Varies	Includes RNA-Seq, Microarray data
Open targets	Drug targets	20,000 +	Integration of genetic, and clinical data
	Disease association	Varies	Links to immunology pathways

11.4 Evaluation Metrics

Assessment of the different methods is essential for determining analysis outcomes using benchmark datasets [74, 75]. Many crucial medical applications lack evaluation metrics and methodologies that could be used to measure the effects of applications based on the outcome analysis using benchmark datasets. These metrics are necessary for evaluating the application from different perspectives such as predictive ability analysis, interpretable representational analysis, and model application with real-world apparatus. The predictive performance is estimated for the analysis models on the basis of appropriate metrics such as accuracy, precision, and F-score. Similarly, by the application of such evaluation metrics, numerous real-world applications might be estimated in different evaluations [76].

The evaluation might be performed by different methodologies such as statistical tests, validation, accuracy, precision, sensitivity, specificity, F-score, geometric mean, C-index, and bootstrapping [77, 78]. The assessment of predictive analysis is necessary for the determination of model applicability to the healthcare domain. Different model comparison or evaluation methodologies should be used to validate predictive utility for various benchmark datasets. Alternatively, for various model comparison methodologies, different datasets might be applicable. The dataset choice is also critical to check different model generalizability because the testing process readily overfits some easy and small datasets. Even though this may be protective for the avoidance of false positive findings, it is necessary to consider the balance between the complexity of the methodology [79, 80].

11.4.1 Accuracy and Precision

Accuracy and precision define the effectiveness of the derived model from ground truth data and thus provide a basic evaluation of the benchmark dataset for the analysis. Accuracy, as the term itself suggests, defines the correctness of the model predictions, i.e., the true positive and the true negative, without any bias. Therefore, if accuracy is used as the model evaluation metric, the importance given to the false positive and false negative is neglected. Precision defines the closeness of the true positive with the model's positive predictions. The model was wrong when it predicted a sample to be in class A or B, when actually it belonged to class B or A, respectively. Precision aims to optimize for prediction probability, so that the achieved threshold could be placed closer to one. However, as we place the threshold closer to one, the false negative rate begins to increase, thereby causing the false negative rates to get equally large [81].

The performance of the predictive model is generally measured by the combination of both metrics, i.e., the optimal downsizing of the model's scale by assuming it to have optimal accuracy without affecting precision. This phenomenon could be analogous to a doctor performing surgery, where he has to have a good model for both precision and accuracy. If he works with a model of only accuracy, he may end up having patients with good values being dead and vice versa. These two metrics have conflicting relationships. Accuracy gets higher if the number of false predictions reduces, but in precision, the false negative prediction can lead to a sharp decrease in the precision score. Currently, the advancement of new techniques such as real-world analytics and histopathology in vitro results in improved model generalization and may be used to boost the reported metrics. However, with the advent of more sophisticated measurement techniques available in the market, and the enhancement of their applications, they could be utilized together with the old ones to improve the algorithms further and give them the accuracy as well as the precision necessary for regulatory approval [82].

11.4.2 Cross-Validation Techniques

There is a set of best practices for technique evaluation using benchmark datasets, which is necessary for proper analysis in the medical system. Cross-validation evaluates the model's ability to generalize to unseen datasets by preventing model overfitting on the training dataset. Cross-validation can reduce the dependence on a single data split by providing multiple train-test splits to the research process. The partitioning of the dataset can make various algorithms less or more effective by comparing the results of different partitioning strategies. Various techniques of cross-validation, like hold-out, k-fold, stratified cross-validation, and leave-one-out, can be used effectively in medical data analysis [83].

However, the ability to predict the performance of the model better than the training and validation datasets can depend on the partitioning strategy. The selection of a suitable partition, the dataset's size, and a target problem can play a principal role in the development of robust data partitioning. Partitioning bias can introduce overfitting during the model selection process. However, the cross-validation technique offers the best evaluation strategy. The possible limitations of this technique include bias in dataset splitting, both toward the model and toward the systematic differences among datasets. More often, proper implementation can make cross-validation an appropriate criterion for internal and external validation of machine learning predictive models [84].

11.5 Summary

A review of recent efforts in creating benchmark datasets for evaluating research objectives in various healthcare domains. As we have witnessed, in mainstream healthcare domains, these datasets have led to improved research and development and can provide better patient care. With the technological advancements and innovations across healthcare domains, it is important to augment research in dataset creation and assessment. A collaborative approach between healthcare professionals, academic researchers, and institutions is mandatory. To make informed decisions when seeking to evaluate new methodologies, the review and quality of existing and available benchmarking datasets are critical. However, currently, efforts related to the sharing of benchmarked datasets are quite limited and in a nascent stage.

Furthermore, from other reviews and our discussions in the paper, it has become evident that the availability of ground truth labels, data privacy, and ethical challenges are some of the barriers to developing a valuable benchmark dataset in healthcare. New approaches and ideas toward extending these datasets to overcome these barriers are needed. Besides that, research in the financial sector has demonstrated the success of benchmark datasets across various domains at a rudimentary level. Benchmark datasets should meet a manifesto that encourages the standardization of dataset descriptions, characteristics, and data views. It should also encourage methodologies aimed at making the sharing of datasets more ethical and motivating end-users to contribute by sharing their datasets. We also propose a decentralized but collaborative approach to creating global benchmark datasets that may transform the development of modern healthcare systems, including intake, analysis, and diagnosis.

References

1. Singh Y et al (2023) Topological data analysis in medical imaging: current state of the art. Insights Imaging 14(1):58
2. Atasever S et al (2023) A comprehensive survey of deep learning research on medical image analysis with focus on transfer learning. Clin Imaging 94:18–41
3. Si-Ahmed A, Al-Garadi MA, Boustia N (2023) Survey of machine learning based intrusion detection methods for Internet of Medical Things. Appl Soft Comput 140:110227
4. Gupta NS, Kumar P (2023) Perspective of artificial intelligence in healthcare data management: a journey towards precision medicine. Comput Biol Med 162:107051
5. Aminizadeh S et al (2023) The applications of machine learning techniques in medical data processing based on distributed computing and the Internet of Things. Comput Methods Programs Biomed 241:107745
6. Goceri E (2023) Medical image data augmentation: techniques, comparisons and interpretations. Artif Intell Rev 56(11):12561–12605
7. Purohit RM et al (2025) FedBlocks: federated learning and blockchainbased privacy-preserved pioneering framework for IoT healthcare using IPFS in web 3.0 era. Cluster Computing, 28(2):139
8. Garcea F et al (2023) Data augmentation for medical imaging: a systematic literature review. Comput Biol Med 152:106391
9. Sachan RK et al (2023) Diagnosis of mental health from social networking posts: an improved ML-based approach. In: International conference on emergent converging technologies and biomedical systems. Springer
10. Cai H et al (2022) A multi-modal open dataset for mental-disorder analysis. Scientific Data 9(1):178
11. Varanasi LVSKBK, Dasari CM (2023) Deep learning based techniques for neuro-degenerative disorders detection. Eng Appl Artif Intell 122:106103
12. Liu G-D et al (2020) A brief review of artificial intelligence applications and algorithms for psychiatric disorders. Engineering 6(4):462–467

13. Surianarayanan C et al (2023) Convergence of artificial intelligence and neuroscience towards the diagnosis of neurological disorders—a scoping review. Sensors 23(6):3062
14. Tyagi A, Singh VP, Gore MM (2023) Towards artificial intelligence in mental health: a comprehensive survey on the detection of schizophrenia. Multimedia Tools Appl 82(13):20343–20405
15. de Lacy N et al (2023) Predicting individual cases of major adolescent psychiatric conditions with artificial intelligence. Transl Psychiatry 13(1):314
16. Bhalla S, Kumar A, Kushwaha R (2023) Comparative analysis on brain stroke prediction using machine learning. In: 2023 14th International conference on computing communication and networking technologies (ICCCNT). IEEE
17. Tandel GS et al (2020) Multiclass magnetic resonance imaging brain tumor classification using artificial intelligence paradigm. Comput Biol Med 122:103804
18. Ranjbarzadeh R et al (2023) Brain tumor segmentation of MRI images: a comprehensive review on the application of artificial intelligence tools. Comput Biol Med 152:106405
19. Saleh A, Sukaik R, Abu-Naser SS (2020) Brain tumor classification using deep learning. In: 2020 International conference on assistive and rehabilitation technologies (iCareTech)
20. Raghavendra U et al (2023) Brain tumor detection and screening using artificial intelligence techniques: current trends and future perspectives. Comput Biol Med 163:107063
21. BRATS. https://www.kaggle.com/datasets/awsaf49/brats2020-training-data
22. https://www.med.upenn.edu/cbica/brats2020/data.html
23. https://www.kaggle.com/datasets/masoudnickparvar/brain-tumor-mri-dataset
24. https://gdc.cancer.gov/about-data/publications
25. https://sites.wustl.edu/oasisbrains/
26. http://brain-development.org/ixi-dataset/
27. https://www.kaggle.com/code/ohbewise/a-rsna-mri-solution-from-dicom-to-submission
28. Noreen N et al (2020) A deep learning model based on concatenation approach for the diagnosis of brain tumor. IEEE Access 8:55135–55144
29. Aamir M et al (2022) A deep learning approach for brain tumor classification using MRI images. Comput Electr Eng 101:108105
30. Köse C et al (2012) Simple methods for segmentation and measurement of diabetic retinopathy lesions in retinal fundus images. Comput Methods Programs Biomed 107(2):274–293
31. Santhi D et al (2016) Segmentation and classification of bright lesions to diagnose diabetic retinopathy in retinal images. Biomed Eng/Biomedizinische Technik 61(4):443–453
32. Franklin SW, Rajan SE (2013) An automated retinal imaging method for the early diagnosis of diabetic retinopathy. Technol Health Care 21:557–569
33. Yadav M, Goel R, Rajeswari D (2021) A deep learning based diabetic retinopathy detection from retinal images. In: 2021 International conference on intelligent technologies (CONIT)
34. https://www.kaggle.com/c/diabetic-retinopathy-detection
35. https://www.kaggle.com/datasets/andrewmvd/drive-digital-retinal-images-for-vessel-extrac tion
36. https://www.kaggle.com/datasets/vidheeshnacode/stare-dataset
37. https://www.kaggle.com/datasets/dreamer07/eyepacs
38. https://www.kaggle.com/competitions/aptos2019-blindness-detection
39. https://www.kaggle.com/datasets/paultimothymooney/kermany2018
40. https://ieee-dataport.org/datasets
41. https://www.adcis.net/en/third-party/messidor/
42. https://www.kaggle.com/datasets/andrewmvd/retinal-disease-classification
43. Saranya P et al (2022) Blood vessel segmentation in retinal fundus images for proliferative diabetic retinopathy screening using deep learning. Vis Comput 38(3):977–992
44. Elsharkawy M et al (2022) The role of different retinal imaging modalities in predicting progression of diabetic retinopathy: a survey. Sensors 22(9):3490
45. Saranya P, Prabakaran S (2020) Automatic detection of non-proliferative diabetic retinopathy in retinal fundus images using convolution neural network. J Amb Intell Humanized Comput
46. Hardas M et al (2022) Retinal fundus image classification for diabetic retinopathy using SVM predictions. Phys Eng Sci Med 45(3):781–791

47. Li YC et al (2021) Automatic detection of atherosclerotic plaque and calcification from intravascular ultrasound images by using deep convolutional neural networks. IEEE Trans Ultrason Ferroelectr Freq Control 68(5):1762–1772
48. Skandha SS et al (2020) 3-D optimized classification and characterization artificial intelligence paradigm for cardiovascular/stroke risk stratification using carotid ultrasound-based delineated plaque: Atheromatic™ 2.0. Comp Biol Med 125:103958
49. Fedewa R et al (2020) Artificial intelligence in intracoronary imaging. Curr Cardiol Rep 22(7):46
50. Yang J et al (2024) Automated classification of coronary plaque on intravascular ultrasound by deep classifier cascades. IEEE Trans Ultrason Ferroelect Freq Control:1–1
51. Masuda T et al (2022) Deep learning with convolutional neural network for estimation of the characterisation of coronary plaques: validation using IB-IVUS. Radiography 28(1):61–67
52. Cho H et al (2021) Intravascular ultrasound-based deep learning for plaque characterization in coronary artery disease. Atherosclerosis 324:69–75
53. Makaryus AN et al (2014) Cardiovascular imaging: current developments in research and clinical practice. Clin Med Insights Cardiol 8(Suppl 4):57–61
54. Johri AM et al (2022) Deep learning artificial intelligence framework for multiclass coronary artery disease prediction using combination of conventional risk factors, carotid ultrasound, and intraplaque neovascularization. Comput Biol Med 150:106018
55. Jamthikar AD et al (2020) Low-cost office-based cardiovascular risk stratification using machine learning and focused carotid ultrasound in an Asian-Indian cohort. J Med Syst 44(12):208
56. Johri AM et al (2021) Role of artificial intelligence in cardiovascular risk prediction and outcomes: comparison of machine-learning and conventional statistical approaches for the analysis of carotid ultrasound features and intra-plaque neovascularization. Int J Cardiovasc Imaging 37(11):3145–3156
57. Fatima N et al (2020) Prediction of breast cancer, comparative review of machine learning techniques, and their analysis. IEEE Access 8:150360–150376
58. din, N.M.u., et al., Breast cancer detection using deep learning: Datasets, methods, and challenges ahead. Computers in Biology and Medicine, 2022. 149: p. 106073.
59. Sengar PP, Gaikwad MJ, Nagdive AS (2020) Comparative study of machine learning algorithms for breast cancer prediction. In: 2020 Third international conference on smart systems and inventive technology (ICSSIT)
60. Nanglia S et al (2022) An enhanced predictive heterogeneous ensemble model for breast cancer prediction. Biomed Sig Process Control 72:103279
61. Gupta P, Garg S (2020) Breast cancer prediction using varying parameters of machine learning models. Procedia Comp Sci 171:593–601
62. Ara S, Das A, Dey A (2021) Malignant and benign breast cancer classification using machine learning algorithms. In: 2021 International conference on artificial intelligence (ICAI). IEEE
63. Abunasser BS et al (2022) Breast cancer detection and classification using deep learning Xception algorithm. Int J Adv Comput Sci Appl 13(7)
64. Basiri R et al (2024) Protocol for metadata and image collection at diabetic foot ulcer clinics: enabling research in wound analytics and deep learning. Biomed Eng Online 23(1):12
65. Reyes-Luévano J et al (2023) DFU_VIRNet: A novel Visible-InfraRed CNN to improve diabetic foot ulcer classification and early detection of ulcer risk zones. Biomed Signal Process Control 86:105341
66. Biswas S et al (2023) DFU_MultiNet: A deep neural network approach for detecting diabetic foot ulcers through multi-scale feature fusion using the DFU dataset. Intell Based Med 8:100128
67. Sathya Preiya V, Kumar VA (2023) Deep learning-based classification and feature extraction for predicting pathogenesis of foot ulcers in patients with diabetes. Diagnostics 13(12):1983
68. Weatherall T et al (2024) The impact of machine learning on the prediction of diabetic foot ulcers–a systematic review. J Tissue Viability
69. Das SK et al (2023) Diabetic foot ulcer identification: a review. Diagnostics 13(12):1998

70. Toofanee MSA et al (2023) Dfu-siam a novel diabetic foot ulcer classification with deep learning. IEEE Access
71. Hudson D et al (2023) Can we predict T cell specificity with digital biology and machine learning? Nat Rev Immunol 23(8):511–521
72. Forlin R, James A, Brodin P (2023) Making human immune systems more interpretable through systems immunology. Trends Immunol
73. MacMath D, Chen M, Khoury P (2023) Artificial intelligence: exploring the future of innovation in allergy immunology. Curr Allergy Asthma Rep 23(6):351–362
74. Kumar A (2023) Visual object tracking using deep learning. CRC Press
75. Kumar A, Walia GS, Sharma K (2020) Real-time multi-cue object tracking: benchmark. In: Proceedings of international conference on IoT inclusive life (ICIIL 2019), NITTTR Chandigarh, India. Springer
76. Kairys A et al (2023) Towards home-based diabetic foot ulcer monitoring: a systematic review. Sensors 23(7):3618
77. Kumar A, Singh D (Eds.) (2025) Revolutionizing Medical Systems Using Artificial Intelligence: A Breakthrough in Healthcare. Elsevier
78. Maindarkar M, Kumar A (2025) Artificial intelligence based disease diagnosis using ultrasound imaging. In Revolutionizing Medical Systems using Artificial Intelligence, pp 147–161. Academic Press
79. Hettinger G et al (2024) An improved clinical and genetics-based prediction model for diabetic foot ulcer healing. Adv Wound Care 13(6):281–290
80. Dubey S, Dixit M (2023) Recent developments on computer aided systems for diagnosis of diabetic retinopathy: a review. Multimedia Tools Appl 82(10):14471–14525
81. Nagaraju S et al (2023) Automated diabetic foot ulcer detection and classification using deep learning. IEEE Access 11:127578–127588
82. Kumar A, Leekha A, Jain R, Sethi A, Pulkit (2024, June) Deep Learning-Based Diabetic Retinopathy Detection Using Enhanced ResNet-152. In International Conference on Data Analytics & Management, pp 101–112. Singapore: Springer Nature Singapore
83. Kavatlawar A et al (2024) Cardio vascular disease (CVD) recognition using convolutional neural networks. Grenze Int J Eng Technol (GIJET) 10
84. Li F et al (2023) Evaluating and mitigating bias in machine learning models for cardiovascular disease prediction. J Biomed Inform 138:104294

Chapter 12
Role of AI and Modern Medical Equipment in Smart Healthcare

Abstract When it comes to healthcare innovation, the Internet of Things (IoT) has a tremendous impact. With the arrival of Medicine 4.0, there has been a flurry of activity in developing platforms, both in terms of hardware and the underlying software. Thanks to this foresight, Healthcare IoT (H-IoT) technologies have been developed. The sensing nodes' ability to communicate with the processors and the algorithms used to process that data are the foundational technologies that allow the system to function. But right now, a number of new technologies are bolstering these enabling technologies as well. Almost every aspect of H-IoT systems has been revolutionized by the usage of AI. By moving processing power closer to the deployed network, the fog/edge concept is reducing the impact of numerous obstacles. However, big data makes it possible to process massive amounts of data. Blockchains are discovering the most innovative applications in H-IoT systems, while Software Defined Networks (SDNs) add system flexibility. Progress in H-IoT applications is being propelled by developments in the IoNT and the TI, or Tactile Internet. This chapter explores how these technologies are changing H-IoT systems and finds out how to use these changes to improve QoS in future.

Keywords Modern medical equipment · Intelligent health systems · Healthcare technology · Digital health · AI in medicine

12.1 Introduction

Artificial intelligence can aid medical devices in adopting newer and better infrastructures, software and algorithms, and data management information. Advanced medical equipment designs and excellent data resulted from powerful research analytics can be used to evaluate and classify patients according to their molecular profiles, their risk of complications, and the most effective treatment. The development of such advanced medical equipment and analysis software with a high degree of accuracy, sensitivity, specificity, precision, and reliability will not only revolutionize the way

A. Kumar and D. Singh, *Artificial Intelligence in Modern Healthcare System*,
Transactions on Computer Systems and Networks,
https://doi.org/10.1007/978-981-96-6703-1_12

medicine is practiced but also improve overall healthcare costs, reduce disparities in care and stimulate economic growth, and improve global well-being [1].

The World Health Organization defines health as a state of complete physical, mental, and social well-being rather than merely the absence of disease or infirmity. Attaining good health involves the synergy of various factors, including research, diagnostics, medication, therapy, and the environment. Researchers workday in and day out to invent and improve new methodologies to alleviate pain, treat diseases, and enhance the quality of life for everyone. Rapid changes in the field of technology make it possible to produce better medical devices, real-time diagnostics systems, highly effective therapy equipment, excellent research methodologies and interpretations, and an environment that decreases risks and stress. The strong relation between medical instrumentation and health outcomes requires real-time health diagnostics and risk reduction with minimal human intervention. Artificial intelligence is now used in a variety of medical fields to provide reliable analysis and interpretations from complex high-volume data sources [2].

Currently, around 10–15% of the world's population suffer from rare diseases, and the healthcare system's interest and patrons are dwindling rapidly. However, the AI-driven Internet of Things (IoT) revolution can make it a little easier to bridge the gap and meet the need for an increase in personalized and efficient healthcare for such patients [3]. The smart healthcare device's AI-integrated design helps in rapid monitoring, assured security, and remote care, and conditions monitored in the comfort of your home instead of visiting a physician. A wide variety of such applications may be provided along with the medically confirmed and recommended algorithms for proven effect. The AI research recommends safer and effective healthcare systems, especially for primary care, control of chronic diseases, preventive care optimization, and remote medicine supervision. In this research, an AI-based health management controller modifies the smart healthcare device, and tests were conducted to detect rare diseases, particularly episodic diseases, irrespective of their affected symptoms and to manage them according to the patients' needs at different levels. Based on the results, the final implementation is expected to have ultra-low power consumption and computability. The results contribute to the development of a future framework that is able to adapt from private and secure healthcare devices, e.g., medical IoT [4].

These AI algorithms were used for intelligent behavior simulation in games and were able to adapt based on gaming experience. The usage of AI is not just limited to games but has started to expand its wings in various fields such as military, finance, analytics, and, as will be discussed further, in the healthcare departments too. Healthcare was considered to be the most sought-after research area in this century, and the transformation of healthcare due to AI is poised to lead to more tech-driven decision and solution implementations with fewer side effects in the patient treatment process. However, areas such as advanced healthcare and smart healthcare are all still easier said than done. The term has all the potential and is expected to be included in people's lives as an extremely special part, as more and more research and commercial activities are shaped up for a better tomorrow. In other words, AI is expected to be the backbone technology for most of the upcoming scientific advancements [5, 6].

Artificial intelligence (AI) is the branch of computer science dealing with the simulation of intelligent behavior in computers. Game AI, derived from the term Artificial Intelligence, is the intelligence exhibited by the characters in video games. Almost all video games implement some level of pre-defined AI to control characters and game objects or events. It had been almost 21 years since the Nintendo 64 created a buzz in the gaming industry with its 3D gaming revolution. The advent of AI and modern-day technology has improved the gaming hardware and software aspect by several folds, offering a breathtaking gaming visual treat. The modern game industries started to alter their game design to include more intellect control for better reception, even in handheld devices [1, 7]. The Key Contribution of this chapter are as:

- In this chapter, we depicted the role of AI technologies in the field of the Internet of Medical Things. We explained the applications of modern technologies in smart healthcare. We also depicted the role of AI in smart healthcare.
- We also demonstrated the role of the Internet of Medical Things in smart healthcare. The healthcare industry plays a vital role in the growth of the economy as it addresses the issues of health services, making them more convenient and efficient.
- This chapter reviews the roles of the IoMT and AI-enabled devices available in the market, enabling superior patient care. The potential applications of smart healthcare are demonstrated with experimental results in computer-aided diagnosis and imaging available services in existing healthcare systems that may result in realizing future healthcare standards.
- The integration of the latest technologies for medical diagnosis, IoT devices, including sensors, wearables, and portables, connectivity, cloud computing, AI, and medical cyber-physical systems, often contributes to patient safety and the quality of care, which in turn makes life-saving decisions possible.

The following chapter is structured accordingly. In Sect. 12.2, The foundation of Modern Healthcare, general medical equipment functions of AI are detailed. In Sect. 12.3, the role of advanced medical equipment backed by AI in diagnostics, repair, and evaluation of considerations linked to smart health care are then presented.

Smart health as an emerging field in health care is aimed at providing personalized, ubiquitous, preventive, promotive, curative, and tailored care and services to citizens in a cost-effective manner. Ethical and legal consideration to create and use of modern healthcare equipment are discussed in Sect. 12.6. In Sect. 12.7, Case studies and Success stories of the modern equipment in smart healthcare system is discussed. Section 12.8 discusses the future of AI-driven care. A summary concludes this chapter in Sect. 12.9.

12.2 Foundations of Smart Healthcare

Corrective diagnostic measures will be deployed to minimize errors. Rich medical imaging services (e.g., birth mammograms, identification of skin cancer) for validated medical types of data that are crucial for breast and skin cancer risk assessment models, by finding risk-minimal expertise and lean team organizational structures, will enable more universal mass-market offerings. Efforts to safeguard patient sensory data can spread satisfaction from government or insurance company regulations to the personal tailored family care plan. Proper use of these integral settings can crucially change healthcare services scalability issues, undertaking mission-critical information (e.g., tort restrictions) out to data in the evolving domain of digital healthcare [1]. Resolved vagueness and imprecise knowledge are the main attributes of precision health as a new "principle of 3P healthcare". The intersection of an infinite number of healthcare long-term and short-term goals, based on each patient's profile, accentuates the importance of index-free healthcare system performance. Also, advanced medical imaging services addressing needs, comprising on-demand visual consultation, complements a non-standard definition of collaborative mental health [2].

Advanced smart healthcare is building an affluent system of present healthcare, comprising the semantization of ubiquitous patient-related personal data, such as wearable campus. In this context, mathematical foundations (neutrosophic sets are loosely defined) have presented a flexible method to describe vague and imprecise, as well as precise information from new engineering. In contrast, concepts can have profoundly ambiguous meanings. The resultant notion of vague sets allows for flexibly modeling optimal preference information while, at the same time, allowing for partially fulfilling minimum conditions, providing a meaningful generalization of traditional fuzzy sets. Performance, security, privacy, cognition, finance, and control are among the specific building blocks. Consumer-oriented healthcare plans will be proposed with strong security measures [8].

12.2.1 Definition and Scope

AI has succeeded in working as a virtual doctor well enough to search for helping patients by scheduling outpatient appointments, telehealth appointments, monitor a patient's vital signs, and manage the patient's electronic health record (EHR) [9–11]. AI has also demonstrated its successful capacity to handle specialized professional medical tasks, such as the use of time-consuming, costly and error-prone manual processes, including the provision of remote intensive care unit services, detecting medication overuse and misuse of opiates, detecting neurologic and psychiatric disorders, and monitoring the life of patients receiving radiation therapy [1].

The term artificial intelligence (AI) means using software and related technologies to learn how to complete tasks that currently require some form of human input. AI is divided into two major subsets: narrow AI and general AI. Narrow AI encompasses systems built to accomplish a narrow range of tasks while ubiquitous general AI systems are accomplished by humans and can accomplish the same variety of tasks as humans. While much narrower AI efforts have yielded significant breakthroughs, AI is currently still narrow. As the development of hardware and applications pushes ever further, it is very unlikely that those breakthroughs will continue to accumulate and that our methods and ideas will continue to develop.

12.3 AI Applications in Healthcare

AI is likely to significantly impact various aspects of the field of medicine and pharmacy, making healthcare more powerful in its role as a social engine. Artificial intelligence, information and communications technology (ICT), big data, the internet of things (IoT), drones, and robotics are developing rapidly and undergoing major technological innovations while also demonstrating the potential to interact and create new value, new ecology, new industry, and new technology [12]. The international development direction of medical diagnosis, artificial intelligence, and smart medical equipment technology was first introduced, with the medical smart device market concept described, high-frequency classification of over-optimization, and primary application layout strategy, and development trends considered. Smart, intelligent technologies have been extensively embraced due to the burgeoning explosion of COVID-19 datasets, which includes datasets of genomics, proteomics, glycomics, metabolomics, epidemiology, at the population level, clinical level, and molecular level, and phenotype data from telehealth. Consequently, a wide array of innovative AI technologies have been chiefly developed to encounter numerous preventive, diagnostic, therapeutic, and predictive detection challenges, such as detecting drugs, repurposing molecules, detecting novel antigenic epitopes, prognosis, clinical stage detection, and new strain detection. The articles and research topics are assessed from preprint publications, top journals, as well as leading research institutes from six renowned companies (Positionly, AiMed, Bootstrap labs, Socialinsider, and Niceprice) around the globe. The credibility of the information is different for the individual websites, but the general statistics are similar, as accuracy is given more priority in studies and at early stages considering the novel COVID-19 pandemic infection where novel strains and mutations kept emerging with repurposed vaccines making vaccination ineffective. Finally, information obtained from academic content and industrial trends has been employed to support the qualitative and quantitative analysis for this article [13, 14].

MN positions in medical scenarios with imaging and in the field of pharma should be taken into consideration. The segment is huge and high because of the lack of public inference on big pharma AI. Drug discovery will benefit from AI and has changed abdominal imaging in hospitals worldwide. It is also widely used in ophthalmology among the subspecialties. MN has implications for software/hardware R&D, and as a function of personalized medicine, AI/ML-guided patient health tracking is

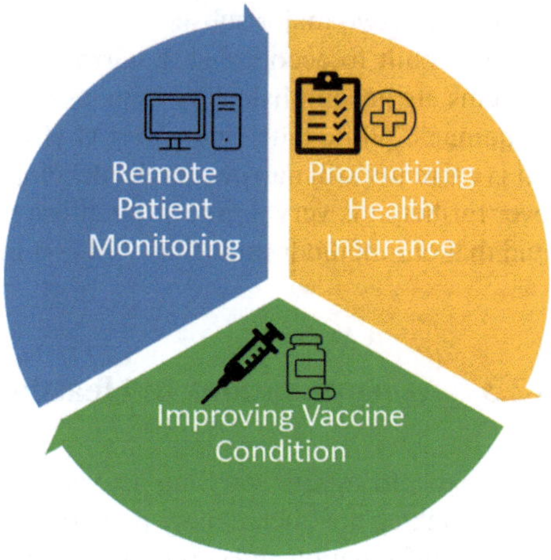

Fig. 12.1 Applications of internet of medical things

growing. Devices and procedures for precision medicine may also benefit as precision values in MN applications increase. The role of MN is less relevant in healthcare, though the subset financial services area for crash department visit predictive analytics is growing and considered very relevant for quality outcomes [1]. AI has already had a significant impact in healthcare and everyday life applications. In the healthcare industry, applications are often categorized into three groups, including personalized medicine, image analysis, and intelligent question–answer systems. An intelligent question-answering system for medical diagnosis has also been proposed, in which a hybrid of a shallow semantic model that leverages syntactic and semantic features and a deep learning model is used to obtain word-level representations and improve diagnostic accuracy. For personalized medicine applications, integrating healthcare with wireless and mobile technologies has gained popularity. In particular, personalized medicine in robots has attracted significant attention in China and overseas. Robotics, including human accompaniment and interaction, has gone mobile, accompanied by the rapid development of artificial intelligence around health centers in China. Facial expression terminal services have already penetrated the market, and discussions on whether virtual or exoskeletons for robots in rehabilitation are more accessible have begun [15] (The Applications of internet of medical things is shown in Fig. 12.1).

12.3.1 Diagnosis and Treatment Planning

There have been remarkable advances of AI in various clinical specialties such as radiology, pathology, and dermatology by the design of AI-flavored software which are primarily developed to aid the clinicians by helping them to detect and diagnose different grades of pathologic abnormalities. Automatic tools for the detection of Tuberculosis and various skin diseases through social media have also been successful in the direction of democratizing medical asset. It is thought that AI research in detection diagnostics will benefit from being addressed as wide decision problems which encompass the posterior tasks like risk assessment, lesion detection sequestration, and lesion characterization, partly in imitation of how humans naturally contemplate upon these problems. We should imagine, in the medical AI environment in which prediction and explanation rather than hard-coded rules is the realistic means for objectivism [16].

Diagnosis and treatment planning are the most critical, as well as intellectual activities of healthcare systems. The significant utilization of AI in the field of medical imaging has opened up several novel directions in this area. A pipeline named Intuition-Enabling will altogether create improvements in the clinical process at every stage and will enable large-scale, opt-in study of the decision-making process.

12.4 Modern Medical Equipment in Healthcare

In fact, this is already happening to the point that misinformation often leaves citizens perplexed about the actual capabilities of modern medicine when faced with known diseases that quickly give ravages. In this context, healthcare is getting smart and refers not only to the use of some tools that characterize the context of the Internet of Things but also to the use of artificial intelligence. In so doing, hospitals and generally health institutions will be able to monitor vital parameters constantly, both in very specific clinical conditions and in situations of chronic disease, with final improvements in clinical risk assessment and variations in the patient's state during the treatment. After diagnosis and therapy have been set out, it is, in fact, possible to continue the dimensional monitoring of the health status of the patient even outside the hospital [17] (Some Examples of smart healthcare devices and tools is shown in Fig. 12.2).

Artificial intelligence or AI is the buzzword in health and medicine. This comprises a wide variety of technologies such as virtual reality, chatbots, and robotics, along with wearable devices and smart machinery [18–20]. All of this medical equipment is without which the modern medical system could hardly operate. These are all the ones that interact with us and monitor our states, to be able to give health professionals the elements they need to make the best decisions. And on these questions is a giant amount of data, from the many tests that we are subject during our lifetime to the vast amount of information generated daily by professional and healthcare professionals. Hence, whenever we talk about the need to develop the future of medicine, we cannot disregard the enormous revolution that AI and modern medical equipment are creating in healthcare. Health systems, and the citizens who need them, can only benefit from this, both in terms of quality and actual efficiency [21].

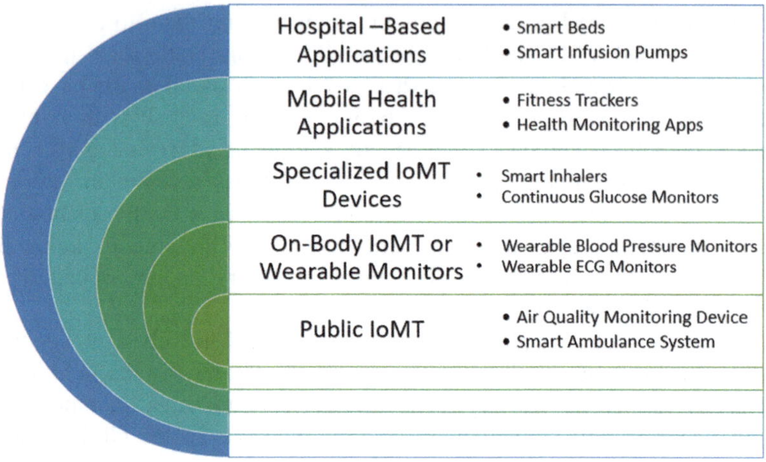

Fig. 12.2 Examples of smart healthcare devices and tools

12.4.1 Types and Functionality

The sensor measures four important parameters for a person where there is a threat of fire: temperature and its change, humidity, the presence of LPG and methane in the air. After processing, the obtained data give the opportunity for AI algorithms to operate efficiently in the emergency cases of a fire hazard.

The reviewed medical parameters contain information about temperature, pulse, oxygen level, and other important vital signs. This information is used to develop an integrated patient information system with smart medical devices. The developed software can store and process data from patients and display set data on a web platform. This functionality increases the ability of remotely operated AI-based lives' residence system.

Currently, SH&MCS in Ukraine is developing actively due to the potential of unmet needs that this system can meet. In the next two sections of this chapter, examples of specific smart healthcare and medical care systems are considered. The technical characteristics of medical equipment that can be used in the healthcare system and the basic principles of the AI algorithms that can be implemented on these devices are also presented. All algorithms are implemented in the LabView graphical development environment [22].

A smart healthcare and medical care system (SH&MCS) typically consists of the following important functional units: an information unit that processes the accumulated data and delivers the specific medical history of medical care to the interested parties, a wireless monitoring unit to track specific patient features, a security unit to control the restricted area, a display unit to provide relevant information, and a medical care unit that provides basic medical assistance using AI algorithms without the participation of the personal physician.

Hospitals, clinics, and other healthcare organizations are considered the second and third most common places for investment in IT and specifically information systems. This is due to the constantly growing needs of the demographic population for healthcare services in recent years. Developing technologies aimed at optimizing these processes and improving the quality dictate the concept of providing smart healthcare services [23].

12.5 Integration of AI and Medical Equipment

Smart healthcare is a semantic fusion of conventional concepts of healthcare, service delivery, location of care (homecare, healthcare, and care), health to be protected or enhanced, and technology integration. It recognizes the need for a system that helps healthcare providers provide care, help more individuals with greater precision, in a timely and cost-effective manner, and help patients take responsibility for their wellbeing. It includes patients involved in their wellbeing and healthcare and needing care. The transition to smart healthcare entails more than the application of digital technologies in healthcare as such. Business processes, governance principles, and services are reengineered as well. Health ecosystems share the joint goal. Digital technology is now at the core of enabling everything to blend together and function as a method of care and support [24].

The realization of Future Healthcare 4.0 is the subject of significant research worldwide, primarily with the objective of reducing preventable illnesses, efficiently delivering care, and having patients take greater responsibility for managing their health. Along with IT concepts, terms such as "smart healthcare," "connected health," and "intelligent healthcare" are used to underscore the vital role of AI. In order to deliver operational cost savings and mitigate some of the stress on the healthcare system, a third tenet is the generation of value to society and organizations.

12.5.1 Challenges and Opportunities

The authors face multiple technical challenges in the design and deployment of modern medical equipment. Furthermore, typically, the clinical community's needs are not recognized or understood. Since AI engineers need to create clinically validated products that physicians would embrace, both early and often, a closer connection with the clinical ecosystem is required. They seek to consider the clinical philosophy behind the processes and devices, and questions where AI and related technologies may be helpful. Regardless of their efficacy and potential to minimize inefficiencies in healthcare and contribute to economic growth, reforming service delivery in the field of healthcare is a markedly difficult feat. In addition, because privacy and vulnerability are often important in the transformation of an organization, industry reforms are complex. Possible benefits and privacy problems can also be contradictory for every technology, and thus partnerships and industry-wide governance become critically necessary [25].

Substandard healthcare provision results in millions of avoidable deaths every year. It is due primarily to the uneven distribution of skilled healthcare personnel and the concentration of expertise and technology in large facilities in metropolitan areas. Smart healthcare marries recent advances in IT and AI to tackle these issues across the healthcare value chain. Appliance firms are now rushing toward the lucrative vertical of smart healthcare. In few other industries, however, do we have the chance to produce a technology framework where older persons can live longer, healthier lives using the AI skills that will support the future of smart healthcare? In this yearbook, the authors share the benefits and obstacles to AI deployment in healthcare that they have seen as AI designers, entrepreneurs, and investors. While there are significant technical and political roadblocks, present groundbreaking breakthroughs and the political climate offer a promise for smooth access to AI-based healthcare solutions [26].

12.6 Ethical and Legal Considerations

We discuss the appropriateness of AI to tackle these challenges and provide examples of how existing AI research can involve these considerations [27]. We discuss access to healthcare, a challenge that AI research can begin to engage with now. We consider how AI innovations can be incorporated into the practice of healthcare from both provider and patient standpoints to promote shared decision-making and not undermine trust. We propose hardware/software co-design as a way to counteract growing costs and bring about a step change in what is feasible. We discuss how AI research should engage with relevant laws and ensure that liability is based on competence, not on professional status. Whatever the future of AI in healthcare, we foresee a future where the clinicians of the future need to understand both the capabilities and the limitations of AI [28].

With every new innovation in science, we believe the benefit must be weighed against potential harm. From explaining the correct construal of data to understanding any counter-intuitive behaviors in popular algorithms, to conducting rigorous evaluations of AI, the committee's vision is that AI for healthcare is a two-way partnership that involves both the AI community and the healthcare community working closely together. Ethical considerations of AI for healthcare are mirrored in other AI contexts and can be represented by impact on different communities, trust and control, laws, liability, risks, access, health disparities, and informed consent. These are some of the unique properties of AI for healthcare.

12.6.1 Data Privacy and Security

Organizations need to have a security mindset to prioritize the privacy of their users and clients and to align their practices to meet the highest security standards. To accomplish this, organizations have to embed privacy in system designs by consolidating data collection to the minimum amount required and concealing sensitive data from all unauthorized users. Any data that is not mandated, any data that presents risk, should not be collected or stored. Blockchain technology is set to change the landscape of secure data sharing. Systems based on blockchain can enhance data reliability and authenticity. Adaptive and proactive security controls should be continuously updated to technological and operational changes in the organization [29]. Specific usage restrictions should be set and enforced by role-based access control mechanisms to manage user behavior in line with their responsibilities. Breach notification is a regulatory requirement to take prompt action in the event of unauthorized breaches of protected data. Data Privacy Impact Assessments are a forward-looking defense mechanism to protect various aspects of data privacy. The IoT presents significant new data privacy risks, including devices that monitor behavior and capture sensitive information about users. Machine learning developed by large-scale data usage, often across borders, requires privacy infrastructure [30].

Data has always been considered as an important asset and it is rightly tagged as "the new oil" in contemporary times. By extension, data privacy and security issues have never been more important than in the present world of automation to prevent misuse. Smart healthcare is completely dependent on large amounts of data in different formats, whether images, texts, audio, or bio-signals. Data privacy regulations impose penalties on organizations that violate them; hence, compliance with such regulations is a necessity. This is a fundamental right for every individual and the protected personal data includes health records and life-threatening treatment constraints. Implementing appropriate data security measures such as data storage, encryption, transmission, and de-identification tools (use of k-anonymity, l-diversity, t-closeness, and user-level preservation and consistency) is necessary to enable the collection, aggregation, and sharing of data. The combination of software, hardware, and network security tools should be properly configured and administered for the highest standard of data security to address privacy and security concerns [31].

12.7 Case Studies and Success Stories

Let's look at a couple of success stories. Philips is a hospital-focused company where clinical expertise and a patient-centric approach go hand in hand. It has always sought to drive innovation in healthcare by partnering with some of the best minds in hospitals and academic institutions globally and applying cutting-edge artificial intelligence. Such collaborations have resulted in some breakthrough solutions and products that today have become vital for the medical community. In the United Arab

Emirates, the government has prioritized investments in the latest healthcare technologies in the interest of the patient. Reflecting this vision, when NMC Healthcare commissioned the 745-bed NMC Royal hospital in Abu Dhabi, they placed a major emphasis on the quality of installation of the medical equipment and diagnostic solutions to be procured. It was not an easy task, demanding exceptionally close working together by the experienced project teams of Philips and NMC Healthcare to identify the latest medical technologies and install them, avoiding the inevitable project challenges [32].

The proliferation of artificial intelligence, machine learning, and the internet of things in smart healthcare is leading to better outcomes for patients while meeting the challenge of increasing demand for medical services. The results are inspiring: more accurate diagnoses, personalized treatment plans, and real-time patient monitoring during every stage of the care process. We are now entering the era of preventive medicine, with AI-driven drugs to halt disease creation at the gene level and predictive medical systems that could remediate healthcare's current reactive philosophy. The impacts also extend to the business performance of healthcare organizations. Physicians are benefiting from the connectivity of smart devices as they can work remotely, and through customized instructions, machines could be performing many routine surgeries [33].

12.7.1 Real-World Implementations

The role of AI is less an automatic decision-making than a decision-support system. Human validation of AI suggestions is therefore an essential component of any medical image analysis pipeline. It requires implementing explanations of AI decisions that are meaningful to human experts. Explanation methods developed in other subfields of computer science, in particular in natural language processing, cannot be straightforwardly transferred to medical imaging, where humans observe localized, high-dimensionality patterns. In the near future, a challenge will be integrating these methods in complex medical imaging pipelines, including pre-, post-, and revisit within and across imaging modalities [34].

Real-world implementations highlight several challenges facing AI for medical image analysis. First, report performance in clinical validation experiments varies widely, depending on patient and disease cohorts and on the type of image analysis. Some studies, in particular on thoracic radiographs, dermatoscopy, and ophthalmic images, show comparable performance to human experts, even in increasingly difficult detection and triage tasks, while others are years behind solving subproblems addressed long ago, in particular in breast and digital pathology. Several commercial solutions show impressive performance on public benchmarks but have not, to our knowledge, undergone clinical validation, revealing a gap between research and real-world impact. This is due to time-consuming and expensive regulatory approval and certification procedures that typically require clinical trials involving hundreds of patients in multiple centers and imitate clinical workflows, including double-reading by human experts, even in subproblems where individual interpretation is fast and easy, and clearly superior human performance is well established [35].

12.8 Future Trends and Innovations

The concept of smart healthcare with AI and modern medical equipment was explained in detail. After monitoring healthcare by modern medical devices and analyzing the results with machine learning techniques, diseases are prevented by AI-enabled systems. Throughout the chapter, many case studies related to smart healthcare were presented. Lastly, future trends and directions for achieving smart healthcare were explained. Nevertheless, future research works proposed include the development of more sophisticated healthcare systems to truly achieve smart healthcare, such as employing advanced data science technology for health information management and utilizing ambient intelligence for embedding intelligence into the environment [33].

Through this study, the potential for AI technologies to enhance healthcare and make truly smart healthcare systems has been explained in detail. AI combined cloud-enabled personalized healthcare with virtual sensors and subsequently applied a neuro-fuzzy method to classify the imaging results from Thailand and developed iCycle, an intelligent ICU telemedicine monitoring and diagnostics system. We have explained modern medical devices applied for treating and diagnosing various diseases and monitored health, and shown how modern medical equipment is able to lower healthcare costs by providing high-quality supportive care. Finally, the current issues and future trends related to smart healthcare and AI were described [36].

12.8.1 Emerging Technologies

The healthcare sector is witnessing significant advancements using structured as well as unstructured data to improve the overall quality and cost of healthcare by using powerful care delivery models. The possibilities of personalized healthcare services make smart healthcare services highly indispensable. The main contributions of this chapter aimed at encouraging the acceptance of smart healthcare practices by presenting a comprehensive overview of AI models/techniques that play a vital role in healthcare and how the application is enabled through modern healthcare techniques are as follows: – Identifying and discussing the direct patient/customer benefits, peer user benefits, healthcare professional benefits, healthcare industry benefits, and healthcare service provider benefits of employing the proposed smart healthcare solution set. – Establishing the essential components and innovative AI solution models in healthcare. – Providing the latest information and trends from research undergoing worldwide by deploying advanced AI models for smart healthcare [34].

The combination of AI and machine learning in concert with big data analytics is proposed as a solution to make healthcare smart. These smart healthcare solutions possess substantial potential in both early detection and personalized treatment. AI-based personalized medicine using deep learning for mass data processing and machine learning algorithms for optimal treatment pathway recommendation are some common applications. Big data-driven technologies can be widely used with promising results as diagnosis and treatment tools for smart healthcare. The key purpose of this chapter is to highlight the significance of smart healthcare and its urgent requirement in the post-COVID world with the help of diversified intelligent solutions. Subsequently, the chapter presents the multifaceted dimensions of smart healthcare with various enabling artificial intelligence (AI) models together with modern medical equipment and related techniques. To benefit from these assorted intelligent models, there is a significant need to process multidimensional information. Smart healthcare is an essential blend of diverse models of intelligence, including but not limited to big data [34, 36].

Emerging technologies such as artificial intelligence (AI) and modern medical equipment are taking the healthcare sector by storm. The arrival of sophisticated models of AI has helped transform healthcare from the traditional model to a more effective smart healthcare. This chapter discusses the implementation of various advanced models of AI and machine learning in the field of healthcare. A few more cutting-edge technologies such as the Internet of Things and edge computing are also looked at. The technological progression in the fields of big data and machine learning is cornerstone for smart healthcare, leading to novel ideas and studies. Given the considerable complexities and challenges posed by healthcare, promoting general healthcare and improving healthcare outcomes are of utmost importance. A mix of sophisticated technologies such as AI, mobile computing, social networks, cloud computing, big data, and the Internet of Things can serve multiple benefits to different players in the healthcare system.

12.9 Summary

The plethora of patient data, the potential for new treatment pathways, the long-term impact of treatment instead of the trial period, and a reduced discovery time for drugs are some of the many benefits that smart medical devices present us with. However, being a young and undefined field in health informatics, data is an asset in narrow silos with a low possibility of compatibility. Of course, a standardized data model that can work across different device platforms and new knowledge while discovering pitfalls during clinical monitoring will provide valuable information to all healthcare consumers. As intelligent health systems develop, more benefits are expected in future. With the strong build of both IT and AI, we can expect smart healthcare to diffuse more and more into daily lives. While ensuring privacy, improving decision making in many areas like wellness, prevention, and being ready even if a sudden illness affects a patient, early accurate diagnosis, swift optimal treatment, and rapid recovery can be a reality. Healthcare delivery today includes the rapidly increasing role of technology in the diagnosis, monitoring, and treatment of diseases. Using AI in diagnosis and disease management is already playing a key role in better and rapid decision making. The challenges of monitoring patients who need constant care can also be better managed using AI and machine learning. The AI tools that we have today could evolve faster and prove to be a much bigger asset if we had better access to data.

References

1. Jayachitra S et al (2023) AI enabled internet of medical things in smart healthcare. AI models for blockchain-based intelligent networks in IoT systems: concepts, methodologies, tools, and applications. Springer, pp 141–161
2. Sankaran KS, Kim T-H, Renjith P (2023) An improved AI-based secure M-trust privacy protocol for Medical Internet of Things in smart healthcare system. IEEE Internet Things J 10(21):18477–18485
3. Bhardwaj V, Joshi R, Gaur AM (2022) IoT-based smart health monitoring system for COVID-19. SN Comp Sci 3(2):137
4. Awotunde JB et al (2022) AiIoMT: IoMT-based system-enabled artificial intelligence for enhanced smart healthcare systems. In: Machine learning for critical Internet of Medical Things: applications and use cases, pp 229–254
5. Khang A (2024) AI and IoT technology and applications for smart healthcare systems. CRC Press
6. Srivastava J, Routray S (2022) AI enabled internet of medical things framework for smart healthcare. In: International conference on innovations in intelligent computing and communications. Springer
7. Alshehri F, Muhammad G (2020) A comprehensive survey of the Internet of Things (IoT) and AI-based smart healthcare. IEEE Access 9:3660–3678
8. Bhambri P, Tripathy B (2024) Role of AI and IoT based medical diagnostics smart health care system for post-Covid-19 world. Smart healthcare systems. CRC Press, pp 135–145
9. Kumar A et al (2023) 6G-enabled IoT and AI for smart healthcare: Challenges, impact, and analysis. CRC Press
10. Kumar A, Mamgai R, Jain R (2023) Application of IoT-enabled CNN for natural language processing. In: IoT-enabled convolutional neural networks: techniques and applications. River Publishers, pp 149–177
11. Maindarkar M, Kumar A (2025) Artificial intelligence based disease diagnosis using ultrasound imaging. In Revolutionizing Medical Systems using Artificial Intelligence, pp 147–161. Academic Press
12. Ghazal TM et al (2021) IoT for smart cities: machine learning approaches in smart healthcare—a review. Future Internet 13(8):218
13. Saif S et al (2021) Data science and AI in IoT based smart healthcare: issues, challenges and case study. In: Hassanien A-E, Taha MHN, Khalifa NEM (eds) Enabling AI applications in data science. Springer International Publishing, Cham, pp 415–439
14. Mohapatra S, Swarnkar T (2021) Artificial intelligence for smart healthcare management: brief study. Springer Singapore, Singapore
15. Bharati S et al (2023) Explainable artificial intelligence (XAI) with IoHT for smart healthcare: a review. In: Kose U et al (eds) Interpretable cognitive Internet of Things for healthcare. Springer International Publishing, Cham, pp 1–24
16. Xie Y et al (2021) Integration of artificial intelligence, blockchain, and wearable technology for chronic disease management: a new paradigm in smart healthcare. Curr Med Sci 41(6):1123–1133

17. Sharma A et al (2024) IoT and AI-based smart healthcare monitoring system. In: Singh PK et al (eds) Data science and artificial intelligence for digital healthcare: communications technologies for epidemic models. Springer International Publishing, Cham, pp 195–209
18. Bhalla S, Kumar A, Kushwaha R (2023) Comparative analysis on brain stroke prediction using machine learning. In: 2023 14th International conference on computing communication and networking technologies (ICCCNT). IEEE
19. Purohit RM, Verma JP, Jain R, Kumar A (2025) FedBlocks: federated learning and blockchain-based privacy-preserved pioneering framework for IoT healthcare using IPFS in web 3.0 era. Cluster Computing, 28(2):139
20. Sachan RK et al (2023) Diagnosis of mental health from social networking posts: an improved ML-based approach. In: International conference on emergent converging technologies and biomedical systems. Springer
21. Chen X et al (2023) Information fusion and artificial intelligence for smart healthcare: a bibliometric study. Inf Process Manage 60(1):103113
22. Ahmed I, Chehri A, Jeon G (2024) Artificial intelligence and blockchain enabled smart healthcare system for monitoring and detection of COVID-19 in biomedical images. IEEE/ACM Trans Comput Biol Bioinf 21(4):814–822
23. Sujith AVLN et al (2022) Systematic review of smart health monitoring using deep learning and artificial intelligence. Neurosci Inf 2(3):100028
24. Umbare R et al (2024) A smart healthcare application using artificial intelligence (AI) and machine learning (ML). In: 2024 5th International conference on image processing and capsule networks (ICIPCN)
25. Rani S et al (2023) IoT equipped intelligent distributed framework for smart healthcare systems. In: Rishiwal V et al (eds) Towards the integration of IoT, cloud and big data: services, applications and standards. Springer Nature Singapore, Singapore, pp 97–114
26. Yigit Y et al (2024) Machine learning for smart healthcare management using IoT. In: Namasudra S (ed) IoT and ML for information management: a smart healthcare perspective. Springer Nature Singapore, Singapore, pp 135–166
27. Kumar A et al (2023) Security and privacy issues in smart healthcare using machine-learning perspectives. 6G-enabled IoT and AI for smart healthcare. CRC Press, pp 41–56
28. Singh R et al (2024) Securing smart health care systems and infrastructure using artificial intelligence. In: 2024 4th International conference on innovative practices in technology and management (ICIPTM)
29. Bibhu V et al (2023) AI model for blockchain based industrial application in healthcare IoT. In: Bhushan B, Sangaiah AK, Nguyen TN (eds) AI models for blockchain-based intelligent networks in IoT systems: concepts, methodologies, tools, and applications. Springer International Publishing, Cham, pp 163–184
30. Rajawat AS et al (2022) AI and blockchain for healthcare data security in smart cities. In: Piuri V et al (eds) AI and IoT for smart city applications. Springer Nature Singapore, Singapore, pp 185–198
31. Khalid N et al (2023) Privacy-preserving artificial intelligence in healthcare: techniques and applications. Comput Biol Med 158:106848
32. Nasr M et al (2021) Smart healthcare in the age of AI: recent advances, challenges, and future prospects. IEEE Access 9:145248–145270
33. Kothamali PR et al (2023) Smart healthcare: enhancing remote patient monitoring with AI and IoT. Revista de Inteligencia Artificial en Medicina 14(1):113–146
34. Pradhan B et al (2023) An AI-assisted smart healthcare system using 5G communication. IEEE Access
35. Kumar A, Singh D (Eds.) (2025) Revolutionizing Medical Systems Using Artificial Intelligence: A Breakthrough in Healthcare. Elsevier
36. Navaz AN et al (2021) Trends, technologies, and key challenges in smart and connected healthcare. IEEE Access 9:74044–74067

Part IV
Artificial Intelligence for Healthcare Digitization

Chapter 13
Evolution of Traditional Healthcare to Modern Healthcare—Benefits, Opportunities and Challenges

Abstract Several AI methods can be applied in the healthcare field. A simple but effective area where AI has already been used for some time is the image recognition on radiological and ophthalmological problems. In some cases, AI software outperforms the experts in disease diagnosis, such as breast and lung cancer detection. Retinography tasks, such as the understanding of retina images or the detection of blind spot injuries due to diabetes, have also been widely studied. For instance, some publicly available software has been built to make diagnoses or even infer diseases such as macular degeneration compared to human performance for some datasets. The most obvious use of AI software for this is to provide faster and more efficient screening systems, as the proposed software provides faster solutions with great benefits associated with this. The use of artificial intelligence in healthcare has been rapidly growing and is gaining more importance progressively. AI has opened novel approaches in the area and enabled us to achieve solutions that were impossible to think about or implement before. The rapid growth of data in the healthcare industry allows researchers and healthcare systems to explore and build innovative methods that could be used in diagnosis, early disease detection, prognosis systems, and many other healthcare tasks. The rapid growth of healthcare data has promoted an opportunity to gain new insights, develop new tools and software, and extract new patterns to build more accurate models and methods that ultimately turn the healthcare systems smarter.

Keywords Traditional healthcare · Modern healthcare · Healthcare evolution · Healthcare transformation · Healthcare system development · Healthcare innovation · Health service delivery

A. Kumar and D. Singh, *Artificial Intelligence in Modern Healthcare System*, Transactions on Computer Systems and Networks,
https://doi.org/10.1007/978-981-96-6703-1_13

13.1 Introduction

A large number of the population in the world do not receive adequate medical care. This has raised the interest of many companies and researchers to work toward enhancing healthcare quality, access, and services through technological innovations. The combination of technological advancements in healthcare areas such as medical devices, electronic health records (EHR), wearable sensors, artificial intelligence, robotics, and telemedicine, with digital components such as cloud computing, mobile communication, and big data, has the potential to revolutionize healthcare [1]. The goal of smart and connected healthcare is to create a ubiquitous and adaptive ICT infrastructure that can integrate unobtrusively the needs, activities, and surroundings of individuals and healthcare staff with the available medical knowledge and the healthcare service processes [2]. Such integration will enable the delivery of evidence-based personalized healthcare to patients based on their unique conditions, characteristics, and the feedback provided during the care period, and will trigger dynamic adaptations of care plans and service processes [3].

Healthcare around the world is going through a digital transformation. This transformation is aimed to enhance the quality of care, increase access to healthcare services, and reduce the cost of such services. Over the last few years, many technological developments have enabled the transformation of traditional healthcare to smart and connected healthcare [4]. Smart healthcare is an integration of traditional healthcare with information, advanced communication, and technologies, offering higher quality, more personalized, and on-demand services to patients [5].

For several years, healthcare services have followed a traditional pattern which is neither speedy nor secure. The patient has to wait for a long time in hospitals or medical centers to make an appointment with a doctor, which may take time based on the emergency. The major disadvantage of traditional healthcare is that the patient may not have proper documentation from a particular place [6]. In order to ensure secure and speedy healthcare services, the traditional system has to move from paper-based services to a smart healthcare system. Any physician, specialist, and patient should be able to access information from anywhere at any time for security, based on authorization and authentication for the provided services. Since this is a large volume healthcare system with rapid technological advancements, we have come across ways to make the information smart and secure for patients. In this chapter, we present a survey on the transition from traditional healthcare services to smart healthcare services. The key contributions of the chapter are as:

- The delivery, organization, and experience of health services have undergone significant change as a result of the transition from traditional to modern healthcare systems.
- Several noteworthy case studies and success stories demonstrating the influence of contemporary healthcare systems.
- The rise of smart healthcare signifies a substantial change in the management and delivery of healthcare, propelled by developments in data analytics, technology, and patient-centered approaches.

The rest of the chapter is organized as follows. Section 13.2 elaborates the foundation of Traditional Healthcare system. In Sect. 13.3, the emergence of smart healthcare represents a significant shift in how healthcare is delivered and managed, driven by advancements in technology, data analytics, and patient-centered approaches. In addition, some notable case studies and success stories showcasing the impact of modern healthcare systems in Sect. 13.4. Some key future directions and trends shaping the landscape of modern healthcare in Sect. 13.5. Lastly the summary of the chapter is concluded in Sect. 13.6.

13.2 Foundations of Traditional Healthcare

According to Gordon et al., there are three main principles that "should form the basis of a truly effective healthcare system: exceeding patients' expectations regarding the ease of access to and communication with their clinicians, rapid and streamlined access to evidence-based care regardless of ability to pay, and a broadened focus, emphasizing health promotion, disease prevention, and the potential role of healthcare as a tool of economic development and social justice". These factors must be taken into consideration, not only in disease approach and healthcare provision but in a broader sense—in the development of population, cities and rural areas, political choices, and any decisions having an impact on human life. The adoption of such principles could support the birth and maintenance of a healthcare system that is sustainable over time and is strongly integrated into society—a truly smart healthcare model.

In recent years, however, life expectations and other health-specific indices have continued to increase, and the high-grade characteristics of the system have been affected by well-developed challenges such as medical errors, healthcare acquisition costs, burnout and other work-related syndromes, quality decline, standardization limitations, and inefficient, unfair care models. Although a solution to these issues is recognized as fundamental, the roadmaps are different [7].

Standardization: The procedures and practices of medicine, as well as practitioners, are standardized. Medical operations and diagnoses are broadly uniform and are replicated on national and international levels. Research represents a key feature, with results being released to the scientific community. Chronic, iatrogenic condition coverage and prolongation of life cycles: The system focuses not only on acute conditions but also on chronic ones, and it helps to prolong the life of patients. It is staffed by networks of practitioners, with an established hierarchy and effective health policies that favor the reinsertion of invalid citizens [8]. Coexistence with alternative systems: Multicultural societies often have important non-Western medical systems with different historical roots and with a theoretical base separate from current medical science principles. They may be anchored in racial or social fragments or may have a specific link with all citizens, as in the case of Africa's primary healthcare system. Universal access: Although the specific ways in which this comes about differ from nation to nation, modern societies provide universal access to their healthcare systems, at least in some specialized areas.

In modern times, the advancements in human knowledge have led to the establishment of different formalized healthcare systems. Among them, the field of Western scientific medicine (or medical science), founded upon principles of modern science, is likely the most widely known. Its characteristics include evidence-based medical treatments, broad spectrum coverage with a commitment to all members of a community, and a codified form of scope restriction. Such features and quality have led this system to a dominant position in many countries. The following are general characteristics of the Western scientific medical system. Healthcare has always been one of the most important necessities for human beings. In ancient cultures, illness was thought to be the consequence of the anger of gods or spirits. Ancient Egyptians considered the body both material and spiritual and had a well-developed system of physical care that included both secular and sacred dimensions. In Greek civilization, Asclepian temples were established to exacerbate dreams and to thereby make the diagnosis.

13.2.1 Scope of Traditional Healthcare

Despite the widespread development of telemedicine systems across the world, which enable doctors to consult patients that are located at a distance, helping reduce the need for hospital admissions, particularly for patients in remote locations, the vast majority of healthcare interactions in traditional healthcare systems are face-to-face. This is because a large set of healthcare services usually require physical examination or medication, and associated actions are taken in a face-to-face mode. Latest advances in information and communication technology (ICT), combined with increasing external pressures that healthcare systems face, such as increasing demands from an aging population, rising healthcare costs, and shortages of healthcare professionals, have led to the development and deployment of smart healthcare systems, which are also referred to as connected healthcare, e-health, or digital health [9].

Traditional healthcare involves a licensed professional (doctor or nurse) providing services to a patient in an office, clinic, or hospital. It is then essential to list some characteristics of traditional healthcare systems [10]. First, they are typically setting-bound and rely on specialized healthcare settings such as hospitals, clinics, and healthcare staff residences. Second, they are infrastructure-bound, relying on specific healthcare infrastructure elements (buildings, beds, waiting rooms, cabinets) for delivering healthcare services. Third, they are time-bound, necessitating healthcare service provision during working or occasional emergency hours. Time, infrastructure, and setting-bound nature of traditional healthcare all imply physical presence requirements for healthcare staff and patients to properly organize and execute healthcare processes.

13.2.2 Historical Overview

Researchers have discovered anthropological evidence that the earliest civilizations in Mesopotamia and Egypt had already incorporated various forms of medical knowledge. Mesopotamian medical knowledge derived from a number of myths and sagas handed down from previous civilizations [11]. One of the most influential figures in the vast field of Mesopotamian medicine was Esagil-kin-apli, the author of a collection of 36 therapeutic treatments that form the core of ancient Babylonian and Assyrian medical knowledge. By contrast, in the case of ancient Egypt, our primary source is the Ebers Papyrus dating from the second millennium BCE, which contains approximately 700 therapeutic recipes for a large number of diseases [12]. The Papyrus contains a wealth of information on the treatment of infections, medical gynecological practices, suggestions for achieving longevity, and even prescriptions for contraception. The written treatment of diseases was also reflected in the initial Hippocratic corpus of ancient Greek literature that, relating to the field of medicines, has over four centuries established fundamental principles still in force today [13].

The history of the West is steeped in centuries of primitive, medieval, modern, and post-modern health systems. The first primitive age of the relationship between humans and their environment saw climate, hygiene, and social coexistence as prime determinants of health and disease. People subsisted by picking plants and hunting animals, using herbal remedies to treat diseases caused mainly by poor diet and contaminated food and water.

13.2.3 Key Principles and Practices

Each of the following sections discusses in-depth specific healthcare models and practices that can benefit from the use of advanced information technology.

Safeguarding patient safety and ensuring care of the highest quality is a top priority in healthcare. Some of the key principles to ensure patient safety and continuous improvement in healthcare quality include: focusing on the care delivery process, not just the outcomes; continually monitoring care processes and identifying risk factors for adverse events; targeting resources on improving performance in areas that need improvement; using information technology and evidence-based care; and engaging the clinical staff in the process and rewarding their involvement in healthcare management and continuous performance improvement. In many cases, use of advanced information technology products can provide a bridge between the core healthcare practices and the technological advances that are being harnessed for their development [14].

Common healthcare planning and operation management principles include: using evidence-based guidelines and protocols to reduce practice variation; developing and monitoring performance measures; using automated approaches for demand management; providing just-in-time processes to enhance patient flow and reduce inpatient bed length of stay; scheduling healthcare resources based on supply and demand; and optimizing surgical and perianesthesia patient flow, to name a few. Not only do these practices improve the delivery of care, they also make use of the information generated throughout the delivery system [9].

While healthcare systems around the world are unique, there are common principles and practices that can be applied to improve operations as well as the overall safety, quality, and efficiency of care. These healthcare principles and practices cover three primary areas: planning and operations management, patient safety and quality, and use of information technology to improve healthcare processes.

13.3 Emergence of Smart Healthcare

In rapidly aging societies the world over, and with a growing global population that is living longer, we live in an era in which health is considered more important than ever before by a large number of people. For the first time in history, the number of elderly people in countries such as Japan and Italy has now surpassed 20% of the entire population. In addition, this health-consciousness is evident as well in the growing popularity of steady-state bicycle training, jogging, and the changes in food being consumed, which are moving toward more individualized diets that avoid processed foods [15]. Moreover, the development of smartwatches and other monitoring equipment that enable the collection of "lifelog-type" personal biological parameters information, such as steps taken, calorie consumption, kilometers jogged, and heart rate, demonstrate that this trend is increasing. Lifelog-type parameters, combined with daily food logs, would provide a comprehensive database that would enable people to gain a better understanding of their own body's functions [16].

Individualized healthcare that meets the diverse needs of people is required in the rapidly aging society. Such healthcare can be enabled by connecting the various data associated with people's lifelogs and biological parameters and applying both conclusive evidence provided by medical professionals, as well as knowledge from academic principles that are based on experiences and reasoning accumulated as tacit knowledge. We refer to such healthcare as "lifelog-based healthcare" (LHC). We have been developing LHC techniques that quantify diverse human activities using accumulated field experiences, and bio-physiological state sensing technology that acquires condition information as medical professionals review over very long periods. By combining and using these types of healthcare technologies, it is possible for individuals to have their health monitored in daily life. This study is a review of recent results associated with LHC so far achieved by our group. It also contains a discussion of the key research challenges that should be addressed in future to further Lifelog-based Healthcare (LHC) [17].

13.3.1 Concepts Behind Smart Healthcare

It is the concept that encompasses the integration of smart systems in the health field, incorporating intelligent elements that implement communication functionality between environments and the people who use it, being proactive and preventive in nature [18–20]. The possibility of complete self-diagnosis and treatment initiating communication with a medical professional, avoiding unnecessarily overloading the system, has made smart healthcare systems increasingly necessary [21]. The figure shows the concept by illustrating the whole process. First, a physical parameter is needed, acquired thanks to a set of intelligent sensors that capture useful information which is converted through data mining, data analysis, advanced analysis and software. In the presence of certain data, the intelligent system suggests a series of actions. Finally, these can be both presentations to the patient according to the medical diagnosis provided, as well as concepts that help the person to continue taking care of his health. The decision-making process flows in two directions, one vertical and the other a bidirectional concept.

Like other activities, the knowledge and use of information systems is of great importance in healthcare. Along with advances in medical technology, it is transforming traditional healthcare into intelligent or smart healthcare. While traditional healthcare is based on the principles of diagnosing disease and providing effective medical care, its smart predecessor not only involves prevention and early detection of diseases, but also gives a greater role to the patient or citizen and their social environment. This section will define the concept of a smart hospital, together with the main challenges and benefits of integrating information systems in healthcare. In general terms, smart healthcare combines data acquisition (for medical staff and patients), processing (management of care processes), and decision-making to provide timely and professional patient care, providing the highest possible safety to both patients and caregivers with technological support. Its main objective is to prevent patients from entering the hospital and, if they have to, discharge them quickly. This allows hospital resources to be focused only on acute patients [9, 22].

13.3.2 Technological Enablers

Preventive care has been made easier by techniques like the monitoring and analysis of the biologic parameters of the body [23–25]. Researchers are some way forward, using non-intrusive devices that monitor parameters of the body, such as heartbeat, and electrical potentials that denote neurologic activity. They employ "sensorized clothing," which has small sensors that detect warped patterns of emitted light. These devices and several others are capturing much of the biologic information, and the harnessing and analysis of their data are enabling predictions of the individual's physical health and detection of the occurrence of diseases. Capturing and analyzing data from these devices, along with understanding a human's life pattern, may support future medical diagnoses from a purely digital approach, providing substantial benefits to society and making healthcare cheaper [26].

Several technologies are enabling the concept of digital healthcare and promoting faster and more efficient healthcare experiences. These can be divided into different types, such as the Internet of Medical Things (IoMT), smartphones and mobile apps, Big Data, optimization, and data discovery. These technologies create high expectations because they increase our awareness about health, individualizing the way we look at our bodies with the support of wearables, smart cosmeceuticals, and other devices. They help with self-surveillance by tracking several body parameters, leveraging our control over the evolutionary pace and path of our bodies. These technologies are the world's first preventive medical care service, enabling us to think about our health each and every second [27].

13.3.3 Technological Foundations

Digital, connected, and smart healthcare is the result of reducing to practice advanced concepts, models, and theories of information, automation, and intelligent sciences applied to the domain of healthcare. Central to modern healthcare are, for example, the models and theories of health informatics, e-Health, and telemedicine, which have, in fact, enabled some degree of healthcare ICT implementations [28]. Recent advances of connected and smart healthcare are built upon and extend these existing domains of healthcare informatics and technologies. They are defined, for example, by enhancing their connectivity with physical and logical linkages that enable new forms of information, knowledge, and intelligent capabilities. The foundational concepts, models, and theories of healthcare of these advanced developments, however, are often buried and invisible. In the remainder of this section, we therefore reveal the technological foundations of smart and connected healthcare, uncovering its basic building blocks and infrastructure [29].

Wireless communications technology enables untethered connections that are essential for mobile and wearable devices, sensors, and actuators to be integrated with the human body, and embedded and cyber-physical systems to be installed in the physical environments of healthcare facilities and communities. The internet technology allows diverse components, systems, and subsystems of healthcare to be logically connected, enabling information and knowledge about healthcare to be seamlessly integrated and shared. Computing and data/information processing technologies provide the brains of connected healthcare systems, enabling distributed intelligence to be effectively implemented so that data and information are transformed into useful knowledge and wisdom for decision-making and action by healthcare stakeholders [30].

The centerpiece of recent advances in digital, connected, and smart healthcare is information and communication technologies (ICTs). Prominent among these ICTs are wireless communications, the internet, computing, and data/information processing technologies, fundamental to the design and operation of the ICT infrastructure of smart and connected healthcare systems.

13.3.4 Internet of Things (IoT) in Healthcare

However, IoT in the healthcare domain faces several challenges. Firstly, data privacy and security are the most serious concerns for stakeholders. Secondly, the huge amount of data generated by IoT devices can exceed the capacity of existing networks and systems. Thirdly, the lack of interoperability and standards and the high cost of IoT technology hinder the deployment and operation of IoT systems in the healthcare domain. To address these challenges, research, testing, and the collaboration of regulators, policymakers, and industry stakeholders are needed. The ultimate result of the effort to overcome these challenges is the creation of IoT systems that reduce the cost of healthcare, improve the quality of medical services, increase accessibility, and make personalized medicine a reality. This offers considerable opportunities for various industries to create innovative IoT applications and services in the healthcare domain [31].

The Internet of Things (IoT) is a key enabler of the smart healthcare paradigm. IoT represents a network of uniquely identifiable interconnected devices that communicate without human interaction using standard and proprietary internet protocols. These devices, which can be specialized (sensing, actuating, data processing) or non-specialized (smartphones, tablets, laptops), work in concert to create added value for services created in the interest of individuals and society. The IoT growth trend is clear as new applications and services are constructed by various industries [32]. In the healthcare domain, IoT technology connects medical devices and equipment, enabling them to communicate diagnostic information directly to healthcare management systems in real time. The patients and the caregivers will also be able to exchange information with the systems using the internet in order to report medical status and receive medical instructions.

13.3.5 Data Analytics and Artificial Intelligence in Healthcare

In accordance with the rich volume of literature that investigates the challenges and the opportunities of data analytics and AI technologies in healthcare, such as expert interviews, academic research, industry reports, and public documents, in this section, we provide an elaborative view of the current landscape of the application of data analytics and AI technologies in healthcare, summarize the state-of-the-art projects and companies, develop a taxonomy of healthcare data analytics and AI technologies, and discuss the major challenges and opportunities in the application of data analytics and AI technologies in the healthcare sector. Finally, we provide research implications and recommendations [33].

Despite the promises and the derived opportunities from data analytics and AI technologies, there exist significant challenges in the application of these technologies in the healthcare arena. The challenges span not only the technical concerns, such as the privacy and security of healthcare data, the transparency and accountability of AI technologies, and the vulnerability to adversarial attacks of the intelligence models, but also the overarching concerns related to the perceptions and readiness of the stakeholders in the healthcare ecosystem [34].

In the era of big data, a substantial amount of data in various forms can be generated from diverse sources along the lifecycle of healthcare processes, from the knowledge discovery and data gathering to data curation and annotation, to data modeling and analysis, and to development and deployment of data-driven and personalized health decision and action systems. Data and analytics are both the lifeline and the hidden force of the future health ecosystem [35]. Data analytics and artificial intelligence technologies possess the potential to enhance healthcare in a variety of aspects, such as improving health outcomes, promoting operational efficiency, and advancing the innovation of the healthcare process and system.

13.3.6 Machine Learning Applications in Healthcare

Machine learning is used in personalized medicine, which is designed for customization of healthcare, with decisions, practices, and/or products being tailored to individual patient characteristics. It can also help in the early discovery of the spread of infectious disease. Because of its potential role in transforming the healthcare industry, ML has become an interdisciplinary research area involving computer science, statistics, and healthcare service. Over the last ten years, the applications of ML in the field of healthcare have surged. Aspects of the healthcare system that have benefited from machine learning include risk evaluation, medical imaging, capturing healthcare information, bioinformatics, and general healthcare management [36]. Machine learning can help discover complex interactions between the features and outcomes in these domains.

Machine learning (ML) covers a set of techniques that allow computers to learn from experience and discern complex patterns in vast, possibly high-dimensional data, and make intelligent decisions for prediction or diagnostic purposes. It has huge potential to transform the operation and structure of the healthcare system. Machine learning applications in healthcare are designed to aid in diagnosing diseases, assessing and predicting patient conditions, delivering treatment, and managing chronic conditions, and can help bring about precision and individualized medicine [37] (The Applications of AI in smart healthcare are shown in Fig. 13.1).

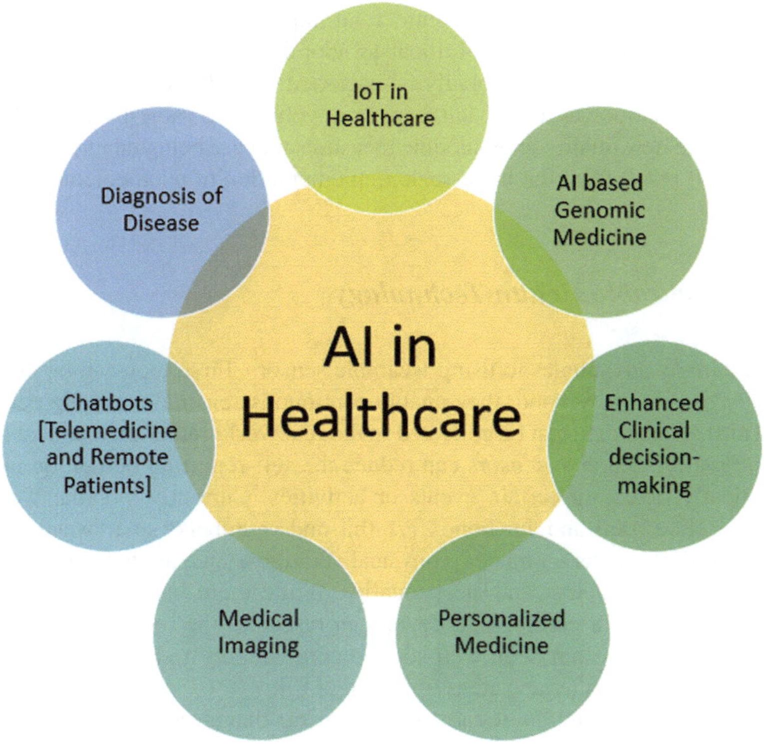

Fig. 13.1 Applications of AI in smart healthcare

13.3.7 Telemedicine and Remote Patient Monitoring

Smart and connected telemedicine services are discussed, starting with the state-of-the-art in specialty telemedicine and the enabling technologies, and then going into the best practices model for building and deploying sustainable telemedicine services. This model, which has been defined by the experience of the UPMC Center for Connected Medicine, is then applied to the outline of a process for rapidly building, piloting, and scaling national or global telemedicine services [38]. The transformation that specialty telemedicine is bringing to healthcare through new quality and outcomes measures and through the use of big data for both research and business intelligence is discussed. The chapter concludes with a vision of the future evolution of specialty telemedicine services.

Telemedicine is a rapidly evolving area in healthcare enabled by the convergence of advanced technologies. It is changing the way care is delivered and presents both new opportunities and challenges. This chapter provides an up-to-date review of the status of telemedicine and remote patient monitoring, including the latest innovations, the challenges that must be addressed to realize its full potential in improving outcomes and access while controlling costs, and the new and emerging business models [39]. Successful commercial implementations in several specialty areas define the factors that are critical to adoption, and scaling of telemedicine services both nationally and globally is discussed. Finally, the transformation that telemedicine is empowering in healthcare from volume- to value-based is addressed in terms of the new quality and outcome measures that are being enabled and the big data that will result from the large-scale implementation of telemedicine services.

13.3.8 Wearable Health Technology

There are many advantages to using wearable sensors. First, they can be integrated into daily use easily. Second, they enable continuous sensing over long periods of time. Third, sensor data can be collected in the real world rather than in a laboratory setting. Finally, wearable sensors can reduce the self-reporting burden on users by automatically detecting certain events or activities. Currently, popular wearables include the Nike FuelBand, Jawbone UP, Fitbit, and a number of smart watches. These devices primarily focus on tracking physical activities, such as steps taken, calories burned, distance traveled, and sleep quality. While there has been an increasing interest in using such commercial devices in research, the reliability and validity of these devices have not been well-studied compared to traditional non-wearable sensors, especially in the new areas of sensing [40].

Today, wearable technologies are usually sensors that are attached to a computing device that people can wear as accessories or clothes. Wearable sensors can easily collect information from the body and the environment in a user-friendly way. People can then utilize this data to gain new insights, improve their lives, and share information efficiently. The concept of the Quantified Self, in which individuals track different types of activities on a regular basis with the help of technology, is often associated with wearable sensors that track body data. It is becoming increasingly popular and new devices are constantly being developed. In the area of emotion and stress research, wearable sensors are also being used to collect physiological data and correlate it with other data modalities, such as location, activity, or user-generated content [41].

13.3.9 Cybersecurity in Healthcare Systems

The growing number of cybersecurity incidents is a clear indicator of the need for a strong, secure posture to protect patient data confidentiality and system availability in the Cyber Physical System (CPS), Internet of Things (IoT), and cloud computing era of healthcare systems. These advanced technologies enhance the quality of care, decrease the cost of care, and improve the efficiency of care. Simultaneously, they create a new vulnerability for malware attacks, in which the consequences can be fatal when an attack successfully targets medical devices performing patient care and monitoring. The consequences of a cyber-attack on patient treatment and diagnosis through Life Critical Medical Devices (LCMDs) are discussed. The attack surface of LCMDs as well as security gaps are analyzed [42].

Cybersecurity in healthcare is extremely important as it ensures patient data is protected and kept confidential. Currently, threats and risks are increasing from various levels, both internal and external, to healthcare systems. Thus, it is necessary to develop and enforce cybersecurity measures through regulation to make sure all healthcare stakeholders are taking it seriously. This chapter discusses the importance of security in healthcare, laws and regulations surrounding protected health information, various types of threats in healthcare, and recommendations for securing healthcare systems [43].

13.3.10 Regulatory and Ethical Considerations

Healthcare is going through a digital transformation with the development of breakthrough technologies, including artificial intelligence (AI), robotics, and the Internet-of-Things (IoT). By connecting patients, caregivers, and medical equipment, data can be collected and analyzed in real-time, which personalizes and improves the quality of patient care and also increases the overall operational efficiency of the healthcare environment. For example, with advanced deep learning techniques, intelligent medical diagnosis systems can achieve an accuracy level that is comparable to human experts by analyzing and learning from a large amount of medical imaging data [44]. With the development of connected healthcare, patients' health conditions can be monitored at home, which reduces the length of hospital stays and the risk of hospital-acquired infections. In addition, the healthcare supply chain can be optimized with IoT technology by tracking and monitoring the conditions of medical inventory and transportation in real-time. It also helps oversee the compound medication production processes with sensor systems.

The transformation of the healthcare sector with innovative digital technologies, including AI, Internet-of-Things (IoT), and robotics, has created breakthrough healthcare applications. Before these data-driven and smart healthcare techniques become widely used, besides addressing the challenges related to cybersecurity, data privacy, and the accuracy of intelligent algorithms, regulatory frameworks and ethical guidelines must be established and implemented. In this chapter, we discuss the promising applications of smart healthcare, review some of the regulations and ethical guidelines that are developed for ensuring the safe and appropriate use of these evolving technologies, and present a few healthcare projects that follow these regulations and guidelines [45].

13.3.11 Impact on Healthcare Delivery and Patient Outcomes

Several areas of healthcare delivery that are or will be impacted by connected health are telemedicine, home healthcare, and medication adherence. It is important for healthcare providers to understand how these changes will affect their role and to proactively address any concerns that may arise during the transition. Moreover, to fully realize the benefits of connected health, the entire healthcare system must be transformed from a system that is reactive and hospital-centered to one that is proactive and patient-centered. This chapter describes promising innovations enabled by connected health within telemedicine, home healthcare, and medication adherence; discusses the challenges that must be addressed in order to realize the full potential of connected health; and outlines opportunities for future research and development.

Connected health holds great promise to transform healthcare delivery and improve patient outcomes. By continuously monitoring a patient's health status regardless of the patient's location and transmitting that information in real time to the patient's healthcare provider, connected health has the potential to enable patient-centered care, improve care team collaboration, reduce hospital readmissions, and expedite recovery. The shift in focus from treatment to prevention and early detection of illness, enabled by connected health, not only improves patient outcomes but also reduces healthcare costs. However, in order for connected health applications to be effective and widely adopted, they must be designed in a way that alleviates the major concerns relating to privacy and security of patient data.

13.3.12 Enhanced Clinical Decision-Making

The formation of clinical informatics as a recognized specialty in the medical field in the United States and the subsequent inclusion of related topics in medical board certifications are significant milestones underscoring the importance of clinical informatics in the healthcare domain. The integration of specialized knowledge and skills related to clinical informatics with those from other healthcare professions provides invaluable support by optimizing decision-making processes that occur within the clinical setting. Such integration routinely results in positive patient health outcome improvements being identified and subsequently addressed through the use of new approaches, or the modification of existing processes, or healthcare policy changes.

Health informatics is the application of informatics in areas related to health. It comprises multiple areas, including those related to the theoretical and practical considerations of acquiring, storing, managing, accessing, and processing information, knowledge, and data. By acting as a catalyst to enhance the decision-making of healthcare professionals, patients, and other stakeholders, health informatics enables the improvement of health outcomes at all levels. Clinical informatics is an applied subdiscipline of health informatics. It focuses on the use of informatics in relation to the specialized knowledge possessed by healthcare professionals, with the overarching goal of enhancing individual and population health outcomes.

13.4 Case Studies and Success Stories

The examination of smart technologies in hospitals and elder care demonstrates that they can support the reorganization of healthcare so that care decisions are made in collaboration with patients and in coordination with professionals across multiple settings, and where hospital-based activity is focused on more acute care that can only take place in a hospital and for which de-hospitalization replaces hospitalization [46]. This transformation recognizes that the hospital is not the endpoint of healthcare. Rather, it is a collection of resources that can combine in many different ways to deliver on the continuum of care that an individual person requires. This continuum does not sit within the hospital. This is despite hospital spending accounting for the vast majority of the overall budget of the health and care system. The cross-cutting effect of smart technologies in supporting a rethinking of healthcare can be observed in the role that they can play in promoting public health and effective behavior change. Smart technologies can provide an important tool to support personalized health care by using tailored information and communications systems, including mobile applications. In addition, the development and implementation of smart cities and smart homes could have significant positive influences on public health by providing safe living conditions, social interaction, and physical activity for residents. The evidence for community-based programs particularly points toward the higher cost-effectiveness of technology-enabled solutions for managing long-term health conditions. In the next section, the chapter moves to discuss the broader implications of the deployment of smart technologies in healthcare [6].

To better understand the potential of breakthroughs in healthcare, this section portrays a number of case studies and success stories of ICT integrated healthcare. Through these case studies, it is proposed to showcase that technology can gradually transform healthcare delivery from almost entirely hospital-based care to packages of different activities, some of which can occur in the home or the workplace. This will require a radical rethinking of the design of healthcare of the future. It is also, however, part of the roadmap to achieving sustainable health systems. It can help mitigate many of the challenges that conventional thinking about the future role of healthcare might generate—capacity constraints for hospitals and other health and social care establishments, and a growing financial burden associated with increasing demand in the face of technological progress. For the transformation to succeed, it will need to take a systems perspective: seeing healthcare not as a series of individual settings and solutions but rather as an integrated ecosystem, encompassing many sectors of the economy. The remainder of this section is structured, in turn, around the deployment of smart technologies in the following four sub-domains of healthcare service provision: hospitals, elderly care, public health, and the management of long-term health conditions.

13.4.1 Implementation Examples

The healthcare-as-a-service concept assumes that the healthcare resources (e.g., medical professionals, equipment, and healthcare services) are managed by a single organization. It increases the efficiency and the quality of the healthcare services provided with the help of a fee-for-service model. Healthcare-as-a-service could both create new businesses and generate massive profits by leveraging new business models in various healthcare services. However, each smart healthcare application has some limitations and implementation issues on certain technical areas such as data analysis, data traffic, user acceptability, data privacy, and retrieval time. In this chapter, several examples are discussed with their strengths and challenges [47].

There are several ways to implement smart healthcare applications. Examples include home monitoring, a smart hospital for patient monitoring, smart healthcare center, and healthcare-as-a-service [48]. Smart homes are equipped with various types of sensors that collect household activity information, lifestyle information, and physiological signals. Furthermore, the smart hospital has been developed with the integration of information and communication technologies (ICT) for inpatients. At smart healthcare centers, various health services are provided to treat patients and maintain health. Physicians in the remote healthcare consultation centers and referral centers perform real-time consultation and patient follow-up implemented through the use of ICT [49].

In telemedicine, IoT-based home monitoring systems are used for continuous health monitoring of a patient to manage chronic disease conditions. It enables sharing health problems remotely and providing advice and possible treatment by an expert. Early detection or risk prediction of diseases is effective for dealing with them through prevention measures or health maintenance services.

Existing smart healthcare applications and implementations can provide realistic insights for creating and implementing real-world smart health systems capable of achieving a paradigm shift in the traditional healthcare domain. Based on its huge potential to meet several healthcare challenges, IoT is being extensively used in healthcare applications and services currently. Smart healthcare applications specific to telemedicine, preventive medicine, healthcare management, and smart living are discussed as follows.

13.4.2 Impact Assessment

The final result confirms that—when applied to specific population-based healthcare problems, such as the prevention of hypertension and diabetes, the management of chronic patients or health guarantee services for critical population typologies—the widespread technological e-health impact on community services might make a substantial difference in the lives of patients living in rural mountain areas and in the savings in care costs for local healthcare services.

To deepen the analysis on how healthcare shifts with the introduction of remote healthcare delivery models and what the actual impact of these initiatives on healthcare costs (broadly conceived) and scope (improving efficacy), we adopted the evaluation framework used by Arora et al.: the Economic Interaction Model (EIM), which includes effects both vertical (on hospital and healthcare facilities management) and horizontal (on markets for goods and services and the living conditions of regional economies) of a service investing in a region. In light of the recent methods in the CBA field, a methodology to fill out a project evaluation form has been proposed and used.

13.5 Future Directions and Trends

The future direction of this field includes, specifically, (1) health data integration, (2) research in constrained settings, for example, (3) discovery for rare disease and personalized medicine, (4) health metric shifts, (5) targeting cost/benefit ratios, and (6) reinforcement learning for health. Although a large fraction of healthcare data is still not directly tied to a single patient or single event, intensive research is crucial to filling the gap by utilizing improved prediction models, programming platforms, or health data representation [50]. Significant different characteristics of healthcare data impose great challenges to building prediction or optimization models. Various real-world problems may refer to research in constrained settings, including a small number of learning data, specifically fewer usable training samples and clinical constraints such as electronic medical records, and laborious rechecking or diagnosing. The future of healthcare research should target cost/benefit ratios coupled with patient outcomes, drug effectiveness, medical-service optimization, and administrative overhead reductions. Reinforcement learning provides a more flexible framework to model sequential patient trajectories to treat patients and refine patient treatment policies based on collected data while avoiding potential biases [51, 52]. The authors can anticipate that reinforcement learning will contribute to policy development or alignment, resource management, and value-based care in the long term.

13.5.1 Artificial Intelligence and Machine Learning

In healthcare, the use of AI may be particularly powerful but also raises uniquely challenging ethical concerns. A demanding discipline: mesoscale models. One of the most challenging aspects of this opportunity space is where simulation and mining become increasingly ambitious as we move from traditional decision optimization space to predictive machine learning models. Management of multimodal data in different conditions from multiple sources in the healthcare domain and frequent fluctuations of decision boundaries in uncertain data products have been very challenging for existing AI technologies and their deployments. Traditional AI algorithms do not yet outperform skilled human clinicians, nurses, and care workers in most of the wellness tasks, and they are generally treated as a supplementary tool for decision-making [53]. Even though with these considerations, various smart healthcare topics are on the edge of creating a pivotal role in influencing patient wellness in the future.

Artificial intelligence (AI) and machine learning, along with deep learning, are propelling diversification in personalized healthcare. A large amount of data is collected daily, and this collection prompts the research of personalized healthcare and artificial intelligence [54]. Machine learning provides a methodology to winnow through the noise: the myriad of data and models are cherry-picked to model trial protocols, candidate agents, and stratify patients for various therapies. Various machine learning algorithms, such as support vector machines, nearest neighbors, neural networks, decision trees, generalized linear models, regularized linear models, ensemble methods, are commonly used in healthcare in the prediction of diseases, healthcare operations, patient health monitoring, and many other applications [55].

13.5.2 Internet of Medical Things (IoMT)

Petabytes of health-related data are, and will be collected, on patients. This incredible medical record powerfully contributes to the data-driven treatment of patients. Nonetheless, respecting the central role of patients' data, ethical and legal issues must be tackled: the topic is so relevant that it deserves performing the collection, analysis, and interpretation of data in agreement with specified guidelines [56]. Researchers should ensure that evaluators, participants, and readers understand the ethical implications of research and information presented. Measures to address the ethical components are indicated; a potential cost to both researchers and healthcare is associated with not following ethical measures. In any case, data-driven medicine lends itself even to personal-elevating organisms, ethically improper evaluations. Finally, development is fundamental too: the new paradigm of IoMT shall not be held hostage by data challenging [57].

The Internet of Medical Things (IoMT) generally employs wearable or implantable devices, sensors, and other smart gadgets to collect and analyze patient data at a large scale. While such data used to be collected within the hospital, with IoMT tools, a lot of pertinent clinical markers and diagnostic modalities may be performed continuously and even beyond hospital premises. The enabling factor of such "disruptive technology" is represented by miniaturization of electronics, biosensorics, and the impressive development of low-power wireless communication. Furthermore, as some sophisticated datasets require even computational power to be processed in the order of teraFlops, computation power available from cloud services may be employed [47]. Nonetheless, very strict constraints need to be fulfilled to make such devices trusted devices, particularly when they are used to control treatments. The confidentiality, integrity, and authenticity of data collected, stored, and elaborated for the health of a patient must be warranted from intrusion and alteration. This is particularly relevant when the health data of a patient represent a pattern, and public or even worse for-profit companies may seek at acquiring such a pattern for many unethical aims [58].

The Internet of Medical Things (IoMT) refers to the network of medical devices, applications, and systems that are connected to the internet. With IoMT, healthcare providers and patients can access and share real-time data, leading to improved diagnoses, treatment, and overall patient care. IoMT devices include wearable sensors, remote monitoring systems, medical robots, and smart implants, among others. These devices collect and transmit data, enabling healthcare professionals to monitor patients remotely, provide timely interventions, and make data-driven decisions. IoMT has the potential to revolutionize healthcare by increasing efficiency, reducing costs, and enhancing patient outcomes [59].

13.6 Summary

However, to move from traditional to smart healthcare, the policy has to protect privacy and meet the necessities of quality and knowledge management and be efficient and affordable—a smart healthcare system. Smart information management is an interactive and iterative loop which needs to be looked beyond. An alliance of skills is needed to reach the smart healthcare system. Indeed, the smart healthcare model indicates the patient and family as one of the stakeholders in care with specific roles and responsibilities. This implies implications that have to be considered in policy design and in professional education. In this connection, medical researchers, health service providers, and policy designers must then make this model a reality to transform a new smart healthcare model for widespread international adoption. The traditional mode of healthcare management is reactive in nature—it waits for diseases to occur and then takes corrective measures. This approach has led to skyrocketing healthcare expenses all over the world. There is a dire need for the transformation of traditional healthcare to smart healthcare, which is an intelligent innovation with data-driven digital-based platforms between patients, healthcare managers, and stakeholders, to provide proper personal and family healthcare management. The smart healthcare system refers to digital solutions that provide high-quality healthcare for individuals. Reportedly, a smart healthcare system can reduce the management costs of chronic diseases and improve patient outcomes in terms of treatment adherence, quality of life, symptom monitoring, and promoting early prevention. As a user, smart healthcare also reduces wasted time through waiting for unaffordable and long periods. It can also lead to improved workflow and support the interests of patients as central stakeholders in healthcare.

Smart healthcare is here to empower us to live healthier lives, mitigate chronic conditions, manage disease at home or on a journey, and improve communities' healthcare public services. Continuously advancing, the speed and cost at which new products and services are now developed can help transform traditional healthcare into smart healthcare, from point of care to community healthcare, hospital care, primary care, clinics, general practices, and individual healthcare, weaving us into an aggregate model of precision healthcare. Incorporating disciplines such as healthcare AI, machine learning, medical IoT, healthcare analytics, wearable technologies, e-Health, healthcare robots, telemedicine, smart hospitals, smart clinics, and patient and public participation, smart healthcare enhances our quality of life, promotes well-being, supports patients at different stages of their care journey, and enables enhanced services to be delivered by healthcare providers. The digital transformation of a hospital through e-Health, enabled by procuring medical-grade routers and networking gear, as well as conducting well-architected e-Health carries excess benefits. There are three revolutions in smart healthcare: digitization, connectivity, and data-driven intelligence.

In the face of numerous promising but still risky technologies, it is incumbent upon the government and professional regulatory bodies to design a light but strong responsive governance mechanism and process that will allow for scientific and technological advancement in healthcare. A substantial increase in public and private investment in digital health and smart healthcare is necessary for developing further solutions that are affordable, reliable, interoperable, timely, inclusive, and human-centered. Legislative, regulatory, legal, and ethical concerns must be addressed. There is still a tremendous amount of research and education to be done. In particular, it is vital to increase awareness and promote training in the unique aspects of digital health among healthcare professionals and students. The importance of customer-friendly design, participatory design, and co-creation of value between the users and the new digital health system must not be underestimated. It is vital to develop and offer user-centered, efficient, ethical and technical expertise and expert support in order to ensure the success, impact, and the much-needed transition to smart healthcare of this Health 6.0 paradigm shift. The health needs of the users should be the real winner over this development. We recommend the establishment of a new organization, competent to take into account the legal, ethical, social, and technical aspects of digital health—to discuss, promote, help when necessary, and enforce principles and governance, quality assurance, and ethical issues such as privacy and data protection.

The transformation to smart healthcare will profoundly and holistically affect every aspect of healthcare. All healthcare stakeholders, whether at the level of health system administration, center of excellence for smart healthcare to be established in every hospital, professional association, and individual healthcare providers such as physicians and pharmacists, and including insurance companies, medical devices and pharmaceutical companies, should join hands and form public–private multistakeholder partnerships to facilitate healthcare system transformation. Research and innovation in medical science and engineering, leveraging artificial intelligence, internet of things, big data analytics, even brain-machine interface and computer–human interaction, must be emphasized and well supported. Ethical, legal, and regulatory considerations have to be addressed as well. Pre- and in-service education curricula for all healthcare professionals should be reformed to include basic knowledge of smart healthcare, so that tomorrow's healthcare workforce will be mindset-ready for the transformation. The last but not least, the end users of the service-patients and citizens-should form a partnership with healthcare systems that enable care to empower them to health and to lead their daily activities.

References

1. Haleem A et al (2022) Medical 4.0 technologies for healthcare: features, capabilities, and applications. Internet of Things Cyber-Phys Syst 2:12–30
2. Awad A et al (2021) Connected healthcare: improving patient care using digital health technologies. Adv Drug Deliv Rev 178:113958
3. Thomason J (2021) Metahealth-how will the metaverse change health care? J Metaverse 1(1):13–16
4. Ahmad KAB et al (2022) Emerging trends and evolutions for smart city healthcare systems. Sustain Cities Soc 80:103695
5. Kumar A et al (2021) A comprehensive study on the role of advanced technologies in 5G based smart hospital. Alex Eng J 60(6):5527–5536
6. Saba Raoof S, Durai MS (2022) A comprehensive review on smart health care: applications, paradigms, and challenges with case studies. Contrast Media Molec Imag 2022(1):4822235
7. Min S et al (2023) Clinical validation of a wearable piezoelectric blood-pressure sensor for continuous health monitoring. Adv Mater 35(26):2301627
8. Niu H et al (2023) Advances in flexible sensors for intelligent perception system enhanced by artificial intelligence. InfoMat 5(5):e12412
9. Mihaescu I, Giroux C (2023) The ways of teaching medicine: whisperings of the silent curriculum. In: Gasping for air and grasping air in medicine: equity, diversity, and inclusion on the frontlines, pp 66
10. Çamlı ŞE et al (2024) Embracing different languages and local differences: co-constructive patient simulation strengthens host countries' clinical training in psychiatry. World J Psychiatry 14(1):111
11. Busch U (2023) Claims of priority–rhe scientific path to the discovery of X-rays. Zeitschrift für Medizinische Physik
12. Serwer D (2024) Science discovers, medicine applies, protection lags, 1896–1902. Strengthening international regimes: the case of radiation protection. Springer, pp 37–71
13. Obodovskiy I (2023) X-ray mania. From radio-phobia to radio-euphoria: low radiation doses: safe, useful, and necessary. Springer, pp 11–28
14. O'Malley CD (2023) The history of medical education: an international symposium held February 5–9, 1968, vol 12. University of California Press
15. Kim J et al (2024) Wireless breathable face mask sensor for spatiotemporal 2D respiration profiling and respiratory diagnosis. Biomaterials 309:122579
16. Shaik T et al (2023) Remote patient monitoring using artificial intelligence: current state, applications, and challenges. Wiley Interdiscip Rev: Data Min Knowl Discov 13(2):e1485
17. Na J, Kim SJ, Lim Y (2023) Designing digital therapeutic content using chronic disease data: a focus on improving urinary dysfunction. Int Neurourol J 27(Suppl 2):S51
18. Kumar A et al (2023) 6G-enabled IoT and AI for smart healthcare: challenges, impact, and analysis. CRC Press
19. Kumar A, Bhalla S (2025) Exploring brain tumor detection through artificial intelligence. In Revolutionizing Medical Systems using Artificial Intelligence, pp 93–120. Academic Press

20. Kumar A, Sareen P, Arora A (2023) Healthcare engineering using AI and distributed technologies. Smart distributed embedded systems for healthcare applications. CRC Press, pp 1–14
21. Alshamrani M (2021) IoT and artificial intelligence implementations for remote healthcare monitoring systems: a survey. J King Saud Univ Comput Inf Sci 34:4687–4701
22. Zeshan F et al (2023) An IoT-enabled ontology-based intelligent healthcare framework for remote patient monitoring. IEEE Access 11:133947–133966
23. Kumar A et al (2024) Artificial intelligence bias in medical system designs: a systematic review. Multimedia Tools Appl 83(6):18005–18057
24. Kumar A et al (2023) Security and privacy issues in smart healthcare using machine-learning perspectives. 6G-enabled IoT and AI for smart healthcare. CRC Press, pp 41–56
25. Kumar A et al (2023) AI-assisted special education for students with exceptional needs. IGI Global
26. Philip NY et al (2021) Internet of Things for in-home health monitoring systems: current advances, challenges and future directions. IEEE J Sel Areas Commun 39(2):300–310
27. Di Carlo F et al (2021) Telepsychiatry and other cutting-edge technologies in COVID-19 pandemic: bridging the distance in mental health assistance. Int J Clin Pract 75(1)
28. Sallam M et al (2023) ChatGPT applications in medical, dental, pharmacy, and public health education: a descriptive study highlighting the advantages and limitations. Narra J 3(1)
29. Lamba K, Rani S (2024) A novel approach of brain-computer interfacing (BCI) and grad-CAM based explainable artificial intelligence: use case scenario for smart healthcare. J Neurosci Methods:110159
30. Ramakrishnan S, Jijitha S, Amudha T (2024) Roadmap of AI and IoMT in smart healthcare: current applications and future perspectives. Internet of Medical Things Smart Healthc:137–161
31. Atadoga A et al (2024) Internet of Things (IoT) in healthcare: a systematic review of use cases and benefits. Int J Sci Res Arch 11(1):1511–1517
32. Puri V, Kataria A, Sharma V (2024) Artificial intelligence-powered decentralized framework for Internet of Things in Healthcare 4.0. Trans Emerg Telecommun Technol 35(4):e4245
33. Bag S et al (2023) Big data analytics and artificial intelligence technologies based collaborative platform empowering absorptive capacity in health care supply chain: an empirical study. J Bus Res 154:113315
34. Guo C, Chen J (2023) Big data analytics in healthcare. Knowledge technology and systems: toward establishing knowledge systems science. Springer, pp 27–70
35. Del Giorgio Solfa F, Simonato FR (2023) Big Data analytics in healthcare: exploring the role of machine learning in predicting patient outcomes and improving healthcare delivery. Int J Comput Inf Manuf (IJCIM) 3
36. Kasula BY (2023) Machine learning applications in diabetic healthcare: a comprehensive analysis and predictive modeling. Int Numeric J Mach Learn Robots 7(7)
37. Sarker M (2024) Revolutionizing healthcare: the role of machine learning in the health sector. J Artif Intell General Sci (JAIGS) 2(1):36–61. ISSN: 3006-4023
38. Lichtenstein MR et al (2024) Researcher experience and comfort with telemedicine and remote patient monitoring in cancer treatment trials. Oncologist 29(4):356–363
39. Kayode S, Luz A (2023) Telemedicine and remote patient monitoring: harnessing neural networks to enable remote healthcare services and remote patient monitoring
40. Shajari S et al (2023) The emergence of AI-based wearable sensors for digital health technology: a review. Sensors 23(23):9498
41. Williams GJ et al (2023) Wearable technology and the cardiovascular system: the future of patient assessment. Lancet Digit Health 5(7):e467–e476
42. Choi SJ, Chen M, Tan X (2023) Assessing the impact of health information exchange on hospital data breach risk. Int J Med Informatics 177:105149
43. Jones DN (2022) Understanding and decreasing security breaches in the healthcare industry: a qualitative case study exploring network-connected medical devices in a large hospital. Northcentral University

44. Javaid M et al (2023) Towards insighting cybersecurity for healthcare domains: a comprehensive review of recent practices and trends. Cyber Secur Appl:100016
45. Argaw ST et al (2020) Cybersecurity of hospitals: discussing the challenges and working towards mitigating the risks. BMC Med Inform Decis Mak 20:1–10
46. Hartmann M, Hashmi US, Imran A (2022) Edge computing in smart health care systems: review, challenges, and research directions. Trans Emerg Telecommun Technol 33(3):e3710
47. Dwivedi R, Mehrotra D, Chandra S (2022) Potential of Internet of Medical Things (IoMT) applications in building a smart healthcare system: a systematic review. J Oral Biol Craniofacial Res 12(2):302–318
48. Poongodi M et al (2021) Smart healthcare in smart cities: wireless patient monitoring system using IoT. J Supercomput:1–26
49. Kondaka LS et al (2022) An intensive healthcare monitoring paradigm by using IoT based machine learning strategies. Multimedia Tools Appl 81(26):36891–36905
50. Itani R, Tobaiqy M, Al Faraj A (2020) Optimizing use of theranostic nanoparticles as a life-saving strategy for treating COVID-19 patients. Theranostics 10(13):5932
51. Kumar A, Singh D (Eds.) (2025) Revolutionizing Medical Systems Using Artificial Intelligence: A Breakthrough in Healthcare. Elsevier
52. Zeadally S, Bello O (2021) Harnessing the power of Internet of Things based connectivity to improve healthcare. Internet of Things 14:100074
53. Rasmy L et al (2021) Med-BERT: pretrained contextualized embeddings on large-scale structured electronic health records for disease prediction. NPJ Digit Med 4(1):86
54. Raparthi M (2020) AI integration in precision health-advancements, challenges, and future prospects. Asian J Multidiscip Res Rev 1(1):90–96
55. Si Y et al (2021) Deep representation learning of patient data from Electronic Health Records (EHR): a systematic review. J Biomed Inform 115:103671
56. Monteiro K et al (2021) Internet of medical things (IoMT) applications in E-health systems context: an overview and research challenges. In: Emerging technologies in biomedical engineering and sustainable telemedicine, pp 1–12
57. Razdan S, Sharma S (2022) Internet of medical things (IoMT): overview, emerging technologies, and case studies. IETE Tech Rev 39(4):775–788
58. Ahmed SF et al (2024) Insights into Internet of Medical Things (IoMT): data fusion, security issues and potential solutions. Inf Fusion 102:102060
59. Sadique MA et al (2024) Engineered two-dimensional nanomaterials based diagnostics integrated with internet of medical things (IoMT) for COVID-19. Chem Soc Rev

Chapter 14
Analysis of AI-Bias in Modern Healthcare Systems

Abstract AI (Artificial Intelligence) has provided many predictive algorithms for the diagnosis of many critical diseases. AI has also presented segmentation algorithms which can segment the desired area from the background for better diagnostic results. But AI-predictive algorithms suffer from AI-bias either due to training data or algorithmic design. This AI-bias leads to variability and inaccuracies in the predictive results which may have severe impact on treatment and clinical deployment of the model. Hence, it is necessary to evaluate the accountability of AI-bias in medical systems. Analysis of bias at various levels of AI-models in medical system design can prevent severity in the medical outcomes. In this chapter, we will highlight the bias accountability at various stages of AI-models. We will also review the various reasons and mitigation techniques to minimize the impact of AI-bias in medical systems.

Keywords Human bias · Data bias · Algorithmic bias · Predictive models · Explainability · Generalizability

14.1 Introduction

AI is revolutionizing many fields that includes computer vision [1, 2], education [3], travel [4], and many more. Healthcare industry is also realizing the potential of AI and providing several upgrades to the traditional system [5]. With AI-based decision support system, medical system is extending real-time patient monitoring, quick evaluation of medical imaging and robot-assisted high-risk surgeries [6]. AI has provided several tools for accurate monitoring, diagnosis and prognosis of patient's statistics recovered from structured as well as unstructured data. AI has penetrated deep into the medical system, providing smarter, quicker, and accurate benefits not only to clinical practitioners but also to the patients.

For the last few years, society realized the potential and benefits of AI based predictive methods in healthcare. Medical practitioners prefer to use AI powered systems to support their various medical related activities and ensuring health equity

A. Kumar and D. Singh, *Artificial Intelligence in Modern Healthcare System*, Transactions on Computer Systems and Networks,
https://doi.org/10.1007/978-981-96-6703-1_14

among patients [7, 8]. Medical staff are becoming dependent on AI based tools due to the wide range of options and possibilities provided by AI-based systems. However, it has been observed that these systems are trained on datasets which lack in gathering the critical aspects of a patient. Those critical aspects include information related to patient's socio-demographics such as gender, age, race/ethnicity, kinship, medical history and many more [9]. The patient's socio-economic status such as medical insurance details are also missing in the training datasets.

Apart from the deficiency in training datasets, AI-predictive algorithms are black box in nature. AI algorithms lack interpretability, transparency, and fairness. Hence, it became necessary to evaluate the models for their appropriateness, robustness and accuracy [10]. AI-based predictive models in medical systems should be audited for quality assessment [11]. Guidelines and tools must be provided which specify the workflows that should be adapted to assess the quality of the AI-model over wide range of parameters such as interpretability, robustness, fairness, and bias.

With the advent in technology, it became essential to evaluate AI-based medical models for the impact of Risk of Bias (RoB) [9]. Bias in AI-models can lead to unforeseen, unreliable and discriminatory outcomes which may influence patient care, diagnosis and treatment. The reason is implicit bias, selection bias and training bias in datasets. Also, the weak algorithmic design and its capabilities to interpret the result introduces bias in the model system. Model trained on biased data, can produce misrepresentative results. Implicit bias in AI-model has a negative impact on the relationship between the medical professional and the patient [12]. The outcomes from a biased model are fatal, which raises the reasonable concern for the evaluation of RoB in AI-model before their practical deployment.

There are lot of examples in which AI-model has been realized to be suffered from bias [13]. For instance, amazon AI-based human resource software was discovered to be suffered from gender bias [14]. It showed high paying and better position jobs to male candidates rather than female applicants. The reason could be its training datasets that contains words which more often found on a male applicant resume rather than a female. In criminal justice, AI-based software which used for identifying the sentence term to the criminal's committing crime was found to be suffered from racial bias [15]. The software suggested harsher and stricter sentence against blacks (American African) than whites for the similar crime committed by them. Authors have explored likelihood based on questionnaire to evaluate the RoB in various AI-based recommendation systems [16]. In healthcare domain, algorithms were identified to be suffered from racial and gender bias [17].

The key impact of bias in AI-model is that they lack generalizability, interoperability, interpretability and explainability. The data bias cannot only be introduced during training data selection but also it is highly dependent on data collection, gathering, cleaning, and processing methods as well [18]. To test the data for its accuracy, completeness, diversity, and acceptability is a huge task in itself. In addition, there are legal law and manifestations which are country specific acts as a hurdle to remove the bias in the training data [19]. The incomplete specification details about the algorithmic design of the medical predictive model also introduce discriminatory, implementation and selection bias [20]. In this chapter, we have reviewed the various AI-based predictive for the RoB and their applicability in the real-time scenarios.

The key objectives of this chapter are as follows:

- We have reviewed the various methods that introduce the data bias and algorithmic bias in the AI-based predictive models that affect the decision-making in healthcare. In addition, the propagation of bias in various stages of model development is explored in detail.
- The methods to mitigate various types of AI-bias are analyzed and a few pointers are suggested for reducing the impact of bias in AI-based predictive model. In addition, bias assessment tools are also evaluated for their performance to assess AI-model for their unbiased outcome.
- Legal manifestations in terms of data privacy, modifications and sharing rules are also discussed to analyze the reasons of RoB in AI-models. These manifestations ensure the transparency and robustness in the AI-models exploited in medical systems.
- Lastly, the limitations and barriers in preventing the AI-bias in the medical system are explored to restrict the negative outcomes due to biased model deployment in real-time.

The rest of the chapter is organized as follows. Section 14.2 elaborates the significance of analyzing the AI-bias in medical systems. The reasons for AI-bias in medical systems in terms of data bias and algorithmic bias are detailed in Sect. 14.3. Various bias assessment tools are evaluated for their appropriateness in the medical system in Sect. 14.4. The methods to mitigate bias from AI-models in healthcare are reviewed in Sect. 14.5. The limitations of bias mitigation strategies which restrict the AI-model to be unbiased are also detailed. Finally, the concluding remarks and future directions are summarized in Sect. 14.6.

14.2 Significance of Analysis of AI-Bias in Medical Systems

Researchers have gained interest in developing AI-based predictive models in medical systems for fast, accurate and efficient outcomes [21–23]. These models have found applications in variety of medical applications to support clinical decision making [9]. However, it has been observed that AI-based models suffer from AI-bias and are prone to impact the quality of clinical results significantly. In addition, AI-bias not only introduces error in the model outcomes but also, reduces the trust of the end users [24]. Hence, it is essential to analyze bias in AI-based medical systems. The following are the reasons which emphasize the necessity of analysis of AI-bias in medical systems.

- Big data and need of reliability

Recently, the applications of AI-based predictive models have been increased exponentially in the medical systems [5]. There is a surge in AI-models in healthcare domain due to increase in data and the difficulty in its manual processing. With the increase in size of training data, it became essential to identify the data source for reliability [25]. Data collection and gathering steps should be very careful examined for having reliable information only. Model trained on unreliable training data are more to produce incorrect and biased outcomes.

- Limitations during in-processing stage

Imbalance, incorrect and missing data is processed by the model introduces selection bias and data bias in the model. The model design and architecture when processing this data exacerbates algorithmic bias. However, it is not easy to identity and eliminate algorithm bias during in-processing of information [26]. The model structure is black box in nature and hence, make it tedious to identify the potential source of algorithmic bias in the AI-based predictive model.

- Enhanced cloud storage, power and speed to reduce clinical errors

Cloud storage facilitates large datasets with minimal processing. It has been observed that large and diverse data can significantly reduce the impact of AI-bias in the medical systems and reduce fatal errors in the outcomes. Enhanced cloud storage has promoted personalized medical aid which help doctors to understand patient's medical history in a better way [26]. In addition, cloud storage has automated the treatment through which patient's vital such as blood pressure, SPO_2, pulse and sugar level can be monitored continuously and effectively. But these AI-devices are determined to be suffered from implicit bias which results in healthcare inequalities [8]. Racial and ethnicity inequalities are predicted by DL-model in patient's ECG data [27]. The technological advancement and progress in medical system emphasize to analyze RoB.

- Automated predictive decision-making

DL and ML based models in medical systems provide fast and accurate prediction for various critical health disorders such as breast cancer [28], cervical cancer [29], CVD [30] and many other diseases [31, 32]. These models are automatic and accurate to provide severity of the disease by analyzing various clinical imaging data such as X-ray, MRI (Magnetic Resonance Imaging) and CT (Computed Tomography) scans. However, the sociodemographic information of the patient such as gender, race, age, and other vitals are not exploited while training the models. Due to which these models are identified to be suffered from AI-bias [9, 33]. In addition, the models are not efficient enough to address the false positive and true negative results effectively.

- Developers limited knowledge and expertise

It has been observed that AI-models which are weak in design and architecture are more prone to AI-bias. To address these limitations, the developer domain knowledge and expertise are vital in ensuring the experiment and information reliability of the system [20]. Perceived bias in the system due to pre-existing beliefs leads to inaccurate results. Deep implicit knowledge of the system design, processing and synthesis is important for an unbiased system which highly depends on the understanding and potential of the developer [34]. Developers should be trained with the healthcare explicit and implicit requirements and outcomes to design robust system. Efficiency in processing unbiased knowledge and eliminating unnecessary details are the prime factors which should be analyzed to reduce the RoB in the AI-models.

- Bias in model selection and feature training

Models trained on biased feature are more intended to produce biased results. Bias perceived in model design is amplified at various stages and can produce catastrophic outcomes [25]. It has been emphasized that the choice of model should not have implicit bias in its architecture which produces false predictions. Before the practical implications of these models, it is essential to evaluate these systems for RoB. These models can be used as a recommender system which additionally supports the medical practitioners in their decision making.

To summarize, there are certain limitations, presumptions and weaknesses in the system design and information gathering that leads to produce biased outcomes. AI-model must be examined during its various stages viz. preprocessing, in-processing and post-processing to eliminate the RoB. There are many other parameters such as societal impact, developer expertise, legal laws and many other unseen reasons that misrepresent the model results and produce incorrect diagnosis. As AI is penetrating deep into the medical systems, it is significant to analyze the models for RoB. Concrete steps, measures and recommendations must be followed to ensure trust, accuracy, and fairness of AI-based healthcare models. It is highlighted that disparities in healthcare can only be minimized by generating unbiased, generalizable, interoperable, and explainable AI-models.

14.3 Types of AI-Bias in Modern Healthcare Systems

Broadly, bias in AI-models is categorized into three categories namely, data bias, algorithmic bias and human bias [34]. These biases are most likely to occur in the AI-models at its various stages of design, development, processing, and deployment. Outcome of predictive model is considered to be bias if it produces variables results for different people depending on their gender, race, age, ethnicity and other socioeconomic parameters. In this section, we have discussed the reasons of these biases in AI-based predictive models in medical systems. Table 14.1 tabulates the details of various representative work in medical domain analyzing the RoB along with the utilized risk assessment strategies. Figure 14.1 illustrates the various types of AI-bias in medical systems.

Table 14.1 AI-based representative work in medical domain and their bias assessment strategies

Reference	Type of bias	Medical domain	Bias assessment	Summary
Brault and Saxena [35]	Data bias and algorithmic bias	Mobile health	Questionnaire based assessment	• Bias can be introduced in AI-model during various stages • Stages viz. problem definition, feature selection, model selection, and training
Suri et al. [36]	Algorithmic bias	CVD risk prediction	Mean score based cumulative plots	• Identified critical AI-attributes • Utilized grading and ranking strategy to visualize bias in the predictive models
Celi et al. [26]	Data bias	Clinical medicine	Confusion matrix and ROC	• Evaluated AI-bias in country specific datasets using gender, racial, countries, and author's expertise • Utilized dataset from various countries to access the bias
Gurupur and Wan [25]	Inherent bias due to knowledge	Healthcare	–	• Emphasized the reasons for bias in AI-systems highly depends on limited subject knowledge, and lack in proper expertise • Data bias due to missing details in the training datasets
Gichoya et al. [37]	Data bias and algorithmic bias	Radiology	–	• Exclaimed the pitfalls during various stages of data collection, and curation • Inefficient model design, development and deployment introduce potential bias in the system
Norori et al. [34]	Data bias and algorithmic bias	Medicine	F1-score	• The potential reasons for AI-bias includes, information gaps, lack of common data standards, performance evaluation and interoperability • Model testing to evaluate algorithms for efficiency, performance, and fairness

(continued)

Table 14.1 (continued)

Reference	Type of bias	Medical domain	Bias assessment	Summary
Nazer et al. [38]	Data bias and algorithmic bias	Healthcare	–	• Discussed sources of potential bias at various stages of model development • Highlighted the strategies to mitigate data bias and algorithmic bias
Noseworthy et al. [27]	Racial bias	ECG analysis	AUC	• Evaluated ECG results and inferred those results suffered from racial bias • Highlighted poor generalizability of the DL-models for detecting low LVEF
DeCamp and Lindvall [24]	Latent bias, emergent bias	Medicine	–	• Highlighted the role of adaptive learning in introducing bias in the model • Clinical implementation and evaluating outcomes exacerbate bias in AI-model
Ueda et al. [39]	Data bias and algorithmic bias	Healthcare	–	• Discussed the source of bias in healthcare namely, data, algorithm, clinical and patient interactions • Recommended various strategies to mitigate bias in healthcare
Sousa et al. [40]	Classifier bias	CT-Scan	ACC, AUC and F1-score	• Examined the bias introduce by artifacts and spurious elements in the image dataset • Applied various explainable AI technique to mitigate bias from the AI-Model
Kumar et al. [9]	Data bias, and algorithmic bias	Medical systems	RBA, RBM, RBS and ANA	• Performed bias assessment by categorizing the work into three classes • Analyzed bias from multiple perspectives and visualized correlation between them using VD

(continued)

Table 14.1 (continued)

Reference	Type of bias	Medical domain	Bias assessment	Summary
Kumari et al. [30]	Data bias, and algorithmic bias	CVD risk prediction	RBA, RBM, and RBS	• Categorized the studies into two categories namely, UNet based model and non-UNet based models • Critically analyzed bias in AI-models considering the characteristics for clinical and scientific applications
Das et al. [31]	Algorithmic bias, training bias	Brain tumor segmentation	RBS, ACC, SEN, ROC, and DCE	• Analyzed segmentation bias in ML-based brain tumor techniques • Ranked RoB in studies into three categories low, medium and high
Suri et al. [32]	Data bias, and algorithmic bias	CVD risk prediction	Analytical ranking slope method, and cumulative plots	• Categorized studies into two categories namely ML and non-ML to predict the RoB • Used risk granularity along with five-point recommendation to reduce the impact of RoB in various prediction models
Suri et al. [33]	Data bias, and algorithmic bias	Covid-19 infected lungs	RBA, RBM, RBS, PROBAST and ROBINS-I	• Quantified RoB in hybrid DL-models for predicting the Covid-19 infection in lungs using randomized and non-randomized techniques • Recommended 8-points to minimize the impact of RoB in DL-models

AUC: Area under the curve, *ECG:* Echocardiogram, *LVEF:* Left ventricular ejection fraction, *ACC:* Accuracy, *RBA:* Radial bias area, *RBM:* Radial bias mean, *RBS:* Radial bias score, *ANA:* Analytical analysis, *VD:* Venn diagram, *CVD:* Cardiovascular disorder, *IVUS:* Intervascular ultrasound, *DL:* Deep learning, *ML:* Machine learning, *RoB:* Risk of bias, *SEN:* Sensitivity, *ROC:* Receiver operating characteristics, *DCE:* Dice coefficient

14.3.1 Data Bias

The potential source of data bias in AI-based medical systems are (i) incomplete data collection, (ii) data gathering from unreliable resources (iii) missing and unseen information in dataset (iv) fallacious data processing and (v) inaccurate data analysis. Also, the predictive healthcare algorithms are suffered from historical biases in the dataset [19]. For instance, an algorithm, when trained on false or incorrect past

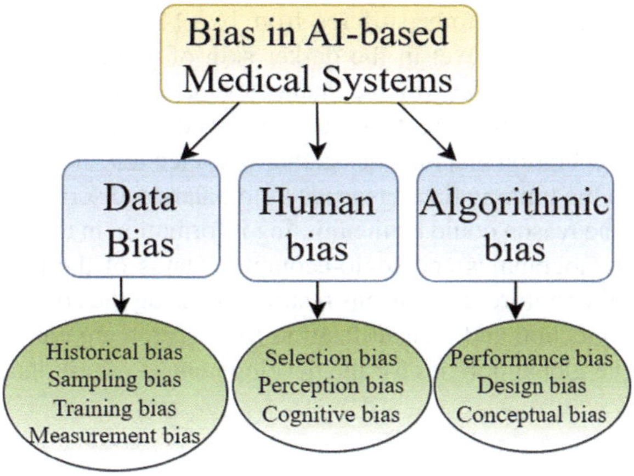

Fig. 14.1 Types of bias in medical systems

information of the patients, is intended to generate incorrect outcomes in future. The missing sociodemographic information such as gender, age, and ethnicity due to societal barrier originates imbalance in the datasets which leads to data bias in the model. In [41], authors have explored that sex-related information in training dataset is essential to generate unbiased outcomes. However, results have inferred that the proposed model is unbiased and has no gender specific results. Model is able to predict covid-19 in chest X-ray neutral to sex attribute.

Oftentimes, training data suffers from historical biases that can be amplified once it is processed through the model [42, 43]. Historical biases lead to racial disparities, health inequities and false outcomes. The reasons for historical bias in the health data are structural barriers such as cultural restrictions or individual limitations. These restrictions influence data collection and can produce under diagnosis outcomes. Historical biases can also arise due misclassification, mislabeling, and missing certain aspects of different segments of population [7]. In addition, underrepresentation of certain group, caste or creed in the training data also lead to potential bias in the outcomes [44]. It has been observed that medical diagnosis is biased toward a specific gender or race as there is a shortage of female gender or darker/African race people in the datasets.

Less diversified training data is relatively more prone to introduce biasness in the model [39]. This kind of training data discriminates between the medical outcomes based on sexual orientation, nationality, and/or socio-economic status and introduces bias in the metadata. This bias is tough to identify and quantify. Sampling bias can be engaged in the model by inappropriate sampling of datasets into training set and test set. This bias is introduced in the model during the initial step of data preparation.

Another potential source of bias in the datasets is classification/ measurement bias [38]. Sometimes, model trained on imbalance datasets are expected to suffer from this type of bias. This bias inaccurately classifies patients based on their demographic features and/or ethnicity and provides different care and incorrect diagnosis. Certain commonly used medical devices such as oximeter, pulse monitor, and thermometer are analyzed to be suffered from bias [45]. Oximeters are recorded to overestimate the oxygen level in the darker skin of people. The reason may be the imbalance training dataset which contains few data from the darker skin people.

To summarize, data bias is induced in the model during the initial steps of data preparation that includes gathering, selection, classification, and sampling. Data bias in the training datasets leads to imbalance, discrimination and inaccurate outcomes. The reason could be the missing information in the training data that includes socio-demographics and socio-economic status of the person. The societal barriers and other unforeseen reasons restrict gathering the complete details of the patients. Data collection and their utilization for research are country specific and hence, prevent the global policies for the development of robust datasets in the medical systems.

14.3.2 Human Bias

Developer expertise, knowledge and perception introduces human bias in the AI-models. Human bias is one of the typical biases which is hard to detect and mitigate [34]. Societal prejudices lead to human bias in the data and model design which is exaggerated during in-processing stage of AI-model. Incomplete population data, and lack of model understanding propagate human bias in healthcare predictive model which impact the quality of model-decision making.

Human design algorithms and their understanding and perception about model features, processing, and evaluation metrics are crucial in preventing human bias. Limitations in variability and diversity of expert knowledge introduces perception bias in the AI-models [37]. There are cases where the designer selects sensitive features and design models using those features. When this model generates poor quality and false outcomes, it is analyzed that selected features are not only inefficient but also inappropriate to generate expected results. The perception bias is introduced in the model due to insufficient human knowledge and expertise. In addition, humans prioritize the problem with their own perception rather than focusing of necessities and urgency of diagnosis to start the accurate treatment.

Data collection, preparation and processing are also done by human experts. This will act as an input data for training the AI-model to predict accurate results. Training data is selected by developers to generate robust outcomes. Data selection is a very critical step for obtaining correct results from the AI-model. Sometimes, developer is not able to select the correct, diverse, and complete datasets and that introduces selection bias in the decision-making [7]. Also, the choice of appropriate AI-model for designing accurate and unbiased model is highly dependent on the experience and knowledge of the AI designer. Fairness in algorithmic design and completeness in training data are essential to prevent the selection bias in clinical predictive algorithms. In addition, the expertise and experience of the developer is significant to minimize the impact the selection bias in the model by neglecting and adjusting the corresponding data that may induce bias.

Limitation of human mind to think critically for generating reliable models introduces cognitive bias in the system [19]. Human reasoning, human-AI interaction, and human past decisions are essential to augment correctness in clinical outcomes from AI-based decision-making systems. Human ability, over-reliance, response, behavior, and human collaborative efforts introduce trust in AI-models by perpetuating human bias in decision-making. Human behavior can recommend strategies to reduce the impact of bias explicitly. Human reliance on AI-models in healthcare system inherits racial and gender bias in the outcome [25].

14.3.3 Algorithmic Bias

The combination of human bias and incomplete data introduces algorithmic bias in the healthcare predictive model. When the model is trained on biased data that will lead to algorithmic bias in the model. Weak design and architecture of the AI-model exaggerate this bias which produces false, different, and inaccurate diagnosis for different group of people. Algorithmic bias introduces systematic errors in the model and hence, impacts the reliability, trust, and fairness in the model. The black-box design of the AI-model prevents the interpretability and generalizability of the results and causes algorithmic bias. The information about the number of hidden layers, optimization function, loss function, weight adjustment and classifier specification are not specified explicitly in all the research papers. Classifier bias can produce different predictive results and biased outcomes for different datasets [40]. The missing information about the model's hyperparameters prevents the reproducibility of the results and prevents to evaluate the interoperability of the model to reduce the impact of algorithmic bias.

Also, feature engineering plays a vital role for inducing algorithmic bias in the model. Identification of feature importance, sensitive feature and feature selection is critical for robust design of the predictive model. DL methodology is used for robust feature extraction from the medical imaging such as X-ray, CT, and MRI images. These extracted features contain quality information suitable for unbiased algorithmic design with accurate outcomes. One of the another potential reason of algorithm bias in AI-based predictive model is lack in description about its contextual specifications [46]. Medical systems are designed with general specifications which vary in design and architecture accommodating the diversity in accordance with the input sociodemographic, lifestyle and medical history.

Robust architecture of AI-based predictive models is crucial for accurate medical prediction. Authors have emphasized that casual reasoning is important for algorithmic fairness [47]. Casual reasoning defines the technology to recognize ethical and social biases. Selection bias can introduce unfairness and biasness in the algorithms if casual knowledge is not formally used for bias inferences and understanding. In [7], authors have defined three potential sources of algorithmic bias namely, direct (model design), training data variance and noise. In addition, incorrect algorithm performance comparison, validation and monitoring phases also introduce bias in implementation, clinical workflows, and other resource usage.

To summarize, the potential reasons of algorithmic bias in AI-models are its weak design, black box architecture, unreliable conceptualization, inefficient implementation, and deployment strategies. Algorithmic bias in healthcare predictive algorithms leads to incorrect diagnosis which can generate highly risky outcomes. Fairness and trustworthiness are important pillars in algorithmic design to ensure transparency, and accountability of AI-based algorithms.

14.4 Bias Assessment Tools

AI-based clinical models have raised concerns for accuracy and reliability in their outcomes as it is identified that these models are suffered from AI-bias [9]. To ensure fairness and trust in these models, it is essential to evaluate these models qualitatively. To measure bias accountability in AI-models, bias and fairness toolkits such as Aequitas [48], ROBINS-I (Risk Of Bias In Non-randomized Studies of Interventions) [49], PROBAST (Prediction model Risk Of Bias Assessment Tool) [50], CHARMS (Checklist for critical Appraisal and data extraction for systematic Reviews of prediction Modeling Studies) [51] and BIAS (Biomedical Image Analysis challengeS) [52] are recommended by various researchers. The details of each of these bias assessment toolkits is as follows.

14.4.1 Aequitas: Bias and Fairness Audit Toolkit

Aequitas is an open-source audit toolkit used to evaluate bias and fairness in an AI-model [48]. This systematic toolkit is easy to use and can test ML workflows, bias and fairness metrics in various subgroups of considered population. It also supports decision making for scientist, developers, and policy makers by testing the developing and deploying stages of AI-models.

Aequitas defines a step of procedure to audit the AI-system for its biased outcomes for a specific demographic/social group. This toolkit performs bias assessment prior to model selection and evaluate disparities in results neural to the type of training data. The bias and fairness audit are executed by checking the operational flow of the AI-model before proceeding to production deployments. Primarily, the audit toolkit has two main users namely, developers who is designing AI systems (scientists and researchers) and policymakers who defines policies for AI-system acceptance.

In healthcare systems, authors have audited the ML-framework which are used for prediction and diagnosis of critical diseases such as diabetic retinopathy, and Alzheimer diseases using Aequitas toolkit [10]. The quantitative bias analysis is performed on test sets and results are compared with the reference set to predict fairness. The toolkit results represent fairness and unfairness by computing the disparity between the test and reference set. Disparity value between 0.8 and 1.25 denotes high similarity results with better quality and fairer AI-model.

To summarize, there are various reasons for bias in AI-models in healthcare domain. AI-models could not provide similar results to all the patients by performing early prediction of critical diseases due to bias intervention in the model. Hence, it is necessary to audit those models for their fairness in outcomes by using toolkits such as Aequitas. This toolkit analyzes models during its development stage and provides results to prevent the failure of models after deployment. This toolkit is useful for both developers and policy makers. The results of the audited model are effective to ensure fairness and equity in the ML-model for solving the desired problem from paper to practice.

14.4.2 ROBINS-I (Risk of Bias in Non-randomized Studies of Interventions)

In order to describe the strength and limitations of non-randomized studies such as healthcare, a bias assessment tool ROBINS-I is proposed by [49]. To analyze the RoB, seven different domains are identified through which bias can be introduced in the AI-models. Broadly, the domains are categorized into pre-intervention, at-intervention and post-intervention levels. Pre-invention examines the confounding bias and the selection bias due to participants in the study. At-intervention domain assesses bias due to non-differential classification. Post-intervention assesses RoB at four different domains namely, bias due to deviation, bias due to missing data, measurement bias and bias in the selected results.

It is essential to determine the potential and magnitude of RoB so that strategies can be planned for its mitigation. To predict overall RoB in ROBINS-I, each study is categorized into five judgment levels namely, low-bias, moderate-bias, serious-bias, critical-bias and no information within each domain as well as across the domains. If a study performs well within and across all the domains, then it is considered to in low RoB and have high quality to be ready for deployment. Moderate-bias zone studies perform well on training data, but their quality and fairness cannot be ensured. For a study in a serious and critical zone, the outcomes must be evaluated to ensure trustworthiness and fairness. Studies can only be kept in no information zone if the number of available parameters is not sufficient for a judgment.

In healthcare, ROBINS-I is exploited for qualitative risk assessment in various AI-based prediction techniques [33, 53]. In [33], authors have analyzed RoB in hybrid DL studies utilized for predicting COVID-19 CT scan data of infected lungs. The three domains of ROBINS-I define seven features as confounding parameters, participant selection, intervention classification, intended deviation, missing data, outcome measurement and reported results. After this, the studies are categorized into three bias zones as low, moderate, and high. Similarly, authors have exploited ROBINS-I tool for risk assessment in AI-model utilized for acute respiratory distress syndrome

[53]. Seven features in three intervention factors are examined for categorizing bias in three zones namely, low, moderate, and high.

14.4.3 PROBAST (Prediction Model Risk of Bias Assessment Tool)

To determine the applicability of medical diagnosis and prognosis studies, RoB is examined using PROBAST tool [50]. PROBAST tool is utilized for investigating the limitations in design, conduct and evaluation that may lead to bias under certain circumstances or RoB in future when certain event will get triggered. Basically, this tool has four domains namely, participants, predictors, outcome, and analysis. Participant domain covers the concerns related to data-driven sources of bias. Predictor domain covers the concerns related to the design, definition, and measurement of the AI-model. Outcome domain covers the concerns about the results produced and measured by the model. Lastly, analysis domain covers the RoB related with the statistical measurement and considerations by the model.

AI-based predictive models are designed to validate and provide prediction scores based on their analysis. In healthcare, these predictions are utilized for diagnosis of critical diseases such as cancer, cardiovascular disorder, COVID-19 and many more [33, 53]. Before the practical applicability of these prediction models, it is essential to evaluate the RoB to prevent their failure in real-time. In order to evaluate the quality and applicability of AI-based models, PROBAST tool had defined certain guidelines to assess RoB. For this, this tool has presented four development stages namely, scope and definitions, review of evidence, web-based Delphi procedure and piloting and redefining the tool.

PROBAST tool is exploited by researchers to estimate the RoB in various AI-based predictive studies in medical systems [33, 53]. Authors have utilized PROBAST risk assessment tool to evaluate hybrid DL studies proposed to predict COVID-19 in infected lung data [33]. The four domains are represented to evaluate the presence and absence of crucial features in AI-models. Participant domain comprises of radiologist validation, data source type and demographics data. Imaging features, pre-processing, data augmentation and optimizers are included in the predictor domain. Outcome domain contains performance evaluation parameters and details about RT-PCR (Reverse transcription-polymerase chain reaction) test. Lastly, the analysis domain includes data partitioning, clinical validation, benchmarking procedures, patient count and statistical evaluation. After this, studies are ranked and categorized into three bias zones namely, low, moderate, and high. Similarly, authors have analyzed RoB in AI-studies used for predicting acute respiratory distress syndrome using PROBAST risk assessment tool in three bias zones [53]. The four domains contain attributes such as participants (data source and radiologist verification), predictors (demographics and imaging features), outcomes (multiple datasets and

RT-PCR test details) and analysis (patient size, optimizer, validation, and design innovation).

14.4.4 CHARMS (CHecklist for Critical Appraisal and Data Extraction for Systematic Reviews of Prediction Modeling Studies)

Mostly, AI-based prediction models are utilized for diagnosis and prognosis of a specific disease. The design and strategies used in development of these models play a vital role in determining the quality and applicability of these models in real-time. To validate these prediction models a checklist known as CHARMS is designed containing questionnaires to review these models [51].

CHARMS assessment tool has defined the key items to review the usefulness and potential application of work. The key items include current and future events of the models, intended scope, prediction modeling type (with or without external validation), target population, prediction outcome, prediction duration and the intended moment of using the model. Further, eleven key domains are specified to review the RoB and applicability of AI-based prediction models. The relevant items which are reviewed are data source, participants, predicted outcome, candidate predictors, sample size, missing data, model development, model performance, model evaluation, results, interpretation, and discussion.

In [54], authors have utilized CHARMS checklist to review the COVID-19 dataset for RoB. The key parameters namely, data source, participant description, outcomes, sample size, missing data and predictors are used for risk assessment. Participants' description includes the methodology for their selection, inclusion, and exclusion from the analysis. Outcome checks the purpose of intention by the model. Overestimation and underestimation by the model for a certain prediction is reviewed in outcomes. Missing data ensures sufficiently large datasets are used for avoiding overfitting and confounding the model. Predictors analyze the source of data acquisition devices and protocol. Finally, the sample size is associated with several aspects of the models such as predictor's size, model preprocessing and importance of effect to be predicted.

Apart from these tools, BIAS (Biomedical Image Analysis challengeS) checklist is also defined to improve the transparency and applicability of the biomedical images for its application and imaging modality [52]. In [54], authors have reviewed biomedical dataset by preparing a checklist of questionnaire. This checklist has key questions about the dataset source, purpose, distribution and intended applications. These reporting guidelines are highly effective for performing the in-depth analysis of the model.

To summarize, the focus of risk assessment tools is to standardize and facilitate the model functionality, and applicability. These tools estimate the RoB at the development stage of the AI-model to prevent failure and unseen outcomes during clinical deployments. These tools evaluate the potential of model design, development, interpretability, generalizability, and interoperability. The AI-based models provide good accuracy and results on training datasets. These tools provide checklists which ensure that these models are suitable for real-time deployment and provide the correct diagnosis and prognosis of the intended disease.

14.5 Approaches to Mitigate AI-Bias in Modern Healthcare System

AI-based models in medical systems must be unbiased, transparent, open, and fair in decision-making for fast and accurate outcomes. It has been observed that a biased AI-model generates discriminatory outcomes for a marginalized subgroup of population and has a striking implication in healthcare. Hence, it is essential to analyze the potential source of bias in AI-based medical system and design strategies to mitigate it. The following are the key steps which help in addressing the bias in the medical system to a great extent. Figure 14.2 illustrates the development stages of AI-model and bias mitigation strategies for a robust and trustworthy prediction model.

- *Data collection and preparation*: The strategies and methodology during the initial step of data gathering and preparation are very crucial to eliminate the impact of data bias in AI-model. For this, protected attributes such as gender, ethnicity, age, smoking history, kinship and insurance status must be collected and considered while preparing the training dataset [9].
- *Large and diverse training dataset*: The size of the training dataset must be large enough for effective training/validation/testing. In addition, datasets can contain details from various subgroups of a population to ensure diversity. Selection of large and diverse size training dataset is essential to avoid selection and sampling bias due to missing/unseen data [38]. Model trained on multiple datasets are

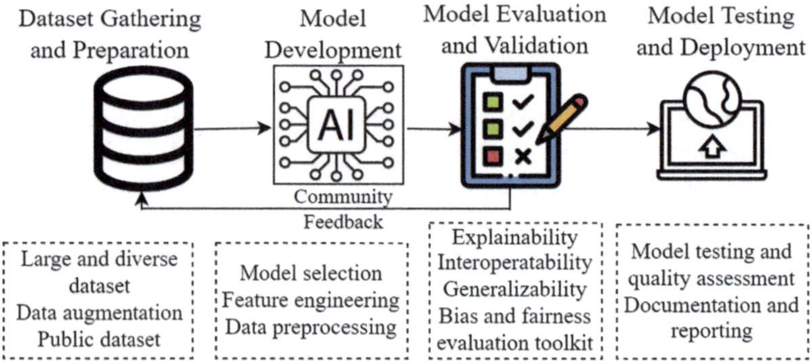

Fig. 14.2 Development stage of AI-model and bias mitigation strategies

effective to address the imbalance and discrimination that may exist in single datasets.

- *Public datasets*: To avoid the impact of societal bias in the model outcomes, it is suggested to train models on public datasets. The public datasets provide clear settings and details which are open and trustworthy. Model trained on public datasets can avoid historical bias that may be present in a self-generated data due to the embedded biasness in its collection [9].
- *Data augmentation for smaller datasets*: Sometimes when the dataset is small in size, data augmentation techniques such as flipping, rotation, and shifting are applied to generate synthetic and surrogate data. Surrogate data not only increase the dataset size but also efficient in handling dataset imbalance [46]. However, it has been observed that surrogate data is insufficient and biased toward a specific subgroup of population due to less diversity. In addition, if the dataset from which the synthetic dataset is generated is suffered from bias, then the bias is propagated in the whole system design and can predict false outcomes.
- *Algorithm selection*: Selection of a robust and reliable algorithms ensured that the algorithmic predictions are unbiased and fair [27]. Clear understanding of the problem statement determines that the algorithms will solve the desired problem accurately by engaging the diverse domain experts and community members. Diverse expertise in the development team can minimize the impact of human-bias in the model.
- *Data preprocessing*: Feature engineering for identifying the sensitive and important feature can prevent the model accountability for bias. Feature bias can be avoided by understanding data and defining input variables effectively during pre-processing stage [34]. Dependent and sensitive features should be utilized in model architecture and their impact should be reviewed consistently for biased and inaccurate outcomes.
- *Model development*: To ensure fairness in the model design and architecture, model methodology and architecture should be transparent, fair, and interpretable. Details of loss function, optimizer, activation function, learning rate and other hyperparameters must be mentioned explicitly [37]. These settings are helpful in verifying the model architecture for its incorrect predictions.
- *Model validation*: Model must be validated internally and externally to avoid implicit and explicit bias in the model. Multiple performance metrics can evaluate the model prediction outcomes from different and wider perspective [25]. This can eliminate evaluation bias that occurs in the model due to disparities in model monitoring and assessment.
- *Risk assessment tools*: Risk assessment tools such as PROBAST [50], ROBINS-I [49] and others [51, 52] are effective in risk assessment during development stage. These tools can not only assess the risk involved in realistic deployment of the model but also prevent catastrophic results that can occur due to biased and unfair outcomes. Tools are impactful for generating trust in the clinical prediction by various AI-model in healthcare.

- *Mixed model strategy*: In [20], authors utilized mixed logistic regression models to combat the impact of racial bias and religious bias in discriminatory algorithms. The experimental results for each of the subgroups in a population was estimated to identify the biasness in the model outcome. In addition, extensive model auditing was recommended to check the model deployment for fairness and bias in its clinical decision making.
- *Model understanding to end users*: End users should have a clear understanding of AI model and potential risks that may cause biased output [39]. Clinical staff and the participating patients will be made aware of the limitations, benefits and usage of AI in decision making. It has been observed that educated patients who understand AI and AI-bias in recommendations by AI-model are more satisfied and provide positive feedback to eliminate the biases. This will help in upgrading the system with better accuracy and efficiency that will serve the needs of all patients.
- *Gap between AI and end users*: Healthcare disparities are observed in a population due to unequal distribution of AI-driven medical systems benefits [39]. Certain AI-based algorithms are not available to all the clinical staff and patients. This will cause privilege bias that brings mistrust in AI-systems unintentionally. The reason for this bias could be lack of knowledge, expertise, resources and perspective of a certain group. In order to mitigate this sort of bias in healthcare, it is essential to provide proper training and education to the clinical staff so that full potential of AI-model can be utilized effectively.

To summarize, bias can creep into model design, architecture, training data and outcome at any stage of its development. From the initial steps of data collection and gathering to the final steps of model implementation and validation all are critical and prone to biasness. There is no defined procedure of steps and gold standards that can mitigate the impact of AI-bias in medical systems. For a model to be unbiased in its outcome, it is essential that it should be trained on large, diverse, indiscriminate, multiple, public, and balanced datasets. In addition, the model architecture should be robust, reliable, and transparent to validate predictions. Models must utilize the risk assessment tools before deploying the model from paper to practice.

14.5.1 Limitations to Prevent AI-Bias in Modern Healthcare System

To address the AI-bias in medical systems, there is a requirement to identify the bias accountability in AI-based predictions to prevent the severe errors and harmful outcomes [39]. AI-bias in predictive model raises concern for the trust and fairness in diagnosis and prognosis in healthcare. It is mandatory to take the necessary steps and procedures to mitigate the impact of various types of bias in AI-model. But there are certain limitations and legal manifestations that restrict the mitigation of data bias and algorithmic bias in AI-model.

In order to mitigate data bias, there is a requirement of public, complete and unbiased data set. However, public availability of medical data involves legal and ethical concerns. Data privacy, modifications, alterations, and usage laws are country-specific and deal strictly with compliance and regulations [55]. Protecting patient's data and ensuring privacy is mandatory for ethical foundation of AI. Patients should be made aware and understandable about the potential threats, benefits, and purpose of using their medical confidential information in AI for decision making. Further, an informed signed consent must be taken from the patients for the usage of their confidential information for clinical purposes.

During data exploration stage, synthetic data is generated to increase the dataset size by data augmentation technique. Surrogate data is utilized to address the missing data values in the dataset. Shortage of effective medical data impact the accuracy of the model and introduce selection bias and data manipulation bias [55]. These biases may cause model failure when tested on unseen data. Limitation to data sharing and usage rules prevent the generalizability of the model [14]. In addition, the ethical concerns respect human rights and privacy for using their secret medical information for clinical deployment which hinders the large and diversified datasets in medical systems.

It is important to specify the roles and responsibilities of each stakeholder to prevent misdiagnosis and false predictions in healthcare. Doctors, clinical staff, and AI-scientists should check and evaluate the AI-generated diagnosis before its applicability. AI-scientist should verify and get the results approved by the specialist doctor and integrate the suggestions and recommendations in decision-making [19]. However, the gap between the knowledge and expertise of AI-scientist and clinical staff may not critically evaluate all the AI-outputs generated by the predictive model. In addition, AI-scientist can be held responsible to design accurate, efficient, and fair model [27]. But the role of clinical staff is vital in providing the community feedback and clear guidance to adjust the limitations of AI-model and enhance overall quality of patient's care.

Establishing an unbiased and fair AI-model in healthcare requires to follow policies and procedures defined by various central agencies. The regulatory authorities and policy makers should be proactive in designing the strong and robust policies that can promote and validate the design, architecture, input and outcome of an AI-model [20]. These agencies specify the requirements that all designers and developers must disclose about their training data, implementation methodologies, and evaluation parameters. In addition, the defined policies should be neutral and focus to eliminate the disparities in healthcare. However, AI-developers and health professionals are unaware about the demand and requirements of these policies and regulations. On the other hand, recognizing dark areas that may cause AI-bias in the model is a challenge for both policy makers and regulatory authorities [39]. The comprehensive guidelines should involve all the stakeholders along with policy makers to govern standards and regulations for a fair and trustworthy AI-model in healthcare.

Strong and robust architecture of AI-model can mitigate the impact of algorithmic bias in predictive model to a great extent [8]. But the black box nature of DL model and non-disclosure of model salient details such as number of input layers, hidden layers, loss function, epochs, and optimizers by many researchers restricts to check the full potential and capabilities of AI-model. These models should monitor the clinical workflow to ensure that the model should not degrade and become biased over time. In addition, automated AI model in healthcare are impacted by bias which is tedious to identify and generate conflict decisions in the predictions [55]. The reason could be human reliability more on the automated predictive model and ignoring the conflicting human decisions.

Regular auditing and monitoring of AI-model is the best practice that should be followed to validate the quality of predictive model for fairness, accuracy and effectiveness [39]. Auditing teams must be established, including all the stakeholders, to verify the performance of the system for different populations under multiple conditions. If AI-model is identified to be suffered from bias, then adjustments should be made to rectify the biases and upgrade the model to be adaptive with the varying conditions [20]. However, quality assessment and constant monitoring of the system cannot ensure that the model will maintain high performance after adjusting all the emerging biases. In addition, emergent bias in AI-model may not extend similar treatment to the patients without any discrimination. The reason could be lack of expertise to identify the key indicators that are accountable for providing underdiagnosis and health disparities.

14.5.2 A Special Note on Explainability, and Generalizability in Modern Healthcare

The key reason for algorithmic bias in AI-model is its black box design architecture. It is suggested to design the model explainable and interpretable to minimize the impact of bias in AI-based prediction models [40]. Bias accountability can be evaluated in AI-model by enhancing transparency and fairness, providing comprehensive explanations to model design, input and predictions.

Explainability is a strategy to analyze the methods for its outcome. Explainable AI models in healthcare are able to define the internal logic of the model which help in understanding the methodology the way outcomes are generated [39]. Explainability in AI-model can be introduced by tools namely, SHAP, LIME and GRADCAM [56]. These tools interpret how AI-model generates an outcome and also validates clinical prediction of the model in presence of certain risk factors [40]. In case of false prediction, preventive measures can be taken to eliminate the inaccuracy in model design and assumption before realistic deployment.

Generalizability ensured that the AI-model will predict accurately when tested on a dataset different from the training dataset. The model architecture is generalizable in terms of its prediction is to be free from any racial and gender bias. Generalizability can be assumed in training dataset if it does not have historical and societal bias during its gathering and processing phases [9]. Due to limited data availability in healthcare, generalizability of AI-model is a challenging task. Addressing all the concerns related to data gathering, patient's privacy and clinical data usage, help in developing generalizable dataset [27]. Model trained on generalizable datasets expected to be fair and unbiased. The outcome of these models is more realistic when deployed in real-time.

14.6 Summary

AI-based predictive models have shown superior accuracy and effective results in predicting the critical diseases such as cancer, brain tumor, CVD and many more. However, these models are suffered from AI-bias and predict discrimatory outcomes based on patient's socio demographics such as sex, age, ethnicity, race, and insurance status. There are three potential sources of bias in medical system design namely, training data, human perception and algorithmic architecture. Data bias in training dataset may creep during its initial steps of collection, and gathering. The unbalanced and incomplete datasets lead to misclassification errors in the outcomes. Model trained on such datasets are realized to be suffered from selection bias, and latent bias when tested on unseen data. Human perception leads to human bias which highly dependent on the knowledge and experience of the developer. Human bias can be embedded in the medical system due to incorrect choice of dataset and improper selection of predictive model. Weak design of the AI-model leads to algorithmic bias. The black box nature of AI-model and insufficient information about the model salient parameters such as loss function, number of optimizer, input and hidden layer variables contribute much toward the algorithmic bias.

Various bias and fairness evaluation toolkits such as Aequitas, PROBAST, ROBINS and many other focus to assess the potential risks in healthcare model during its development stage. These tools identify RoB in the considered AI-model so that the predicted risk can be corrected to prevent the harmful results during deployment stage. The model is audited and evaluated on certain parameters to validate the outcome to be free from bias. These toolkit ensure the trust and fairness in the AI-model predictions.

It is essential to adopt the procedure and strategies to mitigate bias in AI-model in healthcare design. It is advisable to utilize public, large and diverse datasets to avoid data bias. In addition, validation of model on multiple datasets ensures the model' generalizabiltiy and interoperability. Strong and robust design of AI-model architecture is essential to mitigate algorithmic bias. Selection of sensitive and important featuring during in-processing of model development stage is critical to avoid selection bias. In addition, explainable and interpretable AI-model are more reliable in their outcome to be deployed in real-time.

References

1. Kumar A (2023) Visual object tracking using deep learning. CRC Press
2. Kumar A, Sarren P, Raja (2023) Deep learning-based multi-object tracking. In: Object tracking technology: trends, challenges and applications. Springer, pp 183–199
3. Kumar A, Nayyar A, Sachan RK, Jain R (2023) AI-assisted special education for students with exceptional needs. IGI Global
4. Kumar A, Vohra R (2023) Impact of deep learning models for technology sustainability in tourism using big data analytics. Deep learning technologies for the sustainable development goals: issues and solutions in the post-COVID era. Springer, pp 83–96
5. Kumar A, Jain R, Gupta M, Islam SM (2023) 6G-enabled IoT and AI for smart healthcare: challenges, impact, and analysis. CRC Press
6. Kumar A, Sareen P, Arora A (2023) Healthcare engineering using AI and distributed technologies. Smart distributed embedded systems for healthcare applications. CRC Press, pp 1–14
7. Abràmoff MD, Tarver ME, Loyo-Berrios N, Trujillo S, Char D, Obermeyer Z, Maisel WH et al (2023) Considerations for addressing bias in artificial intelligence for health equity. NPJ Digit Med 6(1):170
8. Agarwal R, Bjarnadottir M, Rhue L, Dugas M, Crowley K, Clark J, Gao G (2023) Addressing algorithmic bias and the perpetuation of health inequities: an AI bias aware framework. Health Policy Technol 12(1):100702
9. Kumar A, Aelgani V, Vohra R, Gupta SK, Bhagawati M, Paul S, Laird JR et al (2023) Artificial intelligence bias in medical system designs: a systematic review. Multimedia Tools Appl:1–53
10. Oala L, Fehr J, Gilli L, Balachandran P, Leite AW, Calderon-Ramirez S, Jaramillo-Gutierrez G et al (2020) Ml4h auditing: from paper to practice. In: Machine learning for health. PMLR
11. Oala L, Murchison AG, Balachandran P, Choudhary S, Fehr J, Leite AW, Nakasi R et al (2021) Machine learning for health: algorithm auditing & quality control. J Med Syst 45:1–8
12. Dehkharghanian T, Bidgoli AA, Riasatian A, Mazaheri P, Campbell CJ, Pantanowitz L, Rahnamayan S et al (2023) Biased data, biased AI: deep networks predict the acquisition site of TCGA images. Diagn Pathol 18(1):1–12
13. Kumar A, Vohra R, Jain R, Li M, Gan C, Jain DK (2024) Correlation filter based single object tracking: a review. Inf Fusion:102562
14. Hamilton IA (2018) Why it's totally unsurprising that Amazon's recruitment AI was biased against women. Business Insider
15. Bagaric M, Svilar J, Bull M, Hunter D, Stobbs N (2021) The solution to the pervasive bias and discrimination in the criminal justice: transparent artificial intelligence. Am Criminal Law Rev 59(1)
16. Parra CM, Gupta M, Dennehy D (2021) Likelihood of questioning ai-based recommendations due to perceived racial/gender bias. IEEE Trans Technol Society 3(1):41–45
17. Obermeyer Z, Powers B, Vogeli C, Mullainathan S (2019) Dissecting racial bias in an algorithm used to manage the health of populations. Science 366(6464):447–453
18. Naqa IE, Drukker K (2023) AI in imaging and therapy: innovations, ethics, and impact–introductory editorial. Br J Radiol 96(1150):20239004
19. Vicente L, Matute H (2023) Humans inherit artificial intelligence biases. Sci Rep 13(1):15737
20. Adam H, Balagopalan A, Alsentzer E, Christia F, Ghassemi M (2022) Mitigating the impact of biased artificial intelligence in emergency decision-making. Commun Med 2(1):149
21. Singh D, Singh D, Manju, Gupta U (2023) Smart healthcare: a breakthrough in the growth of technologies. In: Artificial intelligence-based healthcare systems. Springer, pp 73–85
22. Tripathy SK, Singh D, Jaiswal A (2023) Multi-layer feature fusion-based deep multi-layer depth separable convolution neural network for Alzheimer's disease detection. In: 2023 International conference on IoT, communication and automation technology (ICICAT). IEEE
23. Tripathy SK, Singh D, Srivastava S, Srivastava R (2023) Multiscale low-level feature fused multilayer convolution neural network for Alzheimer's disease detection

24. DeCamp M, Lindvall C (2020) Latent bias and the implementation of artificial intelligence in medicine. J Am Med Inform Assoc 27(12):2020–2023
25. Gurupur V, Wan TT (2020) Inherent bias in artificial intelligence-based decision support systems for healthcare. Medicina 56(3):141
26. Celi LA, Cellini J, Charpignon M-L, Dee EC, Dernoncourt F, Eber R, Situ J et al (2022) Sources of bias in artificial intelligence that perpetuate healthcare disparities—a global review. PLOS Digital Health 1(3):e0000022
27. Noseworthy PA, Attia ZI, Brewer LC, Hayes SN, Yao X, Kapa S, Lopez-Jimenez F et al (2020) Assessing and mitigating bias in medical artificial intelligence: the effects of race and ethnicity on a deep learning model for ECG analysis. Circ: Arrhythmia Electrophysiol 13(3):e007988
28. Kumar A, Ahluwalia R (2021) Breast cancer detection using machine learning and its classification. Chapman and Hall/CRC
29. Kumar A, Rai RS, Gheisari M (2021) Prediction of cervical cancer using machine learning. In: Cancer prediction for Industrial IoT 4.0: a machine learning perspective, pp 107
30. Kumari V, Kumar N, Sampath Kumar K, Kumar A, Skandha SS, Saxena S, Fouda MM et al (2023) Deep learning paradigm and its bias for coronary artery wall segmentation in intravascular ultrasound scans: a closer look. J Cardiovasc Dev Dis 10(12):485
31. Das S, Nayak GK, Saba L, Kalra M, Suri JS, Saxena S (2022) An artificial intelligence framework and its bias for brain tumor segmentation: a narrative review. Comput Biol Med 143:105273
32. Suri JS, Bhagawati M, Paul S, Protogeron A, Sfikakis PP, Kitas GD, Saxena S et al (2022) Understanding the bias in machine learning systems for cardiovascular disease risk assessment: the first of its kind review. Comput Biol Med 142:105204
33. Suri JS, Agarwal S, Jena B, Saxena S, El-Baz A, Agarwal V, Fatemi M et al (2022) Five strategies for bias estimation in artificial intelligence-based hybrid deep learning for acute respiratory distress syndrome COVID-19 lung infected patients using AP (ai) Bias 2.0: a systematic review. IEEE Trans Instrumen Meas
34. Norori N, Hu Q, Aellen FM, Faraci FD, Tzovara A (2021) Addressing bias in big data and AI for health care: a call for open science. Patterns 2(10)
35. Brault N, Saxena M (2021) For a critical appraisal of artificial intelligence in healthcare: the problem of bias in mHealth. J Eval Clin Pract 27(3):513–519
36. Suri JS, Bhagawati M, Paul S, Protogerou AD, Sfikakis PP, Kitas GD, Saxena S et al (2022) A powerful paradigm for cardiovascular risk stratification using multiclass, multi-label, and ensemble-based machine learning paradigms: a narrative review. Diagnostics 12(3):722
37. Gichoya JW, Thomas K, Celi LA, Safdar N, Banerjee I, Banja JD, Purkayastha S et al (2023) AI pitfalls and what not to do: mitigating bias in AI. Br J Radiol 96(1150):20230023
38. Nazer LH, Zatarah R, Waldrip S, Ke JXC, Moukheiber M, Khanna AK, Ma H et al (2023) Bias in artificial intelligence algorithms and recommendations for mitigation. PLOS Digital Health 2(6):e0000278
39. Ueda D, Kakinuma T, Fujita S, Kamagata K, Fushimi Y, Ito R, Fujima N et al (2024) Fairness of artificial intelligence in healthcare: review and recommendations. Jpn J Radiol 42(1):3–15
40. Palatnik de Sousa I, Vellasco MM, Costa da Silva E (2021) Explainable artificial intelligence for bias detection in covid ct-scan classifiers. Sensors 21(16):5657
41. Álvarez-Rodríguez L, de Moura J, Novo J, Ortega M (2022) Does imbalance in chest X-ray datasets produce biased deep learning approaches for COVID-19 screening? BMC Med Res Methodol 22(1):125
42. Kumar A, Singh D (Eds.) (2025) Revolutionizing Medical Systems Using Artificial Intelligence: A Breakthrough in Healthcare. Elsevier
43. Singh D, Srivastava R (2022) Channel spatial attention based single-shot object detector for autonomous vehicles. Multimedia Tools Appl 81(16):22289–22305
44. Parikh RB, Teeple S, Navathe AS (2019) Addressing bias in artificial intelligence in health care. JAMA 322(24):2377–2378
45. Charpignon M-L, Byers J, Cabral S, Celi LA, Fernandes C, Gallifant J, Ong BA et al (2023) Critical bias in critical care devices. Crit Care Clin 39(4):795–813

46. Panch T, Mattie H, Atun R (2019) Artificial intelligence and algorithmic bias: implications for health systems. J Glob Heal 9(2)
47. Loftus JR, Russell C, Kusner MJ, Silva R (2018) Causal reasoning for algorithmic fairness. arXiv preprint arXiv:1805.05859
48. Saleiro P, Kuester B, Hinkson L, London J, Stevens A, Anisfeld A, Ghani R et al (2018) Aequitas: a bias and fairness audit toolkit. arXiv preprint arXiv:1811.05577
49. Sterne JA, Hernán MA, Reeves BC, Savović J, Berkman ND, Viswanathan M, Boutron I et al (2016) ROBINS-I: a tool for assessing risk of bias in non-randomised studies of interventions. BMJ 355
50. Wolff RF, Moons KG, Riley RD, Whiting PF, Westwood M, Collins GS, Group P et al (2019) PROBAST: a tool to assess the risk of bias and applicability of prediction model studies. Ann Internal Med 170(1):51–58
51. Moons KG, de Groot JA, Bouwmeester W, Vergouwe Y, Mallett S, Altman DG, Collins GS et al (2014) Critical appraisal and data extraction for systematic reviews of prediction modelling studies: the CHARMS checklist. PLoS Med 11(10):e1001744
52. Maier-Hein L, Reinke A, Kozubek M, Martel AL, Arbel T, Eisenmann M, Onogur S et al (2020) BIAS: transparent reporting of biomedical image analysis challenges. Med Image Anal 66:101796
53. Suri JS, Agarwal S, Gupta SK, Puvvula A, Viskovic K, Suri N, Fatemi M et al (2021) Systematic review of artificial intelligence in acute respiratory distress syndrome for COVID-19 lung patients: a biomedical imaging perspective. IEEE J Biomed Health Inform 25(11):4128–4139
54. Santa Cruz BG, Bossa MN, Sölter J, Husch AD (2021) Public covid-19 x-ray datasets and their impact on model bias–a systematic review of a significant problem. Med Image Anal 74:102225
55. Geis JR, Brady AP, Wu CC, Spencer J, Ranschaert E, Jaremko JL, Shields WF et al (2019) Ethics of artificial intelligence in radiology: summary of the joint European and North American multisociety statement. Radiology 293(2):436–440
56. Jain R, Kumar A, Nayyar A, Dewan K, Garg R, Raman S, Ganguly S (2023) Explaining sentiment analysis results on social media texts through visualization. Multimedia Tools Appl:1–17

Chapter 15
Examining QoS for Modern Healthcare Systems

Abstract Smart healthcare is revolutionizing healthcare delivery by integrating the advantages of IoT, mobile technology, and cloud computing. Cloud computing has greatly facilitated the integration of healthcare institutions, caregivers, and patients in the health business to exchange information. Low latency and quicker reaction times are the primary factors driving efficient healthcare systems' implementation. Therefore, prompt communication across healthcare institutions is crucial in general, but considerable delays among many stakeholders might lead to catastrophic consequences during an emergency. Therefore, innovative methods such as edge computing and artificial intelligence (AI) can effectively address these issues. For a packet to be transmitted from one point to another, it is necessary for the "quality of service" (QoS) requirements to be fulfilled. QoS, or Quality of Service, pertains to the level of performance and reliability that a service provides to its consumers. QoS metrics like as throughput, bandwidth, transmission delay, availability, jitter, latency, and packet loss are essential in this context. We prioritize the individual devices that exist at various levels of the smart healthcare infrastructure and the quality of service (QoS) needs of the healthcare system as a whole.

Keywords Quality of service (QoS) · Modern healthcare systems · Smart healthcare · Digital health · Healthcare service quality · Health system performance · Healthcare communication networks · Health information technology (HIT)

15.1 Introduction

Health is a very precious possession for all individuals, and healthcare is the service that can assist and advise in maintaining this possession. Presently, there is a pressing demand for improved and cost-effective healthcare services, driven by the rapid expansion of the population and the prevalence of various diseases. The health industry has notably benefited from the integration of Internet of Things (IoT), mobile technology, and cloud computing, which have facilitated the connection between

A. Kumar and D. Singh, *Artificial Intelligence in Modern Healthcare System*,
Transactions on Computer Systems and Networks,
https://doi.org/10.1007/978-981-96-6703-1_15

health facilities, caregivers, and patients to exchange information [1]. The implementation of smart healthcare, which involves the transmission and reception of medical information, is very cost-effective for stakeholders. Health devices provide substantial volumes of data and need processing by the system's objectives. The proliferation of smart healthcare systems has resulted in a significant increase in the number of IoT healthcare devices, which is anticipated to exceed 162 billion worldwide as of 2020 [2]. Hence, given the substantial amount of data in an energy-limited setting, modern communication systems are becoming increasingly inefficient. Similarly, current computing approaches are also falling short in meeting the performance needs of smart healthcare applications. Moreover, medical data is subject to time constraints, and medical data that is delayed provides minimal assistance to caregivers, particularly in urgent situations. Efficient healthcare systems may be achieved by prioritizing low latency and improving reaction time. Consequently, prompt reactions from healthcare organizations are crucial, yet during crises, delays for different parties might result in disastrous situations.

Low latency and short reaction times are essential in healthcare services for rapid data access, facilitating precise diagnosis. The relevance of the following real-world circumstances is highlighted:

- If the streaming videos between the doctor and patient function flawlessly, a patient residing in a distant place with few medical resources can access the necessary healthcare. The physician will possess the capability to assess the patient's symptoms and establish a precise diagnosis.
- Low latency is beneficial for X-rays, MRIs, and other medical imaging because it enables fast loading for doctors and several viewing angles for rapid interpretation of the provided results.
- In medical emergencies, prompt access to a patient's medical data, without any obvious delays, might save their life and ensure that they receive appropriate care.

Cloud computing offers extensive computational and storage capabilities to healthcare equipment integrated with IoT technology. However, it suffers from significant latency and sluggish reaction times because of its distance from the end devices. Therefore, to handle such circumstances, cutting-edge methods such as edge computing and artificial intelligence can effectively address these problems [3]. Edge computing is the processing of data in devices positioned at the network's edge [4]. This approach reduces latency and improves energy efficiency. Edge-assisted IoT solutions facilitate the timely delivery of medical services. Furthermore, the integration of these two technologies has the potential to offer answers to several complex issues in healthcare systems. Utilizing AI approaches can significantly enhance the analysis of medical data and decrease reliance on human involvement for decision-making. Artificial intelligence can forecast diseases by analyzing medical records and can provide patients with recommendations for preventing or treating the anticipated illnesses. In order to accommodate the computational demands of AI approaches, it is necessary to develop less resource-intensive AI techniques for edge computing [5]. Edge computing extensively use AI techniques, including machine learning (ML) and deep learning (DL), for system training and knowledge acquisition. Edge intelligence, which is the integration of artificial intelligence (AI) with edge computing, is revolutionizing the functionality of smart healthcare apps.

Edge intelligence is the fragmentation of AI services and IoT data, which are then distributed among many edge devices. Hence, edge devices may possess comprehensive or partial artificial intelligence services or Internet of Things data. Therefore, the services are relocated from the cloud servers to the edge-assisted IoT devices, allowing for AI and data storage to be closer to the end-users [6]. Simultaneously, the healthcare system based on the Internet of Things (IoT) comprises a multitude of devices with distinct specifications. The Internet of Things (IoT) devices present several issues, such as increased demands for battery longevity, interference from other devices, signal weakening in different environments, and reduced dependability caused by increased latency.

High-quality healthcare services are often characterized by precise diagnosis, timely treatment, and exceptional patient care. From a technical standpoint, the efficient functioning of medical monitors and equipment guarantees that the patient's information will be swiftly and seamlessly transmitted to the doctor's computer. This will ensure that the patient receives timely and suitable medical care. This improves the quality of healthcare services by enhancing reaction time and reducing waiting periods for patients and physicians at advanced medical facilities, both on-site and remotely. The wireless link's service characteristics are of highest importance as they directly contribute to enhanced signal receptions, reduced packet loss ratios, and minimized power consumption. Furthermore, the utilization of distributed AI services and IoT data gives rise to many quality of service (QoS) difficulties, such as battery longevity, delay variability, and so on (Role and advantages of AI in modern healthcare system is shown in Fig. 15.1).

The key contributions of this chapter are as:

- This chapter explores the issues and challenges concerning the quality aspect of smart healthcare, particularly the telecare service.

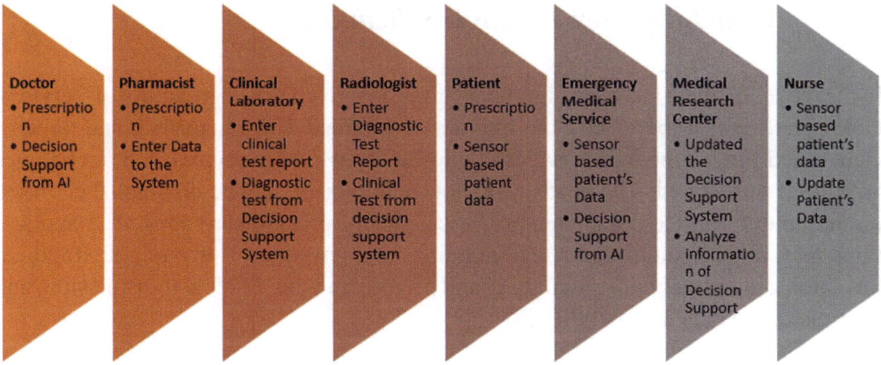

Fig. 15.1 Role and advantages of AI in modern healthcare system

- Given the advancement of technology, improved healthcare could be delivered in a much personalized manner based on individuals' profiles and the associated environmental contexts.
- There are emerging healthcare systems built on the convergence of computing, communications, and other smart technologies. However, no matter how good a healthcare system is in terms of sophistication and cleverness, the quality of service it provides must be close to the services offered in a traditional hospital setting.
- What are the key services or events that should be offered in such a smart hospital setting? How would the quality of services differ when the service delivery/ performance depends on not only health professionals (i.e. doctors, nurses in a conventional hospital) but also on the general patients/their carers?

The rest of the chapter is organized as follows. Section 15.2 elaborates on the AI-based research for QoS in modern Healthcare. In addition, Smart medical services for quality evaluation system are discussed in Sect. 15.3. Technological challenges that emerging the obstacles in the digital era for modern Healthcare system are discussed in Sect. 15.4. Future trends and Innovation of the work is mentioned in Sect. 15.5. The Summary of the chapter is mentioned in Sect. 15.6.

15.2 Related Work

This part focuses on the examination and evaluation of past strategies for ensuring quality of service (QoS), quality of experience (QoE), and cost-effective scheduling. Additionally, it explores metaheuristic algorithms that are relevant to the healthcare field based on existing research.

15.2.1 Optimizing Quality-of-Service (QoS) and Achieving Cost-Efficiency Through Scheduling

In recent years, there has been significant attention given to QoS (Quality of Service), QoE (Quality of Experience), the Internet of Medical Things (IoMT), and the efficient scheduling of medical services [7]. Several Quality of Service (QoS) techniques have been introduced. The needs of various techniques may be distinguished based on the unique service parameters and measures of e-healthcare apps. The applications may include multimedia conferencing, transmission of physiological indicators, high-resolution medical imaging and picture transfer, clinical transmission, and administrative data accessibility [8].

15.2.2 Effective Calculation of Medical Data Processing and Optimization of Service Delivery Solutions in E-healthcare Applications

Recently, the issue of ensuring high-quality service and efficient data processing in the healthcare sector has become a significant challenge in various research fields, such as meta-heuristics, machine learning, artificial intelligence, and deep learning [9–12]. These techniques are applied to medical data and records stored in hospitals [8, 13]. The collection and pre-processing of medical data in healthcare is a crucial task. It involves gathering data records from three main sources: clinical trials, medical research-related records, and organizational data operations [14]. The latest trend in smart healthcare is the advanced development of computer-based assistant aids and real-time platforms for examining, analyzing, and utilizing acquired data. This includes the assessment of quality-of-services [15, 16]. The authors comprehensively analyzed medical-data computation and service-delivery optimization in healthcare in [17]. The study analyzes essential healthcare data, including patients' medical histories, illness prediction, preventive measures, health guidelines, and medical assistance for the elderly. These records enable decision-making based on emergency situations, cost-effectiveness, and improved efficiency. The present study has introduced and implemented diverse probabilistic and adaptive Quality of Service (QoS) frameworks, which efficiently schedule medical data at a lower cost. These frameworks have been applied in various medical settings [17, 18]. A novel approach to health analysis and prediction has been recently proposed. This method utilizes certain quality of service (QoS) characteristics and quality of experience (QoE) mechanisms in real-time. In their study, the authors of reference [19] introduced a novel approach called IoT-fog enabled multi-route for processing and computing medical data. This approach aims to improve the efficiency of real-time medical delivery and optimize the management of healthcare records logs [20]. A multitude of academics have employed metaheuristics to optimize medical services that involve multi-channel pathways and service delivery via healthcare applications [21]. Zhao and Huang [22] proposed a novel design for a fog-based microservice container system. This architecture aims to efficiently execute sensitive applications and accurately measure transmission delays, while minimizing costs [22, 23]. Furthermore, this work examined the challenges and constraints associated with cost-effective job scheduling, specifically focusing on heterogeneous fog servers [22, 24]. In order to achieve this objective, several specialists have put forward various novel adaptive techniques. One potential solution that is of concern is a cost-aware computational offloading and task scheduling architecture. This architecture offers task scheduling solutions through a series of processes, such as task scheduling.

15.3 Smart Medical Services Quality Evaluation System

The utilization of intelligent technology has led to significant divergence in the evaluation criteria for service quality between smart medical care and traditional medical care [25–27]. To the best of our knowledge, there are few methods available to assess the efficacy of smart medical devices. To address this research deficiency, this study conducted a thorough examination of existing literature to identify key indicators for assessing the effectiveness of smart medical services. Additionally, a comprehensive methodology was developed to evaluate the quality of these services. This research initially formulates six aspects after reviewing current literature and assessing the service quality of smart medicine. These dimensions are smart appointment, smart consultation, smart diagnosis and treatment, smart nursing, smart settlement, and smart healthcare.

15.3.1 Smart Appointment

A smart appointment is a method by which a patient may plan a visit with a certain physician, choose a convenient time, and find the hospital. This is done by creating a personal profile using their ID card or medical insurance card on a smartphone or computer. In contrast to the conventional procedure, patients are not required to wait in line at the hospital for registration. Consequently, the implementation of smart appointment systems can greatly reduce the time patients spend waiting and decrease the labor costs for the hospital. This, in turn, enhances the efficiency of hospital management and improves the overall experience and satisfaction of patients.

The utilization of the "mutual health data bank" in the outpatient process enables the implementation of intelligent outpatient procedures. Internet appointment scheduling should be considered a crucial factor in assessing the effectiveness of smart appointment systems [28]. In order to ensure the information security of residents using the smart medicine service platform in Turkish hospitals, it is necessary to establish a service platform information security guarantee system that complies with the relevant national security guarantee standards [29]. An authentic identification system need to be implemented, requiring citizens to undergo real-name registration and documentation using their ID cards or health insurance cards. Patients using a smartphone application to schedule appointments should also be given the GPS coordinates of the hospital and map directions. This would help them save time by avoiding unnecessary diversions. Real-name archiving and smart navigation are two crucial aspects to consider when evaluating smart appointment systems. A study found that sending a short message service (SMS) reminder before a physical examination can successfully decrease the number of missed appointments and increase the rate of real examinations and satisfaction with appointments. According to the findings given above, there are four indicators established for smart appointment:

real-name registration and archiving, Internet appointment, appointmen reminder, and intelligent navigation.

15.3.2 Smart Consultation

Smart consultation is a system that allows patients who have scheduled an appointment online to check their place in the waiting list, the current calling status, and the average waiting time before arriving at the hospital. Patients who have not made an online appointment can use a multi-functional self-service terminal to make an appointment on-site. SMS reminders have been found to enhance appointment attendance, medication adherence, and behavior modification for a range of healthcare concerns [30]. An appointment queuing call system with a unified serial number database can be established for various appointment services at the hospital, including telephone, SMS, online, and self-service appointments. This system can be developed by analyzing the consultation process and its characteristics. Implemented cutting-edge service models by constructing a 3D reservation service network, establishing an intelligent triage call system, and extensively integrating self-service options. The model included comprehensive window service activities and facilitated the integration of outpatient clinics, resulting in enhanced visiting conditions and experiences. This approach significantly enhanced the quality and efficiency of outpatient services [31] (The Smart medical services quality evaluation system is shown in Fig. 15.2).

According to the literature analysis provided, smart consultation involves patients participating in both online and in-person consultations. With the aid of advanced technology, the majority of hospitals have developed their own comprehensive information platforms that include self-service terminals. These terminals allow for real-time data updates. Additionally, hospitals have implemented calling systems that are connected to the appointment platform, which assist in guiding patients. The Internet-based consultation model is more efficient and effective than the traditional consultation process, since it improves all aspects of the process, making it more logical, intelligent, and sensitive to patients' requirements. There are five metrics that are established for smart consultation: online waiting order, online call question, average waiting time, self-service registration, and triage call.

15.3.3 Smart Diagnosis and Treatment

Smart diagnosis and treatment involve utilizing advanced technological tools to enable patients to undergo different operations and get services. The use of the one-stop inpatient care paradigm promotes efficient treatment delivery and enhances the patient experience. This approach successfully mitigates extended waiting times caused by preparation issues [32]. By integrating this platform with the current information system, the scope of information creation was expanded to include bedside

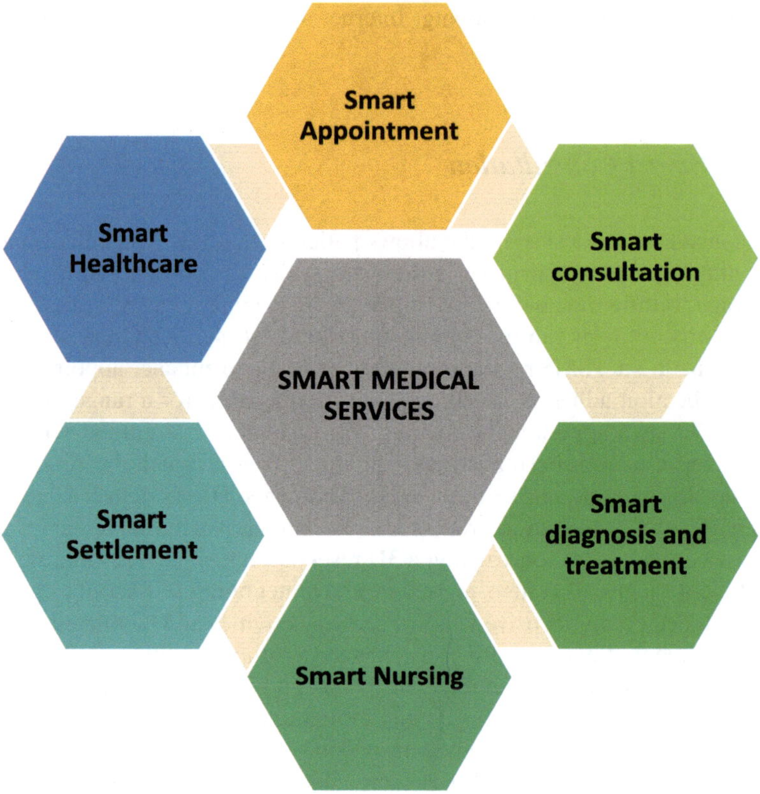

Fig. 15.2 Smart medical services quality evaluation system

access, enabling patients to immediately access information. The technology has significantly enhanced the inpatient experience and the effectiveness of medical care, while also offering data support for hospital administration [33].

According to the literature review provided, smart diagnosis and treatment includes the first consultation with outpatient doctors, as well as the processes involved when a patient has to be hospitalized and after they are admitted. The patients' perception of smart medicine, which integrates medical treatment with modern technology, primarily encompasses doctor-patient communication based on information, timely notification of diagnostic test results, efficient scheduling of non-emergency surgeries, expedited self-check-in for hospital admissions, self-service hospitalization, streamlined referral services, and convenient discharge processes. This study establishes the following indicators for intelligent diagnosis and treatment: doctor-patient contact, prompt notification of test findings, self-scheduling of surgical appointments, self-check-in, self-admission, and self-discharge.

15.3.4 Smart Nursing

Smart nursing encompasses the provision of precise care to hospitalized patients. This involves the use of advanced technology to accurately identify patients, as well as verbal and electronic communication from doctors and nurses. Additionally, it includes the implementation of automatic alerts based on various monitoring methods for infusion and injection management, bedside communication systems, and post-discharge monitoring.

By utilizing 5G technology, it is possible to create a full service platform called a digital ward. This platform includes a smartphone-based electronic medical record inquiry system and a wireless infusion monitoring system that relies on infrared sensing technology [34]. The wireless infusion monitoring system utilizes infrared sensing technology to track the drip rate and progress of infusion, therefore minimizing the need for regular check-ups and enhancing efficiency. Conventional patient identification bracelets that are written by hand encounter several issues [35]. For instance, medication errors may occur when patients are incorrectly identified. However, by using a personal digital assistant (PDA), patients can be accurately identified through barcode-printed wristbands. This method ensures patient privacy and guarantees that the barcode information remains legible, even when patients are bathing or taking medication. Additionally, it has the potential to enhance responsible nursing practices, guarantee nursing safety, decrease nursing time, and enhance patient satisfaction [36].

This study establishes five indicators for smart nursing: patient identity check, inpatient medical order execution, infusion and injection management, patient bedside calling, and post-discharge follow-up.

15.3.5 Smart Settlement

Smart settlement refers to the procedure in which patients may recharge and settle payments using medical insurance cards using self-service machines located in hospital lobbies, consultation rooms, or bedside platforms. There are several methods for recharging and settling medical insurance cards, including at least one of the popular payment platforms such as We Chat, Alipay, and UnionPay.

The use of the "Internet+" settlement method has the potential to assist hospitals in adapting to the advancements in Internet-based care. It can enhance patient experience, streamline settlement processes, and elevate the level of hospital administration [37]. The integrated "Internet+" medical platform aims to enhance the efficiency of appointment scheduling, registration, payment, and inquiries. It also aims to optimize the use of network technology for diagnosis and treatment, procurement, logistics, and follow-up visits. Furthermore, it enables real-time reimbursement from medical insurance companies and facilitates real-time claims for compensation from commercial insurance companies. One of the primary strategies for enhancing the experience is to decrease the number of settlement excursions. In order to enhance convenience, it is necessary to introduce a wide range of payment methods and implement advanced payment technologies. Additionally, the benefits of Internet technology should be utilized to streamline and customize the settlement process, establish flexible payment channels, and minimize the need for physical transactions, all while ensuring the security of funds.

According to the literature research provided, six indicators have been established for smart settlement. These indicators include real-name pre-deposit, medical insurance real-time network settlement, outpatient payment, payment methods, inpatient prepayment, and self-service printing.

15.3.6 Smart Healthcare

Smart healthcare encompasses the range of health services offered by hospitals to both pre-treatment and treated patients. These services include health education, dissemination of health information, provision of health consultation channels, and the creation of electronic health records for patients. Smart healthcare primarily caters to individuals with health needs, offering them the necessary health information.

Patients who test positive during consultation are likely to have a pleasant medical experience [30]. Providing health education on the specific illnesses to patients and their families during outpatient infusion and treatment can significantly reduce their anxiety and enhance their trust and cooperation, ultimately leading to improved patient satisfaction [38]. The adoption of the "Health Education Mobile Course" program resulted in a 3.65% improvement in overall patient satisfaction. This program guides users in preventing or treating ailments such as colds by helping them acquire fundamental medical information [39]. Electronic health records have the potential to provide advantages for patients, providers, and public health [32]. Furthermore, physicians can utilize the information supplied by other healthcare professionals, establish remote access to medical data, send reminders for service requirements, monitor electronic prescriptions and potential medication interactions, and employ clinical information for research purposes.

15.4 Technological Challenges: Emerging Obstacles in the Digital Era

One of the technical challenges of smart healthcare is the hospital information technologies. The second important challenge is the ubiquitous health information system. This is the core enabling technique of smart healthcare, allowing people to receive ubiquitous healthcare services at any time, place, or situation. IoT operated by sensor networks, cloud computing, and mobile device platforms offers infinite patient health benefits [33]. They are pervasive, real-time health monitoring, remote assistance, health self-management, and effective application, among others. The capability of pervasive, real-time health state measurements is directly related to the intelligent algorithm to extract meaningful and effective information. This is also a critical intelligence attribute for telehealth management purposes [34]. Other social systems, such as human mental barriers and public perception-hospital IT system, also influence the utilization of the hospital IT system. An embodiment of this vision by accomplishing improving healthcare efficiency, lowering healthcare cost, enhancing healthcare quality, facilitating a comfortable living environment, and reducing medical staff workloads, and other attributes is the ultimate goal of smart healthcare [35, 36].

Smart healthcare is predominantly enabled by advancements in medical, biomedical, and information technology fields. Ubiquitous computing and pervasive communication media have made it easy to attain high-quality medical care anytime and anywhere through internet-enabled devices [40]. Several challenges should be addressed to enable the promised smart healthcare as an ordinary lifestyle. This study elaborates on the enabling technologies of smart healthcare by creating a vision for healthcare enabled by smart healthcare and understanding potential technological, social, and economic effects of several potentially sensitive enabling techniques [41].

15.4.1 Interoperability and Integration

Hence, a big question arises: "How can we realize the ability of different information systems in traditional e-Health Service, mobile e-Health service, smart elderly-caring service, and smart society service, allowing them to cooperate to such an extent that the different stakeholders can support their different goals?" A way of data sharing algorithm, especially patient's electronic health record, among all stakeholders is described herein as a needed criterion in order to make the different systems, such as different subsystems, a sharing system of e-Health. Proper interoperability becomes a feature of the physical framework's predefined structure and/or serves as a basis for institutional structures in the information sharing process with respect to patient care data and other stakeholders' data, such as private data, counseling data, CDM data, and people's health environment data.

Interoperability and integration are among the main challenges for efficient implementation and management of a smart healthcare environment. In general, we can describe "interoperability" as the ability of systems or products to work with other systems or products without special effort on the part of the customer. It ensures that all systems remain working with minimum conditions to sustain a necessary data flow between the parts. Interoperability serves to create a whole new era of complex systems within the landscape of the smart health area.

15.4.2 Security and Privacy Concerns Explored

Given the constant of change in computer technology, no commitment can ever be forever. Although HTML-based and ID/password computer security has served us for many years, it needs to be upgraded with minimum added user online friction. Minimum adequate security is as relevant in today's covered health systems as it is with smart healthcare systems of the future, with enhanced security mechanisms for enhanced sensitivity data. Tailor-made user-friendly tradeoffs are a realistic approach in the face of all potential compromised security. There can be no absolute security in more personal sensitive health delivery areas, i.e., personal health records held on the cloud, without potential harm in emergency health delivery mode. Furthermore, security and privacy assurance must be predictable, reproducible, measurable, and ultimately authoritative and proven. It can never be completely preventive, given that people will be the weakest links in the security chain.

Security and privacy concerns are by far the biggest and most legitimate concerns which limit the very widespread use and growth trajectory of smart healthcare. When health data availability and utility is maximized, the security and privacy concerns become paramount as precious and sensitive personal health data may be exploited by insider or outsider adversaries. Perfect security cannot be achieved in practice and all security mechanisms, including passwords and physical biometric security measures alone or in combination, can be circumvented. So, tradeoffs must be made which balance security with user convenience, with the nature and sensitivity of the data, and the potential real or perceived public harm from limited data security. Therefore, the very essence in the debate about smart healthcare potential and deliverability and, by inference, the best strategy to be pursued, are health data security degrees and privacy protection measures which could be implemented.

15.4.3 Enhancing the Measures for Ensuring Data Security

Ensuring the security, confidentiality, and integrity (and linearity in audit data recording, i.e., knowing who saw what, when) of the information is key for public and academia adoption, as shown by public opinion toward the privacy breach of patient personal health records (PHRs). Types of Data exchanged and examples of multiple types (e.g., EMR and Personal Informatics (PIs) need different levels of security protection). Different types of systems and solutions used to provide health care, which introduces a level of complexity in managing the authorization of an infinite set of types, as well as subjects of data, operations (like read and write) and performing access control, and encryption and decryption [42].

The security of individual records is essential for the proper operation of the system. Patients will not be willing to come and share their problems if they think that their conversation will be available to everyone the next day. The idea is that records should prove useful today for the general public, and be made available, for instance, twenty years from now when it is safe to do so, making sure that any agreements the patient signed, referring to specific disclosures, are observed, but they should not be available today unless it is in the common interest, say, as part of a public health investigation. The challenge is exacerbated due to the abundance and complexity of sources and types of systems involved in a distributed healthcare system that go as far as encompassing existing independently designed systems including: Provider systems (e.g., EMRs managed by individual institutions), Regional Health Information Organizations (RHIOs) or Health Information Exchanges (HIEs) (e.g., connecting different provider systems), Public Health Information Exchange (PHIE) System (e.g., managing information collected by public health agencies at city, state and national levels), and Personal Informatics (e.g., personal wellness data, such as fitness-of-the-individual data like physical location, and temperature sense data) reports [43].

15.4.4 Exploring the Ethical and Legal Considerations Surrounding the Issues

Despite the widespread use of the term eHealth and telemedicine, it is not always easy to compare and evaluate what is on offer as much of it is poorly evaluated with outcomes that are at best of dubious benefit. The quality of the advice provided is almost never evaluated. Would it stand up to scrutiny by an in-depth peer review process? Many contain basic errors. Access to this high-quality medical information may itself exacerbate inequalities and health as web-based information tends to be targeted more to the empowered and the well-educated. Currently, lack of any form of regulation or evaluation can result in great heterogeneity in the quality of service, from the excellent to the truly dreadful. With the ever-increasing specialization of modern medicine and the rapidly changing therapeutic environment, the ability to reflect critically on one's standard of care is ever more an essential part of medical training. Can people utilizing web-based consultations and web-based sources of referrals ever know the level of themselves being offered? Web-based practice is by no means confined to healthcare professionals based solely in middle and high-income countries. It is therefore important to consider a global response to these problems. Underlying these challenges is the commitment to apply the principles of beneficence, non-maleficence, autonomy, and justice into virtual practice as it becomes an integral part of 21st-century healthcare [44].

In any community and healthcare systems, highly competent health professionals and appropriate access to these professionals are essential. In almost all countries, increasingly healthcare systems are facing problems of scarce resources and financial concerns. The web-based communication between patients and healthcare professionals can provide a low-cost tool to deal with such situations. However, online decision making has up to now been kept at a low, asynchronous, information-seeking level. The offer of opportunity for an electronic consultation service to address and resolve clinical cases in real time represents a major ethical challenge. The problem of the wrong use of the service in inappropriate conditions illustrates the difficulties of applying ethics and legal duty in virtual territories. The international tendency for raising the quality of health services according to pre-established protocols further complicates problem-solving in this particular area. The laws providing good clinical practice apply to experts independently of the medium of use and national boundaries. The lack of public standards for giving good quality can influence the quality of services [45].

Solution: Informed Consent and Ethical Obligations

Therefore, 'informed consent' is an important concept for protecting patient privacy in actual smart healthcare services using IoTs to constantly collect personal health data without patient intervention. In smart healthcare, physicians can form a patient's health-related viewpoints only with a patient's consent. Gathering patient consent for data sharing and integration is important not only for preventing disputes but also for building the trust that should be in the relations among patient, physician, and medical data integrator in smart healthcare services. Informed consent belongs to the ethics of personal privacy and self-determination, and it involves transparency, risk administration, and strengthening autonomy. Therefore, transparent methods for obtaining patient informed consent are needed [46].

The IoT can be used to create digital records of patient data that enable physicians to provide medical services. Although smart healthcare solves many problems, it provokes many issues because this unprecedentedly great collectivity and variety of patient health information can be integrated for physicians. Especially, if patient data is integrated and used in a way that patients do not know or be able to control, medical institutions or stakeholders can unilaterally make and hide decisions and all sorts of misuse that infringe on patient privacy can occur. Current medical information privacy legislation applies not only to health information recorded by medical institutions but also to health information gathered by health devices, but often patients sometimes cannot identify who and what data he or she want to control or cannot effectively control data integration and usage due to an information asymmetry caused by the newly appeared stakeholders [47].

15.4.5 Regulatory Hurdles and Compliance Challenges

The fact that the decisions made by the systems used in smart healthcare should be convergent should be evaluated at the individual patient's expense, the interests of the community, and the quality of health. This gray dilemma becomes more vivid when lawyers and ethicists in the courts try to judge the vicissitudes of system failures. However, reaching a more technical level, this problem should also question the reliability of the operating technique used, which must also satisfy technical, scientific, and engineering requirements since it must be recognized as a powerful tool by the interlocutor who listens to it. There comes the problem of concepts related to the risk of genetic and iatrogenic risk that has even a different weight in calculations depending on the assessment of who paid for the so-called comfort treatments. For all these reasons, it is not possible to conceive of the development of applications of computer intelligence in medicine without an analysis of the implications, also on an ethical level, of the cultural implications, which will be faced with the pervasiveness of new technologies in our lives. The economy of the health budget, in reality, imposes it [48].

The smart healthcare concept encounters regulatory hurdles, just like in other healthcare-related businesses. In addition to accountability and closed-loop control, the concept raises questions such as who is responsible if the smart healthcare system makes a mistake or error, who is responsible for unjustified or even failed kidney removal or heart bypass because of smart healthcare technologies. Such questions cannot be easily answered, playing on ethical questions not only on technology but in a broader plan, by challenging the control of physicians, by allowing machines to make suggestions about health problems, lab test results, diagnosis, treatments, and therapy. The situation is exacerbated if the smart healthcare system happens to be made by a private company. This will further push the health system to consider problems of healthcare in the public–private partnership [49].

Solution: Compliance with Healthcare Regulations and Implementation of Regulatory Requirements

Legislation and technical requirements are placed on hospitals and ambulatory centers to ensure the provision of good quality services. These requirements are specific to the nature of patient care processes and patient characteristics. Often, these requirements are open to interpretation by those subject to them, and organizations may be vulnerable to adverse litigation or prosecution. The growth in the number of regulations and pressure for compliance stems from concerns of employees, their families, and society at large regarding the safety, quality, and regulatory compliance of health services. Even with regulatory measures attempting to enforce the quality of healthcare services, the effects have been mixed in terms of achieving improvements. The issue at hand is how these laws and processes achieve their objectives. Wouldn't it be more effective to employ 'softer' approaches in healthcare services? [50]

Huge amounts of administrative and technical infrastructure are required to ensure that healthcare providers operate within a legal framework of satisfactory quality and comply with regulations in healthcare operations. This concentration on compliance increases the cost of healthcare. The challenges to health service regulations are not just from the delivery of health services, but also reflect changes in the law and society. Over recent decades, changes in health services regulations have been evident in several countries. The reasons for these transformations have been varied and include the growth of new technologies, changes in the level of knowledge and awareness among wide audiences, and a re-evaluation of traditional models of medical care, as well as financial and economic tendencies in managing the health system and strengthening state control to the detriment of medical profession autonomy [51].

15.4.6 Challenges Faced in the Management of Data

With the rapid increase in the volume of digital data, such challenges are likely to grow more acute, and in-depth studies on data exchange metrics, security protocols, and possess more appropriate recommendation tools may be needed. Gu and Niakhail proposed the means to reduce the plethora of data, with a possible technique of using ID photographs and recommendation software tools to present some possible pieces of data within the realm of the subject. These include diagnosis, therapy contract, and care of the present patient, spatial resolution, position inside a chest CT exam, radiation dose verification, and casual information that could impact the diagnostic process. They apply data-driven and biological concepts to solve the problems of data noise detection and cleaning in these medical image examples. They showed that the existing data has mislabeled errors and noise, which may cause biased or incorrect conclusions. Their proposed method assigns small masses to missing data labels and then removes these extra categories, and then analyzes the results. However, investigation in these areas is still in its infancy, and Internet of Things (IoT) technologies possess unique prospects and challenges compared to traditional data management tools used in healthcare [52].

One of the most challenging issues in smart healthcare is data integration and data management. The heterogeneous characteristics of the collected data, such as various formats, storage, access interfaces, and authentication mechanisms, must be supported. Moreover, to maximize the useful information out of the complex data and to build quality of services in smart healthcare, data analysis and management techniques are required. This would involve data mining, analysis, capturing, and divergence in larger data systems, as is the case in healthcare big data. However, big data in the healthcare domain poses huge issues and roadblocks, including medical data security and privacy challenges, lack of data quality means to get knowledge and information from this big data, as well as data mining algorithms that require high-performance computing and computing capabilities and machines, to name just a few. The five main big data scenarios are as follows:

Solution: Big Data Analytics

1. Private practice of healthcare: – The increase in data acquisition through electronic health records and monitoring devices. – Implementation of collaborative care methodologies. – Improvement in hospital data tools and utility methods.
2. Artificial Intelligence provides healthcare services: – Deep learning, natural language processing, and computer vision label medical data. – Data-enabled personal health data promotes personalized healthcare.
3. Hospital-based analytics services: – Use of patient data to create timely responses in treatment, diagnosis, and monitoring of patients. – Utilizing the data extensive electronic health record systems for research using cloud-based solutions without releasing the actual dataset of patients.

4. Healthcare research with big data: – Biomedical research data is integrated with clinical data. – Data-driven discovery solutions to accelerate pre-clinical and clinical research.
5. Big data infrastructure to improve patient outcomes: – Improvement in big data strategies and methodologies. – Controls and regulated industry partnership models.

Big data in the healthcare system occurs due to accessibility to a vast amount, wide variety, and volume of data, which is structured, unstructured, real-time data, and stored in various forms. It can benefit greatly from the innovative analysis and evaluation of healthcare issues and data. The vast availability of data storage, advanced technologies, and device connectivity has led to future healthcare industry growth. Every day, a vast amount of data is produced in healthcare organizations, such as patient data, data sets, and clinical monitoring data. In medical practice, electronic health record datasets have become the main source of clinical data that is released on a day-to-day basis, enabling the provision of healthcare services at any time from any hospital, laboratory, doctor's office, clinic, or any telemedicine solution. Responses to large volumes of healthcare data have resulted in the development of several data analysis methodologies that allow the study of the enormous dataset of patient data and the extraction of useful information. These innovative studies have increased the value of healthcare organizations' patient databases without developing the required infrastructure [53].

15.4.7 Integration with Traditional Healthcare Systems

The QoS model in smart healthcare mentioned in previous chapters is defined by referring to the definition of QoS in ITU-T Recommendation E.800, but more problem cases raised in smart healthcare are not explored in detail. Therefore, this study mainly investigates the challenges and open issues for QoS in smart healthcare. Moreover, as more and more smart healthcare systems integrate existing traditional healthcare systems, QoS issues on the integration of smart healthcare systems and traditional healthcare systems must be noted and solved. In recent years, the continuing development of smart healthcare, particularly when coping with the issue of global aging and the related problem of big data, has attracted considerable attention from both industry and research communities and has been an innovative challenge milestone in the Telecommunication Industry Innovation Roadmap (TIIR) in China [54].

In this chapter, we investigated various challenges and issues for QoS in smart healthcare. Key challenges and issues identified include data storage and management, data sharing and collaboration, device-to-device communication, medical resource management, middleware and infrastructure, near-field communication technology, reliability, security and privacy, and service level agreements. For each category, the current status and related works are summarized. In addition, the challenges, open issues, and implications are also given to provide more inspirations to researchers in the field of smart healthcare so that an efficient and reliable data processing and communication system can be developed to satisfy both healthcare providers and consumers.

Solution: Benefits and Challenges

The big data that will be created by IoTs in smart environments will soon (if it is not already) exceed the capacity of the existing hardware systems as much information is required for efficient processing. Moreover, healthcare infrastructures must guarantee data security while facing hack attacks. The main challenge is transferring knowledge and decisions into practice, to develop sensor technologies that are economically viable for healthcare providers. While the rapid increase in information is becoming a pressing problem, the ability to store, process, and analyze huge amounts of data to gain insight is becoming a key competitive differentiator. A scalable architecture that is able to be built quickly and stably to deliver healthcare services over the network is therefore a second major challenge. This crucial requirement arises from the necessity of developing a flexible ICT infrastructure that is capable of adapting rapidly to every medical requirement. The architecture should be scalable and designed to provide the best trade-offs among communication, computing, and storage resources for costs and needs. The scalable architecture can manage large quantities of data and can optimize the volume and structure of the computational power and storage capacity of servers. Additionally, this architecture can protect against IoT risks; this type of cyber attacks arises not only from the possibility of exploiting the communication infrastructure but also from the significance of the devices or equipment [55].

The smart healthcare environment combines IoTs, next-generation networks, cloud computing, and the latest technologies in healthcare to support the knowledge and decisions of doctors and clinical staff and to enable a sophisticated supply of health and wellness. This environment facilitates the coordination of healthcare services that utilize advanced medical techniques and telemedicine. Although smart healthcare services offer valuable benefits, three main issues challenge the supply of a high quality of service (QoS) to medical staff and patients. These challenges are the large and rapidly increasing amount of healthcare information that requires storage and processing, the scalable architecture able to efficiently deliver healthcare services over the network, and the market competition leading to balancing the investment among the required technological infrastructures and services [56].

15.4.8 *Human Factors and User Acceptance*

In conclusion, healthcare depends on the availability of useful devices and instruments. In addition to leveraging the value of medical knowledge and training for a variety of diagnostic, surgical, prescription, safety, and prevention applications, devices and instruments are necessary for monitoring and supporting technical decisions, handling, administration, transportation, and logistics. In this context, much applies to a wide range of activities and emerging medical applications, which may nowadays be undertaken away from traditional medical professionals and facilities. There is nonetheless evidence that patients across the world have developed some ethical, moral, cognitive, and technological responsibilities and like to be 'working with their doctors', who expect a wider role in future patient-doctor interactions. The question is how much effort is needed to help existing and forthcoming technologies establish a clinically valuable role in medical professional and healthcare services [56].

It is interesting to see the impacts that smart healthcare technology can deliver on patients' and the general population's perspectives, as well as healthcare practices and processes. Most people will readily benefit from the highest quality of health and medical services. They will not be suspicious of how healthcare service provision is organized and regulated, as well as the practical implications and technical requirements for service prescription, delivery, monitoring, and evaluation. However, it is also unquestionable that services should be cost-effective and respect users' safety, preferences, and aspirations in all circumstances. The idea of being able to communicate with health services for professional advice and support at any time is not in the mind of everyone, but the delivery of this kind of service reflects an established reality that medical practitioners have.

This chapter brings together some views on human factors and user acceptance of smart healthcare technologies in general. These concepts combine general knowledge and understanding about engineering and technology in developing and delivering products and services to benefit human quality of life across various healthcare domains. Insights are discussed based on literature reviews, empirical case studies, and practical solutions toward human factors and user acceptance in developing and promoting emerging smart healthcare services in the coming years. Smart healthcare application areas to be discussed include patient-centered healthcare service delivery and personalized health and medicine technologies, such as personalized and predictive health, ambulances, bio-sensors, mobile, stationary, body-worn, and implantable medical devices and systems. The chapter concludes by identifying some challenges and recommendations for R&D for human factors and user acceptance of smart healthcare services, suggesting future development in this active area of engineering and technology sectors.

Solution: Training and Education

Training is a critical cornerstone of ensuring quality functionality of any system in general. The training for the employees to use the ICT tools in a smart healthcare system can be critical for their acceptance and the quality of the services. However, there are still seldom educational training programs and teachers involved. The training of such a setting should be able to convey knowledge of how to use electronic medical records to conduct effective and efficient diagnoses, how to access other healthcare-related information from a variety of sources over the internet in a fast and reliable manner, or how to use the alarm services provided by monitoring devices to decrease the problem of attention (or cognitive) overload. Due to the nature of the content and level of specificity needed, an added worry is misinterpretation between participants that are equivalent in knowledge base with explicitly different educational backgrounds [57].

15.4.9 Financial Considerations

To put it simply, in the broad context of the proposed factors affecting the quality of services of smart healthcare systems, financial elements must not be underestimated. On the contrary, they must be taken into consideration at the same level to design, diagnose, treat, and improve health services. In the following subsections, we will expose the main concepts on available mechanisms for assuring the financial considerations at stake.

Quality of services (QoS) must be maintained at a satisfactory level to ensure a secure healthcare application. The healthcare informatics system must fully cover private and confidential communication, data, information, and processing. Addressing these requirements often comes with a cost in terms of available functionality, performance to be met by the QoS, and measures to assure integrity, availability, and security of the information service or smart health application. Moreover, the potentially involved decision makers (i.e. healthcare professionals, patients, healthcare providers, and suppliers of relevant technology) must be aware and able to evaluate and accept these costs in view of their benefits [58].

Solution: Cost-effectiveness

The cost of smart healthcare is of significance to the community. Here, the cost encompasses initial set-up, operation, and recurrent maintenance costs. Commercialization of the service is hard to achieve if the costs are too high, especially in the current financial crisis environment. People or organizations need proof to believe that this new system is more affordable than the traditional way but with the same quality or better. Although the costs cover numerous aspects that other infrastructures have, the legitimacy of the smart healthcare solution somehow becomes a hot issue to be discussed, in which whether it is a luxury or a necessity is the root of argument. A good balance of providing quality services with the pressure from the bottom line has become a virtual inescapable dilemma [59].

Despite the existing interest and prospects for the smart healthcare paradigm, some important challenges that smart healthcare may confront should be carefully considered. These challenges might not only influence the success of its implementation and diffusion but also set obstacles to resolving the key issues raised previously. Major challenges for quality of services in the smart healthcare are discussed in the following section.

15.4.10 Global Adoption and Cultural Differences

It is important to consider various cultural issues before starting the deployment process of smart healthcare worldwide. Otherwise, if only a few countries accept only a few service items of the system, the productivity of smart healthcare may be suboptimal. The implementation study of global health technologies fundamentally involves culture and society, but there has been little scholarly attention given to in-depth discussions of this issue. As mobility increases and healthcare provision becomes a globalized commercial sector, the exchange of services, know-how, and medical personnel across national borders should not be taken for granted. Therefore, in addition to studying medical and clinical engineering issues in global health, which receive significant attention both domestically and internationally, we also need to explore the institutional settings in which different "smart" technologies, and the knowledge and information embedded in social practices of care, are used, owned, and applied.

Cultural differences are a key issue in the global adoption of smart healthcare systems and technologies. Different cultures may have different attitudes and expectations toward healthcare services. For example, different countries not only have various healthcare policies and systems, but they also have different treatment choices for similar diseases and symptoms. This can greatly influence the implementation of healthcare services such as telehealthcare, infrared sensor devices, and e-medicine, as well as the remuneration model of the smart healthcare system. As a result, customizing smart healthcare technologies and applications to meet diverse needs among various countries, and managing them like tourism, may present some challenges. On the other hand, a broad range of strategies should be focused on to encourage faster adoption of smart healthcare in order to shorten the time of implementation. Developing healthcare technology that fits within the cultural scope and aligns with the interests of healthcare service providers can certainly optimize healthcare development and improvement [60].

Solution: Cross-cultural Communication

In fact, for a healthcare provider, understanding the patient's actual complaints itself is a challenging task without cross-referencing their personal medical records. Understanding the differentiating signs, symptoms, and queries posed for interpretation are necessary to provide the most accurate diagnosis. The communication process invokes the facets of social, emotional and psychological factors regulating the relation between patient and healthcare professionals. According to a study carried out in the UK, the communication gap was identified as a factor responsible for complaints against healthcare providers. The smart healthcare communication-related challenges are unlimited as the links between patient and healthcare increases the patient's perception of healthcare and healthcare quality. Though it directly links caregivers and patients, some typical limitations of smart healthcare include communication pattern analysis, patient adaption over the length of clinical relationship, and healthcare customizations as per the cultural, language differences portrayed by the patient.

The seamless communication between healthcare providers and patients is significant for good patient outcome in relation to compliance and satisfaction with their medical treatment. Smart healthcare, by functioning beyond geographical boundaries with no requirement for face-to-face interaction, poses many challenges associated with cross-cultural communication between healthcare providers and patients. The patients' direct contact with healthcare professionals, expertise of healthcare professionals for treatment procedures, and capacity for understanding comprehension while addressing medical terms used for communication are indeed hindered [61].

15.5 Future Trends and Innovations

We do not need to list them all here; everyone is aware. We have gained valuable experience through the coronavirus pandemic. We know that quality of service in health issues is not dependent only on personnel numbers and technical equipment capability. Infrastructure and leadership/strategy are key factors in care for patients with different scales of health problems, and we are supporting that advances in ICT are strategic in providing this care. Today, telehealth is a subject that is in our healthcare plans. The capability to provide a consult through video links, SMS messages and teleconference anytime and anywhere is not futuristic, but it will expand in the care of patients after COVID-19 ends. At other more advanced stages, intangible innovation will exploit AI with Blockchain, leading to personalized and traceable solutions. The advent of a new fifth generation of high-bandwidth mobile communications, not just within the confined space of a hospital or clinic, but for everyone anywhere, is fundamental in providing value added to healthcare in general and especially to chronic disease control [45].

Introduction, Future Trends and Innovations of Healthcare is an area where services have generally been proven to be somewhat slow in adapting to the speed of technological change. Society has high levels of faith in technologies to cure diseases and provide longer and healthier lives, but not necessarily in ensuring that these benefits reach them. As discussed in the opening chapter, and as shown again in Fig. 15.1, society has other important issues that are associated with technological change in the area of health.

15.6 Summary

The integration of cloud and mobile application is a basic need for smart medical treatment. However, the personal data security, privacy protection, full availability during emergency situations, continuous monitoring, and constant technical improvements need to be carefully maintained with security, reliability, low cost, and especially high QoS. The activity in this study will be helpful for various stakeholders and policymakers in the medical field to deepen their understanding of what they should consider when they need to commission and create the IoT smart healthcare system. The results of this proposed strategy and implementation in cloud-based mobile medicine can be used as a reference for related society of public and private smart healthcare projects. The policy alignment and stakeholders' societal safeguards can be employed to meet all the potential human committed goals for equity, ethics, stakeholders' programs, and the urgency of continuous monitoring with major projects that are health-related. Therefore, from the view of the stakeholders, they can gain due planning of the quoted testbed in the day-to-day project or in the self-assessment of a smart medical project. The IoT smart healthcare system program ultimately needs to be meaningful and prioritized as part of the national or district government policies for stakeholders' dedicated hands-on expression.

The expansion of the IoT has affected various sectors, including medical care. IoT-equipped healthcare systems that use wearable devices, various medical sensors, security and privacy systems (RFID, NFC), and communication modules are developed as smart medical systems. The question is how these features are interconnected through cloud computing systems to create smart medical systems. From the user's point of view, features such as power efficiency, security and privacy, user-friendly design, safety standards, social and technical risks, clearness of responsibility, reliability, and data quality are important. Ultimately, the performance of the IoT medical platform in these respects affects the quality of service for the smart healthcare system, and ensuring high quality of medical care is essential. Furthermore, the global populations of elderly people is increasing, and these groups need reliable, effective, and high-quality healthcare services. The challenges and issues in this respect points will be carefully investigated as a part of the IoT medical platform in this study.

References

1. Ashfaq M et al (2020) Quality of service as a predictor of customer satisfaction in healthcare sector. IBT J Bus Stud (JBS) 16(1)
2. Patan R et al (2020) Smart healthcare and quality of service in IoT using grey filter convolutional based cyber physical system. Sustain Cities Soc 59:102141
3. Eman BAS (2021) Quality of service and privacy in internet of things dedicated to healthcare. Avignon
4. Kurstjens S et al (2021) Performance of commercially-available cholesterol self-tests. Ann Clin Biochem 58(4):289–296
5. Jeong S-M et al (2021) Association between high-density lipoprotein cholesterol level and risk of hematologic malignancy. Leukemia 35(5):1356–1364
6. Shah AM et al (2020) Mining patient opinion to evaluate the service quality in healthcare: a deep-learning approach. J Ambient Intell Humaniz Comput 11:2925–2942
7. Laghari AA et al (2023) Retraction note: a review and state of art of Internet of Things (IoT). Springer
8. Chouat H et al (2023) Adaptive configuration of IoT applications in the fog infrastructure. Computing 105(12):2747–2772

9. Kumar A et al (2024) Artificial intelligence bias in medical system designs: a systematic review. Multimedia Tools Appl 83(6):18005–18057
10. Kumar A et al (2023) Security and privacy issues in smart healthcare using machine-learning perspectives. 6G-enabled IoT and AI for smart healthcare. CRC Press, pp 41–56
11. Purohit RM, Verma JP, Jain R, Kumar A (2025) FedBlocks: federated learning and blockchain-based privacy-preserved pioneering framework for IoT healthcare using IPFS in web 3.0 era. Cluster Computing, 28(2):139
12. Kumar A et al (2023) AI-assisted special education for students with exceptional needs. IGI Global
13. Zhang H et al (2020) Active balancing mechanism for imbalanced medical data in deep learning–based classification models. ACM Trans Multimedia Comput Commun Appl 16(1s):Article 39
14. Wang X et al (2020) ADTT: a highly efficient distributed tensor-train decomposition method for IIoT big data. IEEE Trans Industr Inf 17(3):1573–1582
15. Vizitiu A et al (2020) Applying deep neural networks over homomorphic encrypted medical data. Comput Math Meth Med 2020
16. Lu MY et al (2021) Data-efficient and weakly supervised computational pathology on whole-slide images. Nat Biomed Eng 5(6):555–570
17. Tang S et al (2020) A survey on spark ecosystem: Big data processing infrastructure, machine learning, and applications. IEEE Trans Knowl Data Eng 34(1):71–91
18. Pérez-García F, Sparks R, Ourselin S (2021) TorchIO: a Python library for efficient loading, preprocessing, augmentation and patch-based sampling of medical images in deep learning. Comput Methods Programs Biomed 208:106236
19. Singh M, Baranwal G, Tripathi AK (2020) QoS-aware selection of IoT-based service. Arab J Sci Eng 45(12):10033–10050
20. Qiu Y, Zhang H, Long K (2021) Computation offloading and wireless resource management for healthcare monitoring in fog-computing-based internet of medical things. IEEE Internet Things J 8(21):15875–15883
21. Hassan SR et al (2020) Remote pain monitoring using fog computing for e-healthcare: an efficient architecture. Sensors 20(22):6574
22. Zhao X, Huang C (2020) Microservice based computational offloading framework and cost efficient task scheduling algorithm in heterogeneous fog cloud network. IEEE Access 8:56680–56694
23. Saeed W et al (2021) A fault tolerant data management scheme for healthcare Internet of Things in fog computing. KSII Trans Internet Inf Syst (TIIS) 15(1):35–57
24. Iqbal N et al (2021) A scheduling mechanism based on optimization using IoT-tasks orchestration for efficient patient health monitoring. Sensors 21(16):5430
25. Kumar A, Mamgai R, Jain R (2023) Application of IoT-enabled CNN for natural language processing. In: IoT-enabled convolutional neural networks: techniques and applications. River Publishers, pp 149–177
26. Kumar A, Bhalla S (2025) Exploring brain tumor detection through artificial intelligence. In Revolutionizing Medical Systems using Artificial Intelligence, pp 93–120. Academic Press
27. Kumar A, Singh D, Vohra R (2023) Improving learning abilities using AI-based education systems. In: AI-assisted special education for students with exceptional needs. IGI Global, pp 137–155
28. Küçük A et al (2021) Evaluating of hospital appointment systems in Turkey: challenges and opportunities. Health Policy Technol 10(1):69–74
29. Boone CE et al (2022) How scheduling systems with automated appointment reminders improve health clinic efficiency. J Health Econ 82:102598
30. Shaik T et al (2023) A survey of multimodal information fusion for smart healthcare: mapping the journey from data to wisdom. Inf Fusion:102040
31. Hoti K, Chivers PT, Hughes JD (2021) Assessing procedural pain in infants: a feasibility study evaluating a point-of-care mobile solution based on automated facial analysis. Lancet Digit Health 3(10):e623–e634

32. Rahman A et al (2024) Machine learning and deep learning-based approach in smart healthcare: recent advances, applications, challenges and opportunities. AIMS Public Health 11(1):58–109
33. Quy VK et al (2022) Smart healthcare IoT applications based on fog computing: architecture, applications and challenges. Complex Intell Syst 8(5):3805–3815
34. Gautam A, Mahajan R, Zafar S (2021) QoS optimization in Internet of Medical Things for sustainable management. In: Cognitive Internet of Medical Things for smart healthcare: services and applications, pp 163–179
35. Saba Raoof S, Durai MS (2022) A comprehensive review on smart health care: applications, paradigms, and challenges with case studies. Contrast Media Molec Imaging 2022(1):4822235
36. Kumar A, Singh D (Eds.) (2025) Revolutionizing Medical Systems Using Artificial Intelligence: A Breakthrough in Healthcare. Elsevier
37. Wang G, Shao Q (2024) Design a smart medical service quality evaluation system based on hybrid multi-criteria decision model
38. Mohammed BG, Hasan DS (2023) Smart healthcare monitoring system using IoT. Int J Interact Mobile Technol (iJIM) 17(01):141–152
39. Khang A (2024) AI and IoT technology and applications for smart healthcare systems. CRC Press
40. Haleem A et al (2022) Medical 4.0 technologies for healthcare: features, capabilities, and applications. Internet of Things Cyber-Phys Syst 2:12–30
41. Nasr M et al (2021) Smart healthcare in the age of AI: recent advances, challenges, and future prospects. IEEE Access 9:145248–145270
42. Mehmood G et al (2023) An efficient QoS-based multi-path routing scheme for smart healthcare monitoring in wireless body area networks. Comput Electr Eng 109:108517
43. Younas MI et al (2023) Toward QoS monitoring in IoT edge devices driven healthcare—a systematic literature review. Sensors 23(21):8885
44. Goel G, Chaturvedi AK (2024) Multi-objective load-balancing strategy for fog-driven patient-centric smart healthcare system in a smart city. Eng Technol Appl Sci Res 14(4):16011–16019
45. Singh PD et al (2023) BOSS: a new QoS aware blockchain assisted framework for secure and smart healthcare as a service. Expert Syst 40(4):e12838
46. Aminizadeh S et al (2024) Opportunities and challenges of artificial intelligence and distributed systems to improve the quality of healthcare service. Artif Intell Med 149:102779
47. Singh P et al (2023) A task scheduling algorithm for optimizing quality of service in smart healthcare system. In: International conference on IoT, intelligent computing and security: select proceedings of IICS 2021. Springer
48. Pradhan B et al (2023) An AI-assisted smart healthcare system using 5G communication. IEEE Access
49. Arafat MY, Pan S, Bak E (2024) QQAR: a Q-learning-based QoS-aware routing for IoMT-enabled wireless body area networks for smart healthcare. Internet of Things 26:101151
50. Upadhyay S et al (2023) Challenges and limitation analysis of an IoT-dependent system for deployment in smart healthcare using communication standards features. Sensors 23(11):5155
51. Buyya R et al (2023) Quality of Service (QoS)-driven edge computing and smart hospitals: a vision, architectural elements, and future directions. In: International conference on communication, electronics and digital technology. Springer
52. Almas A et al (2023) Context-based adaptive Fog computing trust solution for time-critical smart healthcare systems. IEEE Internet Things J 10(12):10575–10586
53. Wang L, Ni Z (2024) Deep reinforcement learning for QoS-driven cloud healthcare services selection: a framework and performance evaluation. J Intell Fuzzy Syst 2024(Preprint):1–15
54. Zhang H et al (2023) Microservice deployment mechanism with diversified QoS requirements for smart health system in industry 5.0. IEEE Trans Cons Electron
55. Khodkari H et al (2024) Addressing service quality challenges in the cloud and IoT integration: a case study of tourist remote healthcare system. Wirel Pers Commun:124
56. Safa M et al (2023) Real time health care big data analytics model for improved QoS in cardiac disease prediction with IoT devices. Heal Technol 13(3):473–483

57. Sutradhar S et al (2023) A dynamic step-wise tiny encryption algorithm with fruit fly optimization for quality of service improvement in healthcare. Healthc Anal 3:100177
58. Ghoumid K et al (2024) Optimization analysis of average message delivery time for healthcare monitoring using a developed NB-IoT technology in a smart city. Internet of Things 27:101290
59. Golec M et al (2023) HealthFaaS: AI-based smart healthcare system for heart patients using serverless computing. IEEE Internet Things J 10(21):18469–18476
60. Xu G, Xu M (2024) An effective prediction of resource using machine learning in edge environments for the smart healthcare industry. J Grid Comput 22(2):1–13
61. Peralta-Ochoa AM et al (2023) Smart healthcare applications over 5G networks: a systematic review. Appl Sci 13(3):1469

GPSR Compliance

The European Union's (EU) General Product Safety Regulation (GPSR) is a set of rules that requires consumer products to be safe and our obligations to ensure this.

If you have any concerns about our products, you can contact us on ProductSafety@springernature.com

In case Publisher is established outside the EU, the EU authorized representative is:

Springer Nature Customer Service Center GmbH
Europaplatz 3
69115 Heidelberg, Germany

Batch number: 10725823

Printed by Printforce, the Netherlands